Behaviors of Trace Metals in Environment

Hui Zhang

Behaviors of Trace Metals in Environment

The Pollution in Regional and Metropolis Areas

上海交通大學出版社
SHANGHAI JIAO TONG UNIVERSITY PRESS

Springer

Hui Zhang
School of Environmental Science
and Engineering
Shanghai Jiao Tong University
Shanghai, China

Funded by B&R Book Program

ISBN 978-981-13-3611-9 ISBN 978-981-13-3612-6 (eBook)
https://doi.org/10.1007/978-981-13-3612-6

Jointly published with Shanghai Jiao Tong University Press, Shanghai, China
The print edition is not for sale in China Mainland. Customers from China Mainland please order the print book from: Shanghai Jiao Tong University Press.

Library of Congress Control Number: 2019931535

This Springer imprint is published by the registered company Springer Nature Singapore Pte Ltd.
The registered company address is: 152 Beach Road, #21-01/04 Gateway East, Singapore 189721, Singapore

Preface

Trace metal pollution has always been a hot topic in the field of environmental science and engineering. Based on the accumulation from exploration years, the pollution cases of different causes with typical significances are carried out in this work for the environmental effects and the behavior rules of the trace metals in modern social and economic development. These cases include the regional trace metal pollution in the Hetao area of the Yellow River Basin and the pollution of metropolis areas from Shanghai and Nanjing, China. The study was, respectively, funded mainly by the National Natural Science Foundation of China and the international research cooperation funds (the International Collaboration Foundation of the University of Minnesota, USA and the International Collaboration Foundation of the NEDO, Japan), etc. Through the specific research work on trace metals and the pollution in different origin background cases, the characteristics, restrictive factors, genetic process, and the evolution trends of various kinds of the pollution are discussed with a lot of work in order to explore the regular facts of trace metal pollution. The goal of this study focuses on revealing the regulation of the essential facts in the trace metal pollution in environment based on the cases in different causes of formation for their factors, processes, and features. It is emphasized that the figures from analysis, facts from observations, and experiments simulated the natural conditions are obtained from the work, in the discussion and interpreting on the conclusions for us to understand the chemical or geochemical processes, the relationships in the formation, and the control factors in the pollution. The main six sections of the work are represented in this book as follows:

(1) Introduction on the Trends of Trace Metal and the Pollution Research
(2) The Study on the Regional Trace Metal Pollution in Soil and Groundwater
(3) The Study on the Trace Metal Pollution in Metropolis Water Systems
(4) The Experimental Study on the Environmental Behavior of Trace Metals
(5) The Study on the Research Approaches on Trace Metal Environmental Effects
(6) Main Conclusions from the Work

From this work, it tries to reveal and induce the regularity of the behaviors of trace metals in environmental pollution through the study on specific cases, trace metal pollution caused by the development of society and economy in regional soil and groundwater (the Hetao area, Inner Mongolia) and that caused by the development of modern metropolises (Shanghai and Nanjing, China).

Due to the close relation between metals and human daily life as well as the complex behavior and change probability, the trace metal pollution risen with human production and life may occur everywhere at any time. This pollution is more general, complex, durable, and inconspicuous than the pollutions caused by other hazardous substances. Moreover, there is background content for every metal element in natural environmental systems so that it is always difficult to recognize the pollution from the normal status. But the crises of trace metals to the ecological systems such as hazardous to plants, inducing lesions or deformities to animals and human makes the research on it be an important duty for environmental scientists to have to undertake for long time, specially to understand the principle and regulations of this pollution in environment.

The impacts of the pollution of trace metals may generally be manifold because of complex behaviors of them in environment. In a recent study of trace metal pollution, the major work is the research on cases, methods, and new materials to be used for protecting the pollution. There seems to be little systematic study on the mechanism and regularity of trace metal pollution in the environment. This situation reflects that every case study as well as its problem of trace metal pollution is with the frontier nature, and at the same time the pollution cases all are with complex properties to let them be difficult to be carried out. In this book, the impact status, cause of formation of the pollution of trace metals to be probed from common facts existed in the study cases, associated with experimental study in years in my group, are carried out to induce and conclude some issues of the pollution of trace metals. Based on these points above, the Hetao area of the Yellow River Basin as a regional area case and Shanghai and Nanjing, respectively, as metropolis cases are selected to study for the pollution, which are the typical and heavier cases for the processes and phenomenon of the regional and modern metropolis pollution of trace metals. From the research, some rules of the behavior of trace metals controlled by chemical or geochemical habits in regional and metropolis pollution are probed with discussion.

The samples of this work were in solid and liquid phases, respectively, including rock, ore, soil, sediment from rivers and lake, suspended particles of water, resident hair, and water, respectively, from river, lake, drinking well (phreatic water), and mining well. The measurements were done for concentrations of metals, nitrogen, phosphorus, total organic carbon, and organochlorine pesticides, the isotope of Pb and Sr, mineral composition in soil and sediment.

As important contents of this work, the experiments that simulated conditions at the earth surface were carried out to reveal the rules and factors, respectively, for metals absorbed and desorbed by and from soil and metals released from the grains (ore or rock). In addition, a useful assessment approach on environmental impacts or bioavailability of the trace metals is proposed based on a lot of experiments and comparison discussions.

As mentioned above, this work was funded to me, respectively, by the foundation as the following:

The National Natural Science Foundation of China (granted No. 49863001);

The Foundation of the State Key Laboratory of the Research on Deposits (granted No. SKLMD199910);

The International Collaboration Foundation of the NEDO, Japan (granted No. F-04001);

The Foundation of the State Key Laboratory of Environmental Geochemistry (granted No. SKLEG6007);

The International Collaboration Foundation of the University of Minnesota, USA (granted No. UMN-070604);

The National Natural Science Foundation of China (granted No. 41572329), etc.

In the processes of whole of this work, the help and assistants were, respectively, obtained from the following departments: the Center of Analysis of School of Environmental Science and Engineering Shanghai Jiao Tong University, the Analysis Center of the Shanghai Jiao Tong University, the Center of Environmental Monitoring Shanghai, the State Key Laboratory of Environmental Geochemistry of the Institute of Geochemistry CAS, the State Key Laboratory of the Research on Deposits Nanjing University, the Modern Analysis Center of Nanjing University, the Laboratory of Elemental Geochemistry of the Department of Earth Sciences Nanjing University, the Center of Environmental Monitoring Nanjing, the Beijing Center of Analysis of Ministry of Nuclear Industry, the Center of Environmental Research of the University of Tsukuba Japan, the Department of Civil Engineering of the University of Minnesota USA, etc.

In the process of this research years, a lot of work was performed at the Laboratory of Environmental Pollutant Behavior Chemistry, Shanghai Jiao Tong University. The students who were mainly in the work in the laboratory of my group and who took part partially in this work are: Xiaoyi Zhou, Ying Chen, Lina Zhong, Yi Zhu, Jiying Li, Mengchan Jiang, Chenmin He, Nan Zhang, Aaron Sciore (USA), Yiqing Yao et al. Moreover, there are many students who took part in important work in this research can not be listed for their names here. Professor Dongsheng Ma (the Department of Earth Sciences, Nanjing University), Prof. Zhongling Liu (the Department of Ecological and Environmental Sciences, Inner Mongolia University), Prof. Jingsheng Chen (the Department of Urban and Environmental Sciences, Peking University), Prof. Patrick L. Brezonik (the Department of Civil Engineering, University of Minnesota, USA), and Prof. William Arnold (the Department of Civil Engineering, University of Minnesota, USA) took part in some valuable discussion in the work of this book. In the composing process of the text, Prof. Xiaoye Liu (Shanghai University), Prof. Chiquan He (Shanghai University), Prof. Dengxin Li (Donghua University), Dr. Mingce Long, Dr. Xiaojue Yu, Dr. Yaolan Niu, and Dr. Jia Duo offered some important text-sorting helps.

This work is a huge project performed by collectivity. Except persons mentioned above, there are ones who are not listed on and who contributed common things for this research, and it should say that it would be impossible to finish the work without their participation. The names of all persons who contribute their ability and wisdom to this work will be in my heart, and days worked together with these colleagues have been my most interested and enjoyed memories, especially for the time worked in the Laboratory of Environmental Pollutant Behavior Chemistry, Shanghai Jiao Tong University. The helps, including fund assistance, work assists as well as the friendship and encouragements to me from mentioned departments and persons in this research have become my spiritual wealth to inspire self-confidence and let me more enrich. I would like to express my sincere thanks to everyone and each department who funded or took part in this work of the research!

The opinions in this book may only be limited to my knowledge and academic view, the extents that could not reach to will hope to be pointed out and made it perfection by readers and make it perfection.

Shanghai, China Hui Zhang
August 2018

Contents

Chapter 1
Introduction

1.1 The Trace Metals and the Study History

The knowledge or problem of trace metals has existed for a long time. Especially in recent decades, with the rapid development of environmental science and technology, trace metals have been paid more and more attention as one of the most important pollutants in the environment. In fact, there is also a gradual process of cognition in the concept of metals, which has negative ecological effects in the environment.

In this book, the term of "trace metals" means the metal elements with poisoning or negative impacts in environment when the amount of them reaches certain levels, and their effects on ecosystem are always associated with their chemical or geochemical properties. This term sometimes is mixed up "minor metals," "heavy metals" with seem same implication.

"Minor metals" is quoted from "minor element," a concept of geochemistry, and it is an important study object in corresponding subject named "Minor Element Geochemistry." Minor element is named versus major element in geochemistry. Major element means the main elements composing the matter of the earth, generally containing 13 elements orderly as their abundance in the earth's crust as follows: O, Si, Al, Fe, Ca, Na, K, Mg, H, Mn, Ti, C, and Cl. These 13 elements account for 99.67% of the total earth's crust mass. Another 80 elements (natural elements) except the 13 elements are named minor elements, meaning all, respectively, less than 0.1% in the substances in natural world such as rock, soil, ore, aerolite, water body, air, biology, and sediments of ocean, lake, and river. These elements are generally with smaller abundance in the earth crust within 1×10^{-9}—1×10^{-6}, and the total mass of them accounts to 0.33% of the earth crust mass. The minor element is mainly in the form of ion, dissolved compounds or complexes, hydrosol, aerosol, suspended state or gases. Most of minor elements in natural systems migrate as complexes.

It needs to point out that the minor elements are regarded to be the elements that are not considered as major chemical calculating components in any phases in matter system by some scholars, and also are regarded those are with so less concentration in studied geochemical system that should obey the law of dilute solution (Henry's law).

H. Zhang, *Behaviors of Trace Metals in Environment*,
https://doi.org/10.1007/978-981-13-3612-6_1

In biomedicine research, the minor elements generally imply that the concentration in organisms is less than 0.01% of the studied object's total quality. Therefore, the minor element is a describing term for the element relative concentration in the research system, and trace metals mean the metal elements in the minor elements.

A term related to minor element is "heavy metal," and it is a common term in the field of environmental science and engineering to describe the pollutant in environment, which may lead to negative impacts for organisms in ecosystem when its concentration reaches to certain level. This concept has been widely used, especially at present as the environmental consciousness more and more going deep into the public, and it is always associated with pollution and poisoning in the public consciousness. Because it has been with much closed correlation in many environmental pollution events and with important characteristics, it is named "heavy metal" so as gradually to be this concept today. In textbook, heavy metals imply that are with the specific gravity more than 4 or more than 5 (some regards as), and there are 60 metal elements belonging to first definition and 45 metal elements belong to second definition. After this "heavy metal" term having been used about 100 years, it is found that it is no with the scientific validity. International Union of Pure and Applied Chemistry published IUPAC Report to point the meaningless of "heavy metal" term in 2002, "It is clear that we should abandon classification of metals using terms such as 'heavy metals', which have no sound terminological or scientific basis. A classification of metals and their compounds firmly based on their chemical properties is needed" [1]. In this literature, it was emphasized that the chemical property should be the points of the matter for an academic term, and the "heavy metal" could not contain the environmental behaviors for this matter. Hodson M. E. pointed out that although heavy metal concept used was present, it was a no appropriate one, "Heavy metals is a poor scientific term and alternatives exist. As scientists we should use them" [2]. Due to its specific genesis and development as well as the application background, it is still used today and at the same time this term, heavy metal, is with being replaced trend by other term in academic world. In this book, "trace metals" was used to describe or discuss the problems related to these substances or pollutants.

It has become a very concerned problem for trace metals to affect human health. In fact, trace metals have been the significant pollutants to study due to these matters always heavily to harm ecosystem or human health. These problems not only are issues of environmental science and engineering but also are issues of biomedicine fields, and there are many unknown aspects. Some cognition on trace metal characteristics and biological effects should be belonged to periodical knowledge for human to probe them. Because many subjects such as chemistry, geosciences, biology, medicine, and technical sciences might be involved in for one trace metal pollution issue, there are larger difficulties always in research or treatment. Summarily, the research on trace metals is mainly in the following two aspects. One is trace metals in natural system, and mainly study the origin of them to the environment on their behavior and impacts. Another is behavior mechanisms and effects of them in organisms. In fact, these are the two aspects of a problem. One aspect of the work is to solve the trace metals as contaminants from where they come from and how to go. In addition, the work level is mostly concentrated on the behavioral effects of trace metals in the environmental media before and in the process of entry, which belongs to the domain of environmental science. The second aspects of the work are to solve the problem of the biochemical behavior and the magnitude effect of trace metals in the biological machine. The work mainly focuses on the interaction between the organisms and trace metals and the effects of these effects on the life process, which belongs to the focus of biological medicine. Because these two aspects of work all have the same object of study trace metals, and sometimes naturally unnaturally intertwined, it is also difficult to distinguish. Therefore, researchers who often have biological or medical background in the field of environmental science are engaged in the work, and those who have the background of environmental science in the field of biomedicine are studied. In the research and exploration of these problems, the intersection is frequent or basic. However, the overall view is still hierarchical. Any environmental problem is the basic starting point for the protection of human health, so the trace metal content, which focuses on the biomedical field, is the goal of the study of trace metals in the field of environmental science. The content of trace metals, which is focused on the field of environmental science, is the essential source of trace metals in the field of biomedicine. The two natures are the basic process of trace metal pollution and poisoning, which occurs in different stages of natural and biological processes and effects.

Because of the complexity of trace metals, the understanding of the biological effects of trace metals is different at present, which is the case in both environmental science and biomedicine.

Environmental science focuses on the study of trace metal biological effects, mainly from the speciation of metal elements in environmental media, including trace metal compounds, valence state of elements and the binding form with ligands, the entry of metal elements into organisms, or the ways and means of biological uptake. In addition, the influence of the physical and chemical conditions on trace metals, the method of studying the existence of trace metals in the environmental medium, the correlation of trace metal form, and human health (endemic disease) are also the important contents of the study on the biological effects of trace metals in environmental science. The research on these problems has been deepening as the rising of the role and position of metal form in the environmental effect recognition and evaluation since the 1970s of last century. Especially in recent years, the study of trace metal speciation and bioavailability has become an indispensable basic content in the study of trace metal pollution. On trace metals, the speciation fraction, analysis procedure, analysis method, speciation characteristics, and environmental effects in the study of different media, different metal elements, and the physical and chemical conditions in different systems have been greatly developed and extended in the direction and are still hot topics in the field of trace metal pollution.

In biomedical research, the biochemical behavior of trace metals in organisms, the biological nutrition of trace metals, the biological toxicology of trace metals, the clinical manifestations of trace metal poisoning symptoms, and the corresponding sources of biological intake, dose effect, time effect, metabolic process, etc., all are the important research content. Because of the complexity and uncertainties of these problems, research work is often accompanied by greater difficulties. From these angles, the degree of research and knowledge level of each element is different. In spite of what people hope to achieve, the work seems to be shallow. But in fact, the work has been developing rapidly, and it is also a hot spot of environmental science, medicine, and life science. Many of them are the frontiers of human exploration of the unknown world.

The two levels of the study on the biological effects of trace metals mentioned above all are on the most basic object of which are on trace metals, which are restricted by the natural properties of the metal itself at any levels. Trace metal elements are mostly subgroup elements in the periodic table, that is, the so-called transition elements. The most important feature of these elements is that they are not filled with subelectron layers, and most of them are variable valence elements. This determines the complex and changeable characteristics of its chemical behavior in nature and organism. Among these elements, the common characteristics are:

1. **The particularity of atomic structure and chemical properties**

These elements' atomic radii all are relatively large, sensitive to the surrounding matter, electronegativities are relatively large, and the valence states are changeable. Most elements exist in different valence states according to the different physical and chemical conditions in the environment (Eh, pH, temperature) and the properties of the substances in the system. Many trace metals can be used as central ions to accept the only pair electrons of anions and simple molecules. The complex complexes are formed in the form of complex anion groups in the form of complex anions and other complexes, and they can be formed with some large organic polymers.

2. **The universality in the environment system**

All trace metals are constituents of the earth's crust. There are background contents in various environmental media (rocks, soil, water, air, and biology). In addition, these elements are widely used in all aspects of human life and production. With the geochemical cycle of the natural elements and the exploitation, processing and use of underground resources, the spatial location migration, and the transformation of existing forms on various scales are constantly occurring in the environment. As a result, these elements are widely available and can be entered into the human body through various ways (food, drinking water, breathing, and skin contact), and their different chemical or valence toxic effects are often very different.

3. **The negative biological effects are concealed and latent**

Studies have shown that many trace metals have carcinogenic, teratogenic, and mutagenic effects. When these elements enter organisms, they can be enriched thousands of times through the food chain in higher organisms. Its toxicity has accumulative effect. It can cause chronic damage to organism and lead to acute poisoning in organism too. When the environment is contaminated by these elements, there is often no very apparent representation. The toxic effects of some trace metals are usually reflected in a few years or even longer. They are not easy to detect in the short term, but are hidden and latent.

4. **It is difficult to identify and evaluate for the pollution and poisoning.**

Because of the trace metal background content in the natural environment medium, the toxic effect is hidden and latent, and it is difficult to identify the pollution and poisoning.

The biological effect of trace metals is a hot topic in the field of environmental science and interdisciplinary subjects. Besides the complexity and important environmental significance of these problems, there are still many frontiers in these problems. It is objective to say that trace metal elements are toxic or beneficial, and the corresponding doses and forms of action are all the understanding at the stages or at least some are at stages. To be precise, the understanding of these problems is developing. With the deepening of research and the improvement of people's awareness, more important and clear biological effects of trace metals will be revealed. Therefore, the biological effect of trace metals is a very important research content in environmental science, and their understanding is developing rapidly.

1.2 The Objects of This Work

As mentioned above, the geochemical characteristics of trace metals are complex, and their behavior is changeable in the environment. Its pollution may occur in every link of human production and life, which has two characteristics of universality and complexity. Because of the background content of trace metals in natural environmental media, it is difficult to identify and evaluate. However, a large number of facts, such as inhibition of crop growth, animal and human teratogenesis, carcinogenesis, and some pathological changes, and even the gene trait variation of the life entity has confirmed the serious harm of trace metal pollution to biological and human health [3–7]. Nowadays, trace metals have a serious toxic effect on organisms and human body, which has become a common understanding. This consensus has not only been confined to academia but also been increasingly accepted by the public as a whole.

The behavior complexity of trace metals in the environment leads to the pollution and toxic effects in various statuses. At present, most of the research work in this field is pollution case study, pollution methodology research, and new materials and new ideas for pollution control. Systematic studies on pollution mechanisms and regularities are relatively few. This aspect reflects the frontier of their problem in many case studies and is undoubtedly related to the complexity of the trace metal pollution mechanism, and it is necessary to explore systematically the specific cases. In order to solve the cause and evolution of the specific environmental problems in the study area, this study attempts to summarize the geochemical behavior and constraints of the elements related to the trace metal pollution mechanism and regularity from the geochemical perspective, and explore the general case in the case of the typical cases. In fact, this work tries to understand the basic mechanisms of the trace metal pollution and improve some research methods from the systematic work, and regard them as the basic starting point and the goal of research to be expected to reach.

1.2.1 *The Regional Trace Metal Behavior and Pollution Research Case*

Based on the above research ideas, this work is specifically selected as a specific case of regional trace metal pollution related to the background factors and urban trace metal pollution mainly caused by human activities. It is related to the medium of rock, soil, ground water, surface water and sediment, and hair in polluted areas. Elements and isotopes were studied. The location of the study area is shown in Fig. 1.1.

Arsenic poisoning and other trace metal poisoning in the area of the Hetao prefecture in Inner Mongolia involve 11,000 km^2 regions in seven counties. The local residents have palmoplantar keratosis, skin cancer, hair dry and myocardial injury, and other related lesions, the number of victims of up to 200 thousand, attracting the attention of the international environmental science community [8–10]. In 2000, the USEPA had dispatched workers to take samples from the Hetao Area to develop environmental standards for the determination of trace metal elements such as As. The As Internet Site Project (The Arsenic Website Project, run by Harvard University risk), hosted by Professor Richard Wilson, a toxicant expert at Harvard University in 2000, listed the population poisoning in the Hetao Area into the world famous

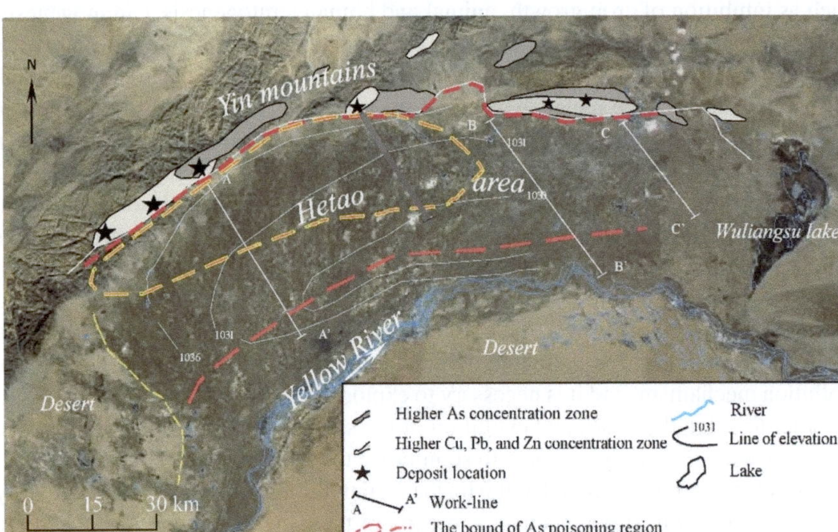

Fig. 1.1 Study area for the behavior and pollution of trace metals in regional scale, Hetao Area, Inner Mongolia

regional cases, such as Bangladesh, the USA, and Canada. A serious case of group As poisoning was reported in [11].

Since ancient times, Hetao Area has been the famous grain silo in China, which is the important commodity grain base in the north, and it is also an important producing area of the economic crops such as oil (sunflower seed), beet, melon, and fruits (apple and pear). The area is advantageous in natural conditions. The area is densely populated and developed, and there are many large- and medium-sized metal and nonmetallic deposits and numerous mineralization sites in the northern margin (upper reaches) near the mountains (1 large pyrite deposit, 2 large Cu polymetallic deposits, 1 large limestone deposit, 1 medium size Pb, Zn polymetallic deposit, 2 small Cu and Pb polymetallic deposits, and 63 mineralization sites). In addition, there are nearly continuous geochemical high background belts of trace elements, such as Cu, Pb, Zn, As, Sb, and Cd, as shown in Fig. 1.1. These unique and superior natural conditions have benefited the people of the area. With the expansion of the production scale and the accumulation of time, the potential pollution sources (the geological bodies rich in metal elements) and the trace metal pollution effects brought to the environment by artificial activities have been also increasingly obvious.

The As content of groundwater in this area is seriously exceeding the standard. The State Council and relevant ministries and commissions of China have attached importance to it in the early 90s of last century. At that time (1992), the general office of the State Council, in conjunction with the department of environmental protection, the departments of geology and mineral administration, and the departments of the health and medical treatment administration of the state and the Inner Mongolia Autonomous Government, carried out a large-scale investigation on the As pollution and the local diseases caused by the groundwater contamination in the area to monitor for the nature wills. According to the statistics at that time (1995), in the local residents' drinking well water As was seriously overweight (at that time the standard was 0.05 mg/L), with an average As of 0.1–0.5 mg/L, up to 0.969 mg/L, and the people number in threats was more than 200 thousand [10].

Arsenic pollution and poisoning has been a hot issue in the international environmental science field for many years. As early as 1998, As and Hg were listed as the most important toxicants in the international toxicology meeting of the International Society of Toxicology in Seattle, which was attended by more than 5000 scientists in the world. On the Science 279 volume, new views and different views on the differences of human body resistance to As and Hg environmental toxicology were specifically reported for different understandings [5]. The earlier typical regional As pollution research was reported by Nickson et al. at Nature (1998) and at Applied Geochemistry (2000) related the Ganges River alluvial plain of Bangladesh on the causes of As pollution in the groundwater [4, 12], the background of As in the northern region of New Zealand reported [14], the As pollution present in the sediments and river water of the Blue River basin in Colorado, USA [13, 14]. In 2000, the Harvard University set up an Arsenic Internet Website Project to report on the Internet pollution and poisoning cases, monitoring and research results, and dynamic [12]. In addition to the Hetao Area in China, a large range of As pollution and As poisoning endemic have appeared in the Baiyingao area of Hexigten County in eastern Inner Mongolia, Kuitun area of Xinjiang, Datong Basin of Shanxi, and Jiayi in Taiwan [15–17], and many works have been carried out in succession. A large number of reports on arsenic pollution and poisoning have also been reported in recent years. The representative work is such as the achievements of Fendorf et al. on the As pollution research reported at Science in 2010, etc. [18].

In fact, many trace metals have similar geochemical habits, which determine their natural characteristics and their regular exercise in the natural environment. The facts of As pollution in the Hetao Area and its unique geochemical background enlighten us that other trace metal elements associated with it have the possibility of pollution in the water. Arsenic and the elements associated with it such as Cd, Sb, Bi, Cu, Pb, and Zn are also likely to be contaminated in the water and soil.

1.2.2 The Metropolis Trace Metal Behavior and Pollution Research Cases

Shanghai and Nanjing are the most economically active cities in modern China. Their development scale, speed, industrial characteristics, and the interference or influences on the natural environment are typical of the environmental impact in the development of modern cities.

Shanghai is located in the estuary of the Yangtze River and the shore of the East China Sea. The river network of the whole area is all across over. It has both the vast land space and the characteristics of the water system. It has a good representative significance for the water environment pollution caused by the urban development, and this study takes it as a case study of the trace metal pollution caused by the urban development. Because Shanghai has always been the leading area of China's economic development, the basic attributes of large population, high urbanization, and industrial ratio determine the special and prominent pollution load that it brings to the environment. In particular, in the recent twenty years in China's rapid development era, the evolution of substance components in the natural environment of Shanghai caused by economic development has characterized the environmental effects of a typical artificial interference to the natural systems such as water, soil, and atmosphere. This study focuses on the relationship between the pollution evolution and the interference factors of human activities and selects the main objects of the surface waters, the Dianshan Lake, the Huangpu River, and the Suzhou Creek, which have a wide coverage of the study area and the stable pollution records. The trace metals and the related substances such as N, P, organochlorine pesticide, and organic carbon in the water and sediment are the main objects of the study. The location of the study area is shown in Fig. 1.2.

The city of Nanjing has a fast development speed, which contains undulating terrain in the urban area, the rule of wind direction is obvious, the soil vegetation is well developed, and the functional division of urban area is clearer. The research on trace metal pollution caused by urban development in Nanjing is better representative.

Fig. 1.2 Study area for the behavior and pollution of trace metals for metropolis, Shanghai (The Dianshan Lake, the Huangpu River, and the Suzhou Creek)

This is a case study of river pollution caused by urban development, as Nanjing is one of the fastest growing cities in China since 1980s, the content and distribution of trace metals in the environmental media (river sediments) have been changed significantly in the process of urbanization. Urban water body and sediment is the record of urban pollution and a comprehensive embodiment of pollution. The pollution accumulation in river sediment in Nanjing has reached a level that cannot be ignored. In this study, the behavior and pollution of trace metals in the sediment of the Yangtze River at Nanjing section were studied. The location of the study area is shown in Fig. 1.3.

As important regions of China's social economy, Shanghai and Nanjing have always been the political and economic centers in the middle and lower reaches of the Yangtze River with superior natural conditions. At present, under the new situation of the rapid development of national economic construction, it is developing with unprecedented scale and speed and has formed the heavy industry supporting system with chemical, mechanical, automobile manufacturing, metallurgy, electric power, and so on. The absolute value of its industrial output value is in the forefront of the same kind of cities in China. Corresponding to the rapid development of industry, the transportation infrastructure construction has been unprecedentedly developed in terms of capital input and scale effectiveness. The development of production and the development of traffic are bound to bring great synthetic influence on the environment, especially heavy industrial factories and traffic lines as trace metal pollution sources, which would bring harm to the environment. These are all the problems that need people to study and deal with in time.

Fig. 1.3 Study area for the behavior and pollution of trace metals for the Yangtze River, Nanjing section

These areas are representative of trace metal pollution research, and they are all facing pollution or become a hot issue in the field of environmental science. In this study, the characteristics of trace metal pollution affected by artificial and geochemical background and the characteristics of the trace metal pollution in the typical urban areas in the northwest of China and the modern cities of Shanghai and Nanjing in southeastern China are studied. Through specific work, the geochemistry of the environmental media in the study area is separately studied. It is studied or probed systematically what was or were the geochemical background of the area, trace metal element distribution (content distribution, speciation distribution, time distribution) characteristics and migration regulation, population disease or environmental effect, regional social and economic development history and trace metal behavior process in water, sediment, and soil, pollution status, cause of formation, mechanism, and so on. On this basis, the trace metal pollution, effect, origin, and development trend of trace metals in the study area are discussed and evaluated. Associated with the experimental data, the general rules of trace metals, such as the chemical properties of trace metals and the environmental behavior dominated by geochemical habits, are summarized.

In addition, in light of the specific problems encountered in the above case studies, the research methods of trace metal pollution, such as the speciation analysis of trace metals, the mechanism, and conditions for the release of trace metals in earth surface solid media, are explored and summarized.

1.3 The Methods of This Work

The research targets include solid medium and liquid medium. The solid medium is rock, ore, soil, river sediment, lake sediment, and human hair; liquid medium is the surface water of the ore field, river water, lake water, and dive (well water).

1.3.1 Sample

All kinds of samples were taken from the original raw substances on site. According to the design plan of the institute, the sampling lines and points were designed separately. The details refer to the relevant chapters below.

1. **Solid sample**

The rock and ore samples were taken from fresh bedrock and ore; the soil samples were taken from the local B layer soil in the depth of 15–30 cm; The samples are dried, grinded, and sifted, and the <50 μm was intercepted to carry out the related project research. The sediment samples of rivers and lake were divided into the silt of riverbank at different depths in the work profile and the underwater sediment. Samples were dried, grinded and sifted, and intercepted for study of <50 μm. The hair sample was directly cutting into the hair of the adult fixed resident in the trace metal poisoning area researched. All kinds of samples were delivered, respectively, to the next stage by special step management.

2. **Liquid sample**

The liquid samples were the river water, lake water, ore field water, and the diving. The surface water samples of river and ore field water were directly sampled from the sampling sites, and the river and lake water samples were sampled at 50 cm below the surface of the water. The underground water samples were mainly taken from the wells of the villages in the study area, the well water was taken from the handle pressing well depth in 15–25 m, and the water was continuously pressurized for 3–5 min before taking. The water samples of the Dianshan Lake, Suzhou Creek, and the Huangpu River were directly sampled at half a meter below the surface of the river water. The sample pails were of polytetrafluoroethylene (Teflon), it was prewashed by sample water five times before loading, and the sample was treated with super pure HNO_3 on site acidification to pH 2–3 to be transferred to next treatment.

1.3.2 Analysis and Test of Composition

The analysis and test items of this research include analysis of trace metal content (total content and speciation content), analysis of N, P, total organic carbon (TOC), polychlorinated biphenyls (PCBs), organochlorine pesticides (OCPs) content (total content and the content of each monomer), Sr, Pb isotopic analysis, and phase analysis of mineral composition. Each item analysis method and related quality control in detail are introduced in the following relevant chapters.

1. **Analysis of trace metal element concentration**

The analysis of trace metal elements was, respectively, completed in the Laboratory Center of the School of Environmental Science and Engineering of Shanghai Jiaotong University, the Shanghai Jiaotong University Analysis and Testing Center, the State Key Laboratory of the Research on Deposits of Nanjing University, the Modern Analytical and Testing Center of Nanjing University, and the Beijing Analysis Center of the Institute of Nuclear Industry Geology. Analysis method details are introduced respectively at the relative chapters below.

2. **Analysis of N, P, total organic carbon, organochlorine pesticides (OCPs), and polychlorinated biphenyls (PCBs)**

The analysis of N, P, total organic carbon, and organochlorine pesticides was completed at the Laboratory Center of the School of Environmental Science and Engineering of Shanghai Jiaotong University and Shanghai Jiaotong University Analysis and Testing Center, respectively. Analysis method details can be found at the relative chapters below.

3. **Mineral composition analysis of soil, river sediments**

The phase analysis of mineral composition was completed at the Modern Analytical and Testing Center of Nanjing University. Analysis method details are introduced respectively at the relative chapters below.

1.3.3 Simulation Experiments on Desorption and Sorption of Trace Metals from or on the Medium

Simulation experiment is one of the key contents in this research work. The simulated experiments of this research have been designed and completed including the experimental simulation for trace metals desorbed and adsorbed from or on soil substances and the experimental simulation for trace metals released from solid particles (ores or rock) to environment.

1.3.4 The Speciation Analysis Experiment of Trace Metals

The chemical speciation analysis of trace metal elements was for the chemical speciation analysis of trace metals in soils in the Hetao Area, in water sediments (lakes, river sediments) and the water suspension particulate in the research areas of Nanjing and Shanghai.

1.3.5 Analysis and Testing Precision and Quality Index

The detailed conditions for the analysis and testing accuracy and the analysis quality of the respective analytical instrument, equipment, and laboratories are described in detail in the corresponding chapters in the following relative chapters.

References

1. IUPAC. "Heavy metals" a meaningless term (IUPAC Technical Report)? Pure and Applied Chemistry, 2002;74:793–807.
2. Hodson ME. Heavy metals—geochemical bogey men? Environ Pollut. 2004;129:341–3.
3. Lee TC, Noriho T, Patricia WL, Tona MG. Induction of gene amplification by arsenic. Science. 1988;241:79–81.
4. Nickson R, Mcarthur J, Burgess W, Ahmedt KM, Ravenscroft P, Rahman M. Arsenic poisoning of Bangladesh groundwater. Nature. 1998;395:338.
5. Kaiser J. Toxicologists shed new light on old poisons. Science. 1998;279:1850–1.
6. Elizabeth PK, Herman JS, Hornberge GM, Mills AL. The effect of distribution of iron-oxyhydroxide grain coatings on the transport of bacterial cells in porous media. Environ Geol. 1998;33:243–8.
7. Miller JR, Plechler J, Desilets M. The role of geomorphic processes in the transport and fate of mercury in the Carson Rover basin, west-central Nevada. Environ Geol. 1998;33(4):249–62.
8. Wu K, Xin CM, Sun TZ. The distribution and effcts of drinking groundwater with higher As to the residents. Inner Mongolia Prophylaxis and Treatment for the Endemia (In Chinese), 1993;19(1):30–32.
9. Wu K, Ma H, Yu G. Approach to the relationship between the poisoning and dosage of As intaking. Inner Mongolia Prophylaxis and Treatment for the Endemia (In Chinese), 1994;19(1):52–54.
10. Inner Mongolia Environmental Protection Bureau. the report of environmental quality on 1995 (In Chinese). Huhhot: Inner Mongolia Environmental Protection Bureau Press; 1995.
11. Kaiser J. Arsenic Dilemma. Science. 2000;288:1.
12. Nickson R, McArthura JM, Ravenscroftb P, Burgessa WG, Ahmedc KM. Mechanism of arsenic release to groundwater, Bangladesh and West Bengal. Appl Geochem. 2000;15:403–13.
13. Craw D, Chappell D, Reay A. Environmental mercury and arsenic sources in fossil hydrothermal systems, Northland. New Zealand. Environ Geol. 2000;39(8):875–87.
14. Apodace LE, Driver NE, Bails JB. Occurrence, transport, and fate of trace elements, Blue River Basin, Summit Country, Colorado: an integrated approach. Environ Geol. 2000;39(8):901–13.
15. Sun T, Wu K, Xing C. Epidemiological Survey of Endemic Arseniasis in Inner Mongolia. Chin J Endemiology (Proceedings of Endemic Arseniasis, in Chinese), 1995:1–4.
16. Wang L, Sun X, Ai H, Xu X, Lin F, Liu H. A investigation on as poisoning of Dzungaria, Xinjiang. Inner Mongolia Prophylaxis and Treatment for the Endemia (In Chinese), 1994;19:37–40.
17. Wang J, Zhao L, Wu Y. The geochemistry of the as poisoning area of the Datong basin, Shanxi. In: Proceedings of fifth national conference on geochemistry (In Chinese);1997. p. 13.
18. Fendorf S, Michael HA, Van Geen A. Spatial and temporal variations of groundwater arsenic in south and Southeast Asia. Science. 2010;328:1123–7.

Chapter 2
The Regional Pollution of Trace Metals—The Hetao Area, China

2.1 The Geochemical Setting of the Study Area

The Hetao Area in this research is located in the western region of Inner Mongolia Autonomous, at 106°30′–109°00′E and 40°40′–41°20′N and with an area of 21,000 km^2. The environmental media in this region related to this research includes the Yellow River, groundwater (<30 m), soil and its parent material—Quaternary Holocene, silty-fine sand and clay. The research area belongs to the alluvial plain of Yellow River, bordering the steep mountains of the Yin Mountains on the north edge of the plain. The Hetao Area appears to be oxbow-shaped in a nearly E–W direction. As shown in Fig. 2.1, the terrain is slightly higher in northwest and slightly leans to the southeast. The elevation is 1200–1500 m with the gradient of 3%–5‰.

The whole area is the quaternary sedimentary coverage area. The lower Pleistocene of the area is divided into two groups according to the lithofacies. The upper layer with low salinity is alluvial layer, which is mainly composed of yellow and grayish yellow fine sand and silty sand. The lower layer with high salinity is lake sediments, which mainly consists of yellow or gray fine sand and silty sand. This set of strata appears a certain sedimentary rhythm features, including the finer lower part and coarser upper part. The Holocene stratum is mainly composed of the fluvial and alluvial layers, and the burial depth of the sediments is 55–250 m. According to the data from S. Li and others in 1994, the quaternary sedimentary thickness is approximately 1200–1500 m, and the thickness in the front of the mountains on the north edge is thinner (20–50 m) [1]. The whole area of the Hetao Area is covered by the farmland and is irrigated from the Yellow River. The soil in this area is mainly yellow powder-based soil [2]. Soil minerals in different areas are generally identical, including detrital minerals (quartz, calcite, feldspar, and dolomite) and clay minerals (illite, kaolin, talc, and chlorite). On the foot of Yin Mountains, there is famous polymetallic metallogenic area (Cu, Pb, Zn, and S) in the northern China. Many metallic and nonmetallic ore deposits and anomaly belts with high content of geochemical setting of the elements (Cu, Pb, Zn, As, Sb, Cd, and et al.) are concentrated on the front of the Yin Mountains, as shown in Fig. 2.1. Four large ore deposits have been

H. Zhang, *Behaviors of Trace Metals in Environment*,
https://doi.org/10.1007/978-981-13-3612-6_2

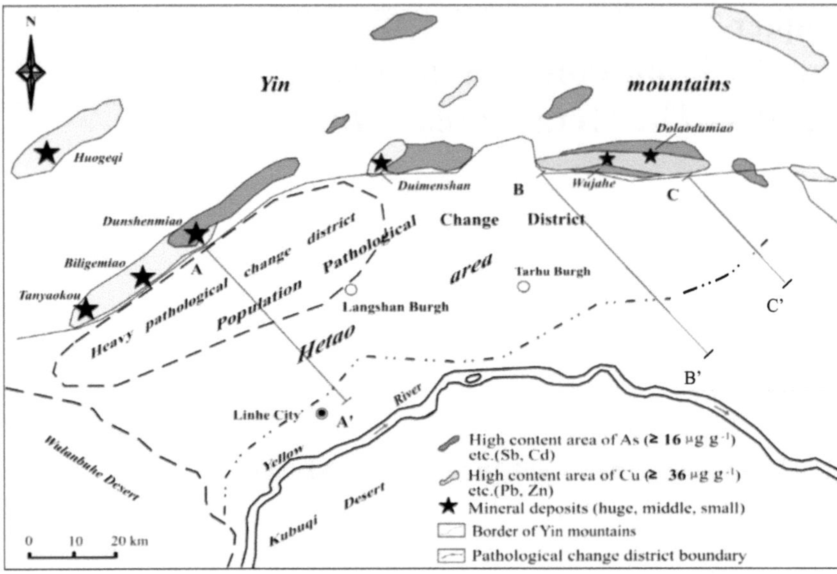

Fig. 2.1 Environmental features of the Hetao Area and the sites of main working lines

mined since years ago, and the longest history of exploration mining has lasted more than 80 years.

2.2 The Distribution of Trace Metals in the Soil

The belt of Yin Mountains in the north of Hetao Area belongs to the groundwater-supplying region. Hydrochemistry of the groundwater of the Hetao Area is mainly the feature with HCO_3–C_lNa and HCO_3–Na/Mg, with the pH of 7.5–8.5 [1]. The geochemical data of trace metals in Hetao Area are shown in Table 2.1.

In the Hetao Area, the research on the trace metal elements in the soil is less than that in the groundwater. However, the high content of the elements in the groundwater in the area had led to pathological changes of the local people. According to the bio-geochemical cycle, the contents of elements in the soil must also be to some extent affected by the groundwater. At the same time, the soil as one of the intermediaries from natural materials to the human body plays a very important role in the circulation of materials in the ecosystem. Therefore, in the studies on the pollution by trace metals and population-based poisoning in the Hetao Area, the research on the trace metal pollution in the soil must be an indispensable part. According to the distribution of pathological changes for local residents and flow characteristics of groundwater in research area, the sampling was carried out on the work-line along the flow direction of groundwater and across the resident poisoning zone in this study, in order to

Table 2.1 Data of geochemical setting in the environmental medium from the Hetao Area (unit: mg/kg)

	V	Cr	Mn	Co	Ni	Cu	Zn
Geochemical setting in soil (2466)	45.85	36.53	384.52	7.32	11.19	12.88	36.96
The concentration in Quaternary strata (15)	130.22	118.64	565.72	17.11	32.68	28.89	62.39
The concentration in groundwater (5)						0.02	0.05
The concentration in Yellow River water (1)						0.01	0.04
Clarke value of soil	90.00	70.00	1000.00	30.00	50.00	30.00	50.00

	Mo	As	Cd	Sn	Sb	Pb	Bi
Geochemical setting in soil (2466)	0.56	5.20	0.06	1.28	0.26	23.48	0.20
The concentration in Quaternary strata (15)	0.45	8.95	0.07	2.35	0.50	17.00	0.08
The concentration in groundwater (5)		0.12	0.00		0.00	0.02	
The concentration in Yellow River water (1)		0.00	0.00		0.00	0.01	
Clarke value of soil	2.00	6.00	0.35	10.00	1.00	35.00	

Note The numbers in parenthesis are the sample numbers used for calculating and in square brackets are the numbers of references

understand and verify the variation characteristics, the average content of trace metal elements, and the relationship of the variation of trace metals with pathological changes in the local residents. The main sampling lines are shown in Fig. 2.1.

Samples were all yellow fluvo-aquic soil, which were taken from B layer (15–30 cm). Samples were dried, ground, and sieved, then the parts (≤50 μm) of which were used in the experiments. The concentrations of trace metals (V, Cr, Mn, Co, Ni, Cu, Zn, Mo, Sn, Sb, Pb, Bi, and Hg) were determined by inductively coupled plasma atomic spectrometry (JY-380ICP-AES, France). The detection limits (mg/kg) are 8×10^{-4} (V), 5.7×10^{-4} (Cr), 2×10^{-4} (Mn), 4.2×10^{-4} (Co), 1×10^{-2} (Ni), 2×10^{-3} (Cu), 4×10^{-3} (Zn), 1×10^{-3} (Mo), 4×10^{-3} (Sn), 3.8×10^{-3} (Sb), 3.4×10^{-3} (Pb), 3.6×10^{-3} (Bi), and 2×10^{-12} (Hg). The concentrations of elements in speciation analysis were determined by TJA-1100 ICP-AES (USA), and the detection limits (mg/kg) are 2×10^{-3} (V), 5×10^{-3} (Cr), 1×10^{-3} (Mn), 3×10^{-3} (Co), 1×10^{-2} (Ni), 2×10^{-3} (Cu), 4×10^{-3} (Zn), 5×10^{-3} (Mo), 1.5×10^{-2} (Sn), 5×10^{-2} (Sb), 2.5×10^{-2} (Pb), and 1.5×10^{-1} (Bi). In addition, the concentration of As and Cd is, respectively, measured by atomic fluorescence spectrometer (AF-610, Beijing Rayleigh Analytical Instrument Co., Ltd., China), and the detection limits (mg/kg) are 8×10^{-5} (As) and 8×10^{-5} (Cd). The analysis qualification was controlled by analyzing the relative deviation limits (5%), and making a statistic analysis (≥95%), the qualification rate was 100%. The mineral composition and phase analysis for the soil was completed by rotating anode X-ray diffraction (D/Max-RA, Rigaku, Japan). The average contents of trace metals and mineral composition in the research area are shown in Table 2.2.

Results show that the average contents of elements in the soil of the Hetao Area are obviously higher than the corresponding background values, and the concentration of As was 1.15–4.81 times of the background value. The concentrations of trace metal elements in the soil were evaluated by the index of geo-accumulation proposed by Muller [3], and the results are shown in Table 2.3.

The index of geo-accumulation ($I_{geo} = log_2[C_n/(1.5B_n)]$), proposed by Muller in 1969, is a quantitative standard through level division of the pollution in order to quantitatively evaluate the pollution degree of trace metals in the sediment, as shown in Table 2.4 [3]. The index of geo-accumulation is actually equivalent to the pollution index in some literatures. Both of them are the quantitative parameters for analyzing the pollution degree of the pollutant based on many work experiences. This index of geo-accumulation was obtained by some objective data with inevitable limitations. It seems with some serious defects and certain limitation, especially for the evaluation of environmental effects of some elements with higher-background values. However, there are many facts too to prove that the index of geo-accumulation has certain significance for the evaluation of the trace metal pollution in soil and sediment. Until now, the index of geo-accumulation is one of the most important parameters of measurements for evaluating the degree of metal pollution and widely used in the research in the field of environmental science. Based on the index of geo-accumulation, the contents of Cu, Zn, As, Cd, Hg, Co, Ni, Cr, and V have reached the weak pollution degree in the Hetao Area, and the concentration of Sb has reached moderate pollution degree in the area (Table 2.3). This is a serious fact for the Hetao Area.

2.2.1 *The Distribution of Total Concentration of Trace Metals*

The distribution of the trace metals in the soil of the Hetao Area is shown in Fig. 2.2. These show that the concentration of the elements (Cu, Zn, As, Cd, Sb, and Pb) in the soil of the Hetao Area decreased gradually from the foot of the Yin Mountains to the downstream of groundwater, with just a few exceptions in partial section. Combined with the local geological and geochemical background, it suggests that: (1) The region with high content of trace metals (e.g., Cu, An, and As) in the soil in Hetao Area was concentrated in the front of Yin Mountains, and the contents of metal elements gradually decreased in the direction of plain and away from the mountain; (2) on the contrary, some other elements (e.g., Bi, Mo, and Sn) which did not appear with higher concentration in the front of the Yin Mountains did not show obvious variation trend in the soil, as shown in Fig. 2.3. These changes seem related to different geochemical background and mining history, and there were differences for theses changes among different trace metals. The ore deposit of Cu, Zn, and S

Table 2.2 Concentration of heavy metals and minerals in soil in the Hetao Area (unit: μg/kg for Hg and μg/g for others)

	Cu	Pb	Zn	As	Sb	Cd	Bi	Mo	Hg	Minerals
Average concentration (51)	24.84	28.14	67.86	13.74	1.52	0.13	0.27	0.37	28.74	Quartz, feldspar, calcite, dolomite
Geochemical setting (2466) [66],	12.88	23.48	36.96	5.20	0.26	0.06	0.20	0.56	13.96	illite, kaolinite talc, chlorite
Clarke value of soil [67]	30.00	5.00		6.00	1.00	0.35			0.06	

Note The numbers in parenthesis are the sample numbers used for calculating and in square brackets are the numbers of references

Table 2.3 Indexes of geo-accumulation and the pollution grades in the soil from the Hetao Area

	Pb	Zn	Cu	As	Bi	Cd	Sb	Hg
I-geo	−0.32	0.29	0.36	0.82	−0.14	0.54	1.97	0.46
Pollution grade		1	1	1		1	2	1
	Mo	Mn	Co	Ni	Cr	V	Sn	
I-geo	−1.17	−0.18	0.55	0.92	0.47	0.02	0.28	
Pollution grade			1	1	1	1	1	

Note ***I-geo*** = $\log_2$ (Cn/1.5Bn), Cn meant the concentration of the metal in the soil, and Bn meant the geochemical setting concentration in soil [3]

Table 2.4 Grades for geo-accumulation indexes in sediments

The index of geo-accumulation(*I-geo*)	Pollution grade	Pollution degree of the sediment
10–5	6	Very serious pollution
4–5	5	Strong—very serious pollution
3–4	4	Strong pollution
2–3	3	Middle—strong pollution
1–2	2	Middle pollution
0–1	1	No pollution—middle pollution
0	0	No pollution

Note I-geo = $\log_2$ (Cn/1.5 × Bn), Cn meant the concentration of the metal in the soil, and Bn meant the geochemical setting concentration in soil [3]

(named Tanyaokou deposit) at the northern–western edge in the research area has the longest history of exploiting and mining (more than 80 years); another large ore deposit of Pb, Zn, and S (named Dongshengmiao deposit near Tanyaokou at the northern–western edge in the research area) has a mining history of 70 years; and the middle-sized deposits named Duimenshan and other smaller deposits at the central front of the Yin Mountains have been exploited later than above deposits. Therefore, for trace metals (e.g., Cu, Zn, and As) in the soil, the variation grads of the concentration distribution curves from the foot of the Yin Mountains to the downstream of the flow are decreased from the section of deposits with longer mining history (Tanyaokou and Dongshengmiao deposits) to that with relative short mining and exploring history (Duimenshan deposit) (Fig. 2.2 AA′, BB′). This corresponding characteristic of the spatial distribution of the element concentration is not accidental, and it may reveal the causal relationships in the long migration processes affected under natural forces.

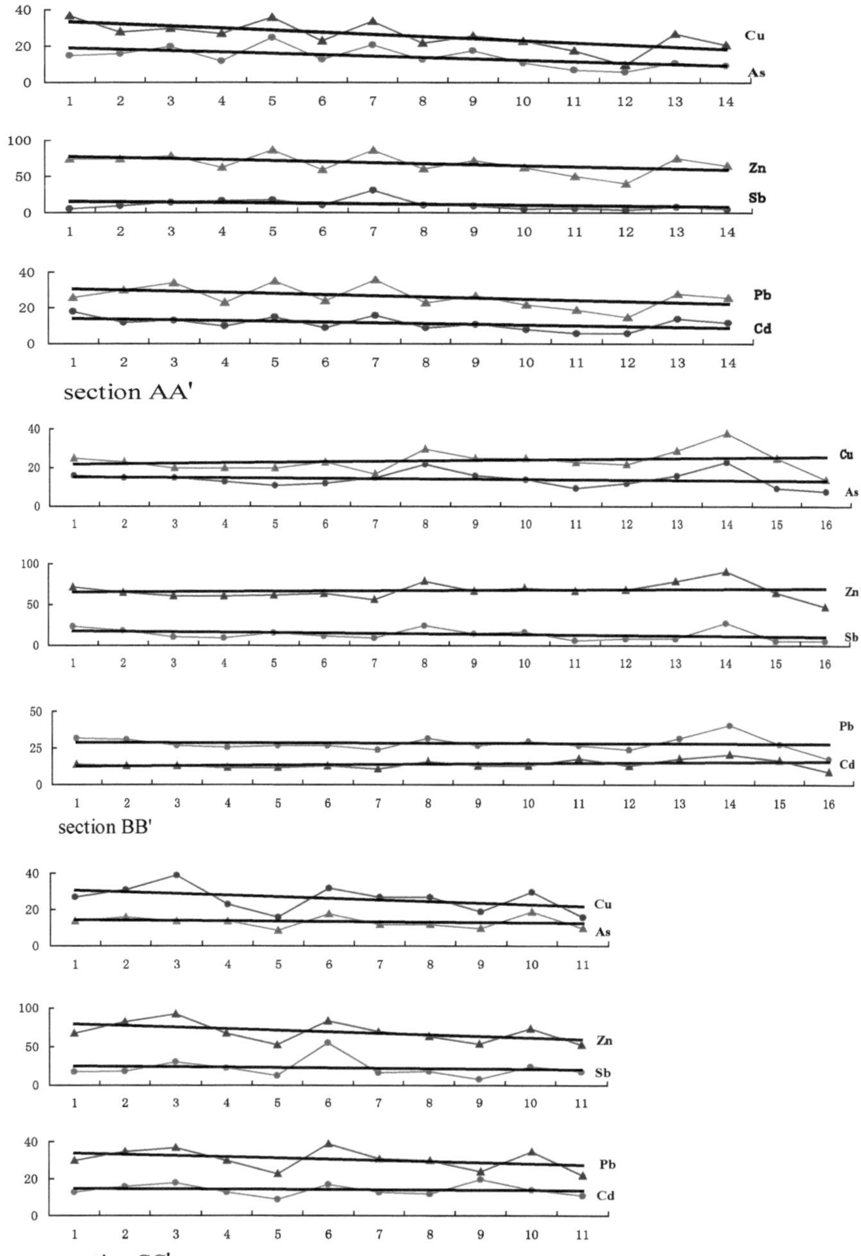

Fig. 2.2 Variations of the concentration of Cu, Pb, Zn, As, and Cd in soil in the Hetao Area. The numerals under *X* axis meant the numbers of the samples and the distance between near samples meant 4 km. The *Y* axis, respectively, was element concentration. The units of Cu, Pb, Zn, and As were mg/kg, and Sb and Cd were $\times 10^{-1}$ mg/kg

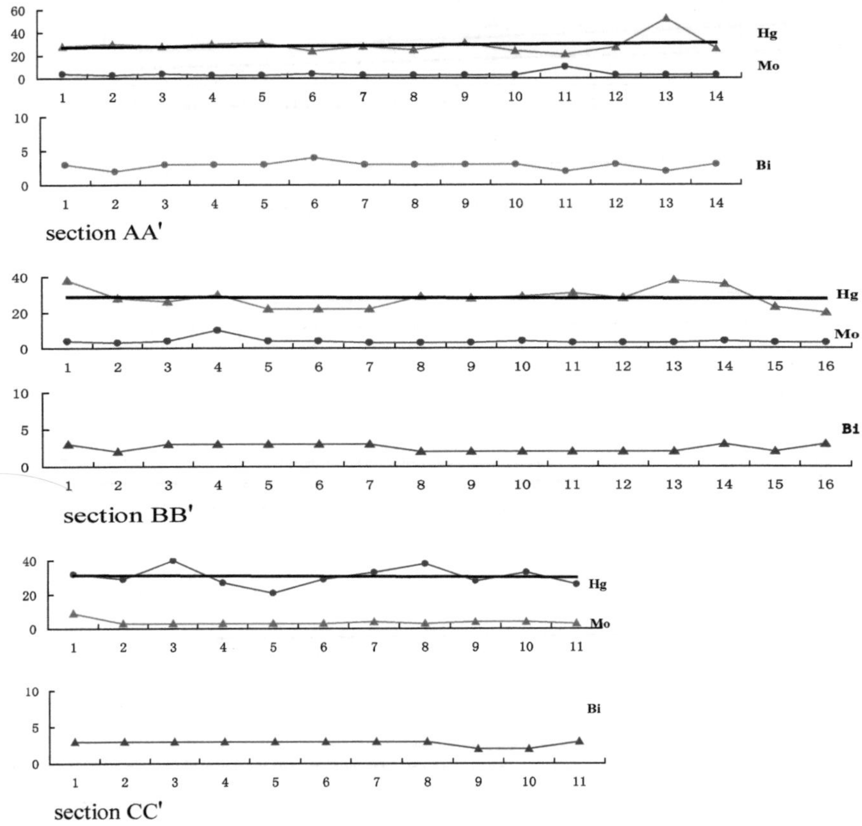

Fig. 2.3 Variations of concentration of Bi, Hg, and Mo in the soil in the working lines (sampling sections). The numerals under *X* axis meant the numbers of samples, and the distance between near samples was 4 km. The *Y* axis meant element concentration, and the unit of Mo and Bi was $\times 10^{-1}$ mg/kg, and Hg was μg/kg

2.2.2 The Distribution of Speciation of the Trace Metals

The chemical forms of elements in environmental media are key parameters to evaluate the ecological environmental effects of the elements in the environment, which are also important and profound to understand the chemical and geochemical behavior of the elements. Therefore, the chemical forms, speciation, of elements have been the important issues which have to be faced and solved in the environmental science research.

To determine the concentrations of the speciation of elements in the environmental media has been paid more attention in the research of environmental toxicology all over the world. For the past few years, the speciation analysis of the trace metals for various environmental media has been developed and research results in this area

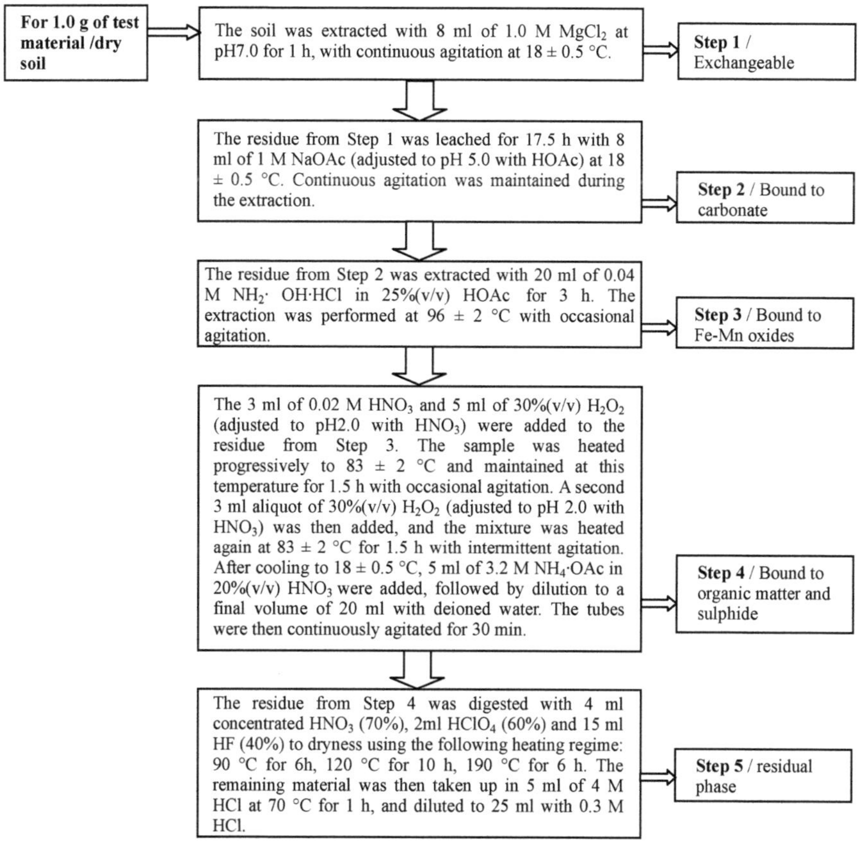

Fig. 2.4 Schedule for speciation analysis of heavy metals in soil

are often reported in the international environmental journals. However, due to the complexity of the speciation itself, the chemical speciation analysis of elements in sediments (water sediments) and soil has not been given a satisfied solution for a targeted and realistic scheme at present [4–12]. Moreover, there are some differences in the analytical results obtained from various analytical procedures which are used in present work because the analysis methods of chemical speciation are only the approximation to the actual species of the elements in the medium from special reagents and operating procedures recommended by the researchers. In particular, in some analysis schedules of the trace metals in common use, today the analysis method recommended by Tessier from University of Quebec in 1979 can be able to consider as a relatively simple and feasible ways. Thus, it has been relatively used widely in the field of speciation analysis applied in the practical work [4].

In this study, the method of chemical speciation analysis proposed by Tessier A. (1979) was used to analyze the chemical speciation in soil of Hetao Area. The work focused on the chemical speciation of seven representative samples for the trace

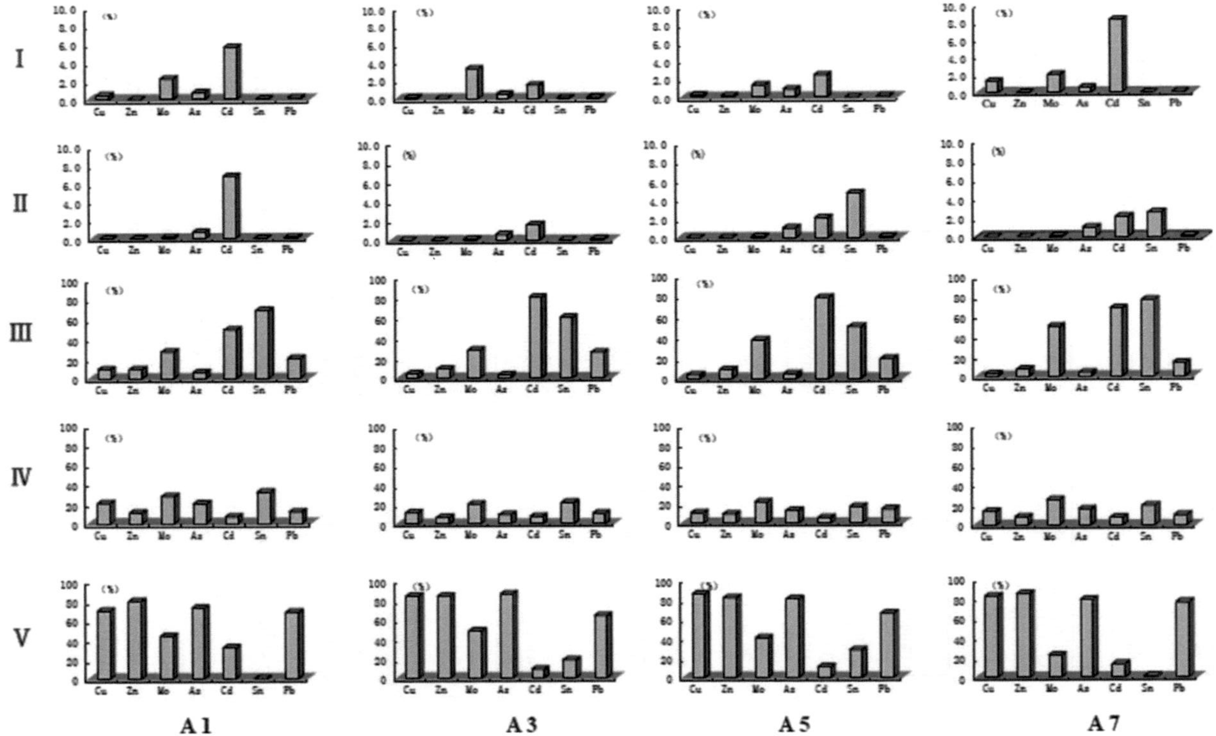

Fig. 2.5 Distribution of heavy metal speciation in the soil from the Hetao Area (1–3). The numerals at the left of the vertical axes of the charts were the concentration percent for the speciation fraction in the total concentration of the heavy metal elements. A1, A3 A5, A7, A9, A11, A14, C1, C5, and C11 meant, respectively, sampling sites and sampling numbers at the working lines. I, II, III, IV, and V meant orderly exchangeable, carbonate, Fe–Mn oxide, organic matter, and sulfide

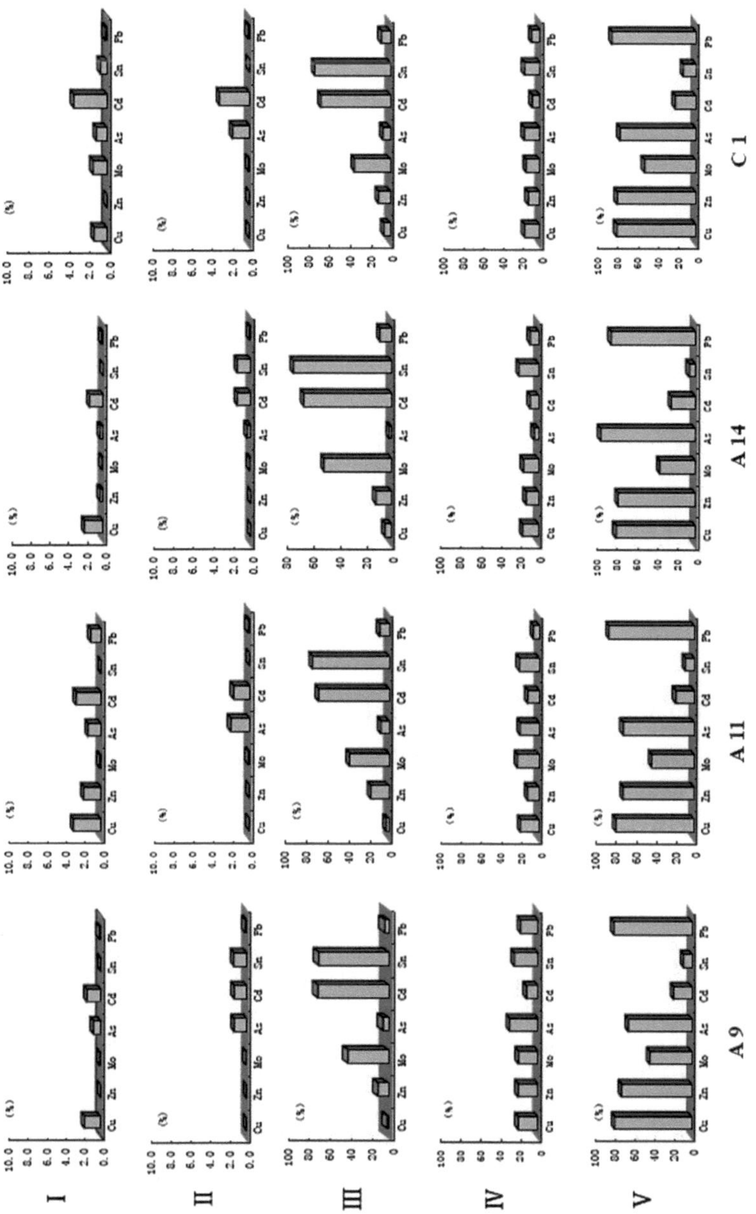

Fig. 2.5 (continued)

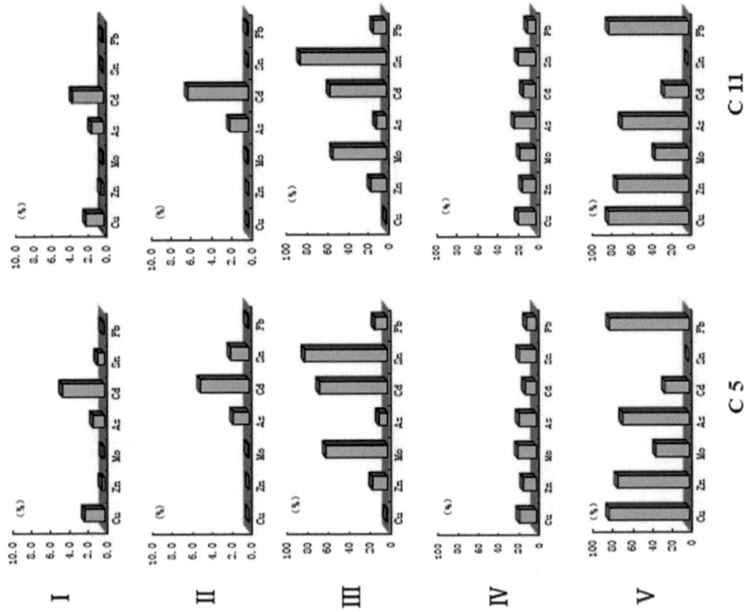

Fig. 2.5 (continued)

metals (Cu, Zn, Mo, As, Cd, Sn, and Pb), and the analysis procedure is shown in Fig. 2.4.

1. **Distribution of the speciation of the metals**

For the soil of Hetao Area, the characteristics on the concentrations of trace metals with different chemical speciation were shown in Tables 2.5, 2.6, 2.7, 2.8, 2.9, 2.10, and 2.11, respectively.

The results showed that the amount of exchangeable trace metals was 0.004–8.272 percent of total concentration of trace metals in Hetao Area, and Cu, Mo, As, and Cd were the common elements with average percentage of 0.855–3.421% and variation range of 0.096–8.272%. Carbonate concentration was 0.003–6.82% of the total concentrations of trace metals, and As, Cd, and Sn were the elements with relatively high concentration in the fraction of carbonates, with the proportion of 1.18–3.07% and variation range of 0.01–6.82%. The concentration of Fe–Mn oxide fraction was 1.04–83.08% of the total concentration, and Mo, Cd, and Sn are the elements with higher concentration of Fe–Mn oxide fraction, with average percentage of 41.29–70.01% and variation range of 21.17–83.08%. Organic concentration was 4.45–28.29% of the total concentration, and the elements with higher concentration of organic speciation were Mo, As, and Sn, which had the average percentage of 16.19–20.02% and variation range of 4.45–28.29%. The concentration of residual species had great difference for different elements in soil of Hetao Area, from the lowest of 0.02% (Sample No. 1 of work-line A, the concentration of residual

Table 2.5 Characteristics of Cu speciation in the soil from the Hetao Area (unit: mg/kg)

	Exch		Carbon		Fe–Mn Oxide		Organ		Residual		Efficient of all	
Sam. No.	C	P (%)	C	P (%)	C	P (%)	C	P (%)	C	P (%)	C	P (%)
A1	0.176	0.438	0.002	0.005	3.78	9.413	8.2	20.420	28	69.728	30.272	40.156
A3	0.048	0.126	0.002	0.005	1.6	4.205	4.4	11.564	32	84.104	15.896	38.048
A5	0.104	0.270	0.002	0.005	1.4	3.636	4	10.389	33	85.705	14.295	38.504
A7	0.528	1.284	0.002	0.005	1	2.431	5.6	13.616	34	82.669	17.331	41.128
A9	0.528	1.707	0.002	0.006	0.4	1.293	6	19.400	24	77.600	22.400	30.928
A11	0.528	2.913	0.002	0.011	0.4	2.207	3.2	17.652	14	77.229	22.771	18.128
A14	0.352	1.961	0.002	0.011	0.8	4.456	2.8	15.597	14	77.986	22.014	17.952
C1	0.352	1.250	0.002	0.007	1.6	5.683	4.2	14.919	22	78.147	21.853	28.152
C5	0.352	2.031	0.002	0.012	0.18	1.039	2.8	16.155	14	80.775	19.225	17.332
C11	0.352	1.891	0.002	0.011	0.06	0.322	3.2	17.193	15	80.593	19.407	18.612
Average	0.332	1.387	0.002	0.008	1.12	3.469	4.44	15.691	23	79.454	20.546	28.894

Note Exch meant exchangeable species, and Carbon meant carbonate, Fe–Mn oxide meant Fe–Mn oxides, Organ meant organic matter, Residual meant residual, Efficient of all meant the sum of efficient species of 4 species ahead; C meant concentration, P meant the percent in total concentration

Table 2.6 Characteristics of Zn speciation in the soil from the Hetao Area (unit: mg/kg)

	Exch		Carbon		Fe–Mn Oxide		Organ		Residual		Efficient of all	
Sam. No.	C	P (%)	C	P (%)	C	P (%)	C	P (%)	C	P (%)	C	P (%)
A1	0.025	0.142	0.025	0.142	3.6	20.455	2	11.364	12	68.182	31.818	17.600
A3	0.025	0.133	0.025	0.133	4.8	25.532	2	10.638	12	63.830	36.170	18.800
A5	0.025	0.137	0.025	0.137	3.6	19.780	2.6	14.286	12	65.934	34.066	18.200
A7	0.025	0.095	0.025	0.095	3.6	13.740	2.6	9.924	20	76.336	23.664	26.200
A9	0.025	0.164	0.025	0.164	0.6	3.947	2.6	17.105	12	78.947	21.053	15.200
A11	0.16	1.130	0.025	0.177	1.2	8.475	0.8	5.650	12	84.746	15.254	14.160
A14	0.025	0.174	0.025	0.174	1.2	8.333	1.2	8.333	12	83.333	16.667	14.400
C1	0.025	0.174	0.025	0.174	1.2	8.333	1.2	8.333	12	83.333	16.667	14.400
C5	0.025	0.164	0.025	0.164	1.8	11.842	1.4	9.211	12	78.947	21.053	15.200
C11	0.025	0.167	0.025	0.167	1.8	12.000	1.2	8.000	12	80.000	20.000	15.000
Average	0.039	0.248	0.025	0.153	2.34	13.244	1.76	10.284	12.8	76.359	23.641	16.916

Note Exch meant exchangeable species, and Carbon meant carbonate, Fe–Mn oxide meant Fe–Mn oxides, Organ meant organic matter, Residual meant residual, Efficient of all meant the sum of efficient species of 4 species ahead; C meant concentration, P meant the percent in total concentration

Table 2.7 Characteristics of Mo speciation in the soil from the Hetao Area (unit: mg/kg)

	Exch		Carbon		Fe–Mn Oxide		Organ		Residual		Efficient of all	
Sam. No.	C	P (%)	C	P (%)	C	P (%)	C	P (%)	C	P (%)	C	P (%)
A1	0.08	2.174	0.005	0.136	1	27.174	1	27.174	1.6	43.478	56.522	3.680
A3	0.16	3.226	0.005	0.101	1.4	28.226	1	20.161	2.4	48.387	51.613	4.960
A5	0.05	1.351	0.005	0.135	1.4	37.838	0.8	21.622	1.5	40.541	59.459	3.700
A7	0.08	2.010	0.005	0.126	2	50.251	1	25.126	0.9	22.613	77.387	3.980
A9	0.005	0.096	0.005	0.096	2	38.462	1	19.231	2.2	42.308	57.692	5.200
A11	0.005	0.132	0.005	0.132	1.4	36.842	0.8	21.053	1.6	42.105	57.895	3.800
A14	0.005	0.125	0.005	0.125	2	50.000	0.6	15.000	1.4	35.000	65.000	4.000
C1	0.08	1.361	0.005	0.085	2	34.014	0.8	13.605	3	51.020	48.980	5.880
C5	0.005	0.147	0.005	0.147	2	58.824	0.6	17.647	0.8	23.529	76.471	3.400
C11	0.005	0.128	0.005	0.128	2	51.282	0.6	15.385	1.3	33.333	66.667	3.900
Average	0.048	1.075	0.005	0.121	1.72	41.291	0.82	19.600	1.67	38.232	61.768	4.250

Note Exch meant exchangeable species, and Carbon meant carbonate, Fe–Mn oxide meant Fe–Mn oxides, Organ meant organic matter, Residual meant residual, Efficient of all meant the sum of efficient species of 4 species ahead; C meant concentration, P meant the percent in total concentration

Table 2.8 Characteristics of As speciation in the soil from the Hetao Area (unit: mg/kg)

	Exch		Carbon		Fe–Mn Oxide		Organ		Residual		Efficient of all	
Sam. No.	C	P (%)	C	P (%)	C	P (%)	C	P (%)	C	P (%)	C	P (%)
A1	23.44	0.699	22.82	0.681	200	5.983	669	19.962	2435	72.676	27.324	3351.04
A3	21.88	0.453	28.16	0.583	159	3.281	457	9.460	4167	86.224	13.776	4832.53
A5	31.17	0.879	35.99	1.015	167	4.707	452	12.741	2860	80.658	19.342	3545.97
A7	21.65	0.556	37.22	0.956	141	3.624	623	15.999	3070	78.865	21.135	3893.34
A9	20.73	0.806	32.71	1.272	142	5.526	728	28.285	1649	64.111	35.889	2572.16
A11	20.55	1.383	28.29	1.903	113	7.587	275	18.503	1050	70.624	29.376	1486.23
A14	21.72	0.301	28.46	0.395	109	1.515	321	4.454	6726	93.335	6.665	7206.73
C1	22.59	1.038	37.22	1.711	142	6.504	341	15.684	1633	75.064	24.936	2175.97
C5	22.78	1.132	31.62	1.572	162	8.068	330	16.380	1466	72.848	27.152	2011.9
C11	22.88	1.300	32.31	1.836	165	9.361	359	20.415	1181	67.089	32.911	1760.27
Average	22.94	0.855	31.48	1.192	150	5.615	455	16.188	2624	76.149	23.851	3283.614

Note Exch meant exchangeable species, and Carbon meant carbonate, Fe–Mn oxide meant Fe–Mn oxides, Organ meant organic matter, Residual meant residual, Efficient of all meant the sum of efficient species of 4 species ahead; C meant concentration, P meant the percent in total concentration

Table 2.9 Characteristics of Cd speciation in the soil from the Hetao Area (unit: mg/kg)

	Exch		Carbon		Fe–Mn Oxide		Organ		Residual		Efficient of all	
Sam. No.	C	P (%)	C	P (%)	C	P (%)	C	P (%)	C	P (%)	C	P (%)
A1	9.36	5.578	11.44	6.818	82.6	49.225	11	6.555	53.4	31.824	68.176	167.800
A3	2	1.409	2.24	1.578	114	80.457	11	7.750	12.5	8.807	91.193	141.940
A5	4	2.430	3.44	2.089	130	78.839	9.4	5.709	18	10.933	89.067	164.640
A7	14.24	8.272	3.6	2.091	117	68.084	13.2	7.668	23.9	13.884	86.116	172.140
A9	2.16	1.460	1.92	1.297	99.2	67.036	17.2	11.623	27.5	18.584	81.416	147.980
A11	2.08	2.651	1.28	1.631	51.8	66.021	8.8	11.216	14.5	18.481	81.519	78.460
A14	1.76	1.435	1.68	1.370	80	65.232	10.2	8.317	29	23.646	76.354	122.640
C1	4.08	3.271	3.76	3.014	81.8	65.576	9.6	7.696	25.5	20.443	79.557	124.740
C5	3.53	4.361	3.92	4.843	52.6	64.986	8.4	10.378	12.5	15.444	84.556	80.940
C11	3.12	3.340	5.6	5.994	50.8	54.378	11.4	12.203	22.5	24.085	75.915	93.420
Average	4.633	3.421	3.888	3.073	86	65.983	11	8.912	23.9	18.613	81.387	129.470

Note Exch meant exchangeable species, and Carbon meant carbonate, Fe–Mn oxide meant Fe–Mn oxides, Organ meant organic matter, Residual meant residual, Efficient of all meant the sum of efficient species of 4 species ahead; C meant concentration, P meant the percent in total concentration

Table 2.10 Characteristics of Sn speciation in the soil from the Hetao Area (unit: mg/kg)

	Exch		Carbon		Fe–Mn Oxide		Organ		Residual		Efficient of all	
Sam. No.	C	P (%)	C	P (%)	C	P (%)	C	P (%)	C	P (%)	C	P (%)
A1	0.015	0.016	0.015	0.016	66	68.750	30	31.250	0.02	0.016	99.984	96.000
A3	0.015	0.013	0.015	0.013	72	60.000	26	21.667	22	18.333	81.667	120.000
A5	0.015	0.011	6.72	4.709	72	50.448	24	16.816	40	28.027	71.973	142.720
A7	0.015	0.011	3.36	2.519	102	76.485	26	19.496	2	1.500	98.500	133.360
A9	0.015	0.012	1.68	1.326	84	66.309	30	23.682	11	8.683	91.317	126.680
A11	0.015	0.011	0.015	0.011	96	71.642	26	19.403	12	8.955	91.045	134.000
A14	0.015	0.012	1.68	1.362	90	72.945	24	19.452	7.7	6.241	93.759	123.380
C1	0.88	0.616	0.015	0.010	102	71.389	22	15.398	18	12.598	87.402	142.880
C5	0.88	0.644	2.48	1.815	108	79.028	22	16.098	3.3	2.415	97.585	136.660
C11	0.015	0.012	0.015	0.012	108	83.077	22	16.923	0.02	0.012	99.988	130.000
Average	0.188	0.136	1.6	1.179	90	70.007	25.2	20.018	11.6	8.678	91.322	128.568

Note Exch meant exchangeable species, and Carbon meant carbonate, Fe–Mn oxide meant Fe–Mn oxides, Organ meant organic matter, Residual meant residual, Efficient of all meant the sum of efficient species of 4 species ahead; C meant concentration, P meant the percent in total concentration

Table 2.11 Characteristics of Pb speciation in the soil from the Hetao Area (unit: mg/kg)

	Exch		Carbon		Fe–Mn Oxide		Organ		Residual		Efficient of all	
Sam. No.	C	P (%)	C	P (%)	C	P (%)	C	P (%)	C	P (%)	C	P (%)
A1	0.025	0.142	0.025	0.142	3.6	20.455	2	11.364	12	68.182	31.818	17.600
A3	0.025	0.133	0.025	0.133	4.8	25.532	2	10.638	12	63.830	36.170	18.800
A5	0.025	0.137	0.025	0.137	3.6	19.780	2.6	14.286	12	65.934	34.066	18.200
A7	0.025	0.095	0.025	0.095	3.6	13.740	2.6	9.924	20	76.336	23.664	26.200
A9	0.025	0.164	0.025	0.164	0.6	3.947	2.6	17.105	12	78.947	21.053	15.200
A11	0.16	1.130	0.025	0.177	1.2	8.475	0.8	5.650	12	84.746	15.254	14.160
A14	0.025	0.174	0.025	0.174	1.2	8.333	1.2	8.333	12	83.333	16.667	14.400
C1	0.025	0.174	0.025	0.174	1.2	8.333	1.2	8.333	12	83.333	16.667	14.400
C5	0.025	0.164	0.025	0.164	1.8	11.842	1.4	9.211	12	78.947	21.053	15.200
C11	0.025	0.167	0.025	0.167	1.8	12.000	1.2	8.000	12	80.000	20.000	15.000
Average	0.039	0.248	0.025	0.153	2.34	13.244	1.76	10.284	12.8	76.359	23.641	16.916

Note Exch meant exchangeable species, and Carbon meant carbonate, Fe–Mn oxide meant Fe–Mn oxides, Organ meant organic matter, Residual meant residual, Efficient of all meant the sum of efficient species of 4 species ahead; C meant concentration, P meant the percent in total concentration

Table 2.12 Ratio of exchangeable speciation in the total concentration of heavy metals in the soil from the Hetao Area (%)

Sam. No.	Cu	Zn	Mo	As	Cd	Sn	Pb
A1	0.4383	0.0040	2.1739	0.6995	5.5781	0.0156	0.1420
A3	0.1262	0.0039	3.2258	0.4528	1.4090	0.0125	0.1330
A5	0.2701	0.1716	1.3514	0.8790	2.4295	0.0105	0.1374
A7	1.2838	0.0034	2.0101	0.5561	8.2723	0.0112	0.0954
A9	1.7072	0.0040	0.0962	0.8059	1.4597	0.0118	0.1645
A11	2.9126	1.8297	0.1316	1.3827	2.6510	0.0112	1.1299
A14	1.9608	0.3250	0.1250	0.3014	1.4351	0.0122	0.1736
C1	1.2504	0.0048	1.3605	1.0382	3.2708	0.6159	0.1736
C5	2.0309	0.2541	0.1471	1.1323	4.3613	0.6439	0.1645
C11	1.8913	0.2616	0.1282	1.2998	3.3398	0.0115	0.1667
Average	1.3871	0.2862	1.0750	0.8548	3.4207	0.1356	0.2481

Table 2.13 Ratio of carbonate in the total concentration of heavy metals in the soil from the Hetao Area (%)

Sam. No.	Cu	Zn	Mo	As	Cd	Sn	Pb
A1	0.0050	0.0040	0.14	0.68	6.82	0.02	0.14
A3	0.0053	0.0039	0.10	0.58	1.58	0.01	0.13
A5	0.0052	0.0039	0.14	1.01	2.09	4.71	0.14
A7	0.0049	0.0034	0.13	0.96	2.09	2.52	0.10
A9	0.0065	0.0040	0.10	1.27	1.30	1.33	0.16
A11	0.0110	0.0070	0.13	1.90	1.63	0.01	0.18
A14	0.0111	0.0054	0.13	0.39	1.37	1.36	0.17
C1	0.0071	0.0048	0.09	1.71	3.01	0.01	0.17
C5	0.0115	0.0064	0.15	1.57	4.84	1.81	0.16
C11	0.0107	0.0065	0.13	1.84	5.99	0.01	0.17
Average	0.0078	0.0049	0.12	1.19	3.07	1.18	0.15

speciation of Sn) to the highest of 86.22% (Sample No. 3 of work-line A, the concentration of residual speciation of As). Cu, Zn, As, and Pb were the elements with relatively high residual speciation concentration, with average ratio of 76.15–79.45% and variation range of 64.11–86.22%. According to the characteristics of speciation and distribution of the trace metals in the researched area, the changes of the relative concentration in different samples generally showed the consistent regularity, as shown in Tables 2.12, 2.13, 2.14, 2.15, 2.16, and Fig. 2.5. In addition, the concentration in exchangeable showed better correlation with the concentration in the carbonate, as shown in Fig. 2.6.

From these features as described above, the facts revealed were as follows:

Table 2.14 Ratio of Fe–Mn oxide speciation in the total concentration of heavy metals in the soil from the Hetao Area (%)

Sam. No.	Cu	Zn	Mo	As	Cd	Sn	Pb
A1	9.41	9.56	27.17	5.98	49.23	68.75	20.45
A3	4.21	9.27	28.23	3.28	80.46	60.00	25.53
A5	3.64	8.77	37.84	4.71	78.84	50.45	19.78
A7	2.43	7.06	50.25	3.62	68.08	76.48	13.74
A9	1.29	9.70	38.46	5.53	67.04	66.31	3.95
A11	2.21	16.89	36.84	7.59	66.02	71.64	8.47
A14	4.46	11.38	50.00	1.51	65.23	72.95	8.33
C1	5.68	10.77	34.01	6.50	65.58	71.39	8.33
C5	1.04	14.29	58.82	8.07	64.99	79.03	11.84
C11	0.32	14.72	51.28	9.36	54.38	83.08	12.00
Average	3.47	11.24	41.29	5.62	65.98	70.01	13.24

Table 2.15 Ratios of organic matter in the total concentration of heavy metals in the soil from the Hetao Area (%)

Sam. No.	Cu	Zn	Mo	As	Cd	Sn	Pb
A1	20.42	10.76	27.17	19.96	6.56	31.25	11.36
A3	11.56	6.76	20.16	9.46	7.75	21.67	10.64
A5	10.39	9.16	21.62	12.74	5.71	16.82	14.29
A7	13.62	8.07	25.13	16.00	7.67	19.50	9.92
A9	19.40	19.60	19.23	28.29	11.62	23.68	17.11
A11	17.65	10.91	21.05	18.50	11.22	19.40	5.65
A14	15.60	12.46	15.00	4.45	8.32	19.45	8.33
C1	14.92	11.48	13.61	15.68	7.70	15.40	8.33
C5	16.16	12.39	17.65	16.38	10.38	16.10	9.21
C11	17.19	13.08	15.38	20.42	12.20	16.92	8.00
Average	15.69	11.47	19.60	16.19	8.91	20.02	10.28

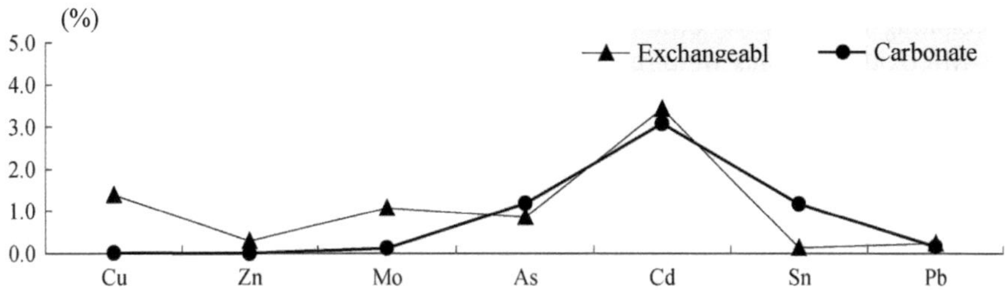

Fig. 2.6 Correlation of the exchangeable and carbonate speciation concentration of the heavy metals in the soil from the Hetao Area. The numerals on the left of vertical axis meant the speciation concentration percent in the total concentration of the element. They were from the analysis for ten samples from the research area

Table 2.16 Ratio of residue in the total concentration of heavy metals in the soil from the Hetao Area (%)

Sam. No.	Cu	Zn	Mo	As	Cd	Sn	Pb
A1	69.73	79.68	43.48	72.68	31.82	0.02	68.18
A3	84.10	83.98	48.39	86.22	8.81	18.33	63.83
A5	85.71	81.89	40.54	80.66	10.93	28.03	65.93
A7	82.67	84.87	22.61	78.87	13.88	1.50	76.34
A9	77.60	70.71	42.31	64.11	18.58	8.68	78.95
A11	77.23	70.37	42.11	70.62	18.48	8.96	84.75
A14	77.99	75.84	35.00	93.33	23.65	6.24	83.33
C1	78.15	77.75	51.02	75.06	20.44	12.60	83.33
C5	80.78	73.06	23.53	72.85	15.44	2.41	78.95
C11	80.59	71.94	33.33	67.09	24.08	0.01	80.00
Average	79.45	77.01	38.23	76.15	18.61	8.68	76.36

A. The forming process of soil and the geochemistry evolution after the soil forming in Hetao Area were roughly the same in all the area. The homogeneous parent material may be thoroughly homogenization in soil-forming process, and the same geochemical process was due to the soil homogenization in all the area.
B. The Hetao Area seems generally in a stable geochemical field, and the conditions of soil geochemical cycle in the area are approximately same in the area. The dominant factor that restricts the exchange and cycling of elements of this area was the geochemical background of the area.

The average concentrations of chemical speciation of trace metals in Hetao Area are shown in Table 2.17. These concentrations can generally represent the total speciation characteristics of trace metals in the soil of the Hetao Area. It demonstrated that the concentrations of Mo, Sn, and Cd were mainly concentrated in the Fe–Mn oxide fractions, accounting for 41–70% of the total concentration, whereas the concentrations in the residual states were relatively lower, about only 9–38%. The concentrations of Cu, Zn, As, and Pb were mainly concentrated in residual fractions, accounting for about 76–80% of the total concentration, and concentrations in the Fe–Mn oxide fractions were relatively lower, with the maximum lower than 13%. The concentrations (around 10%) in organic matter seemed stable, varying from 10 to 16%. It was shown in Fig. 2.7.

2. **The distribution of the efficient fraction of the speciation of the trace metals**

The efficient fractions of trace metals means the sum of the four fractions except for the residual fraction in the five fractions of the Tessier schedule for the speciation analysis, and it is an important part for the bioavailability of the metals under natural conditions in the environment. Therefore, on the current understanding on the issue, the efficient fraction is generally considered as an environmental active part in the environment. The concentration of trace metals in the efficient fraction in the research

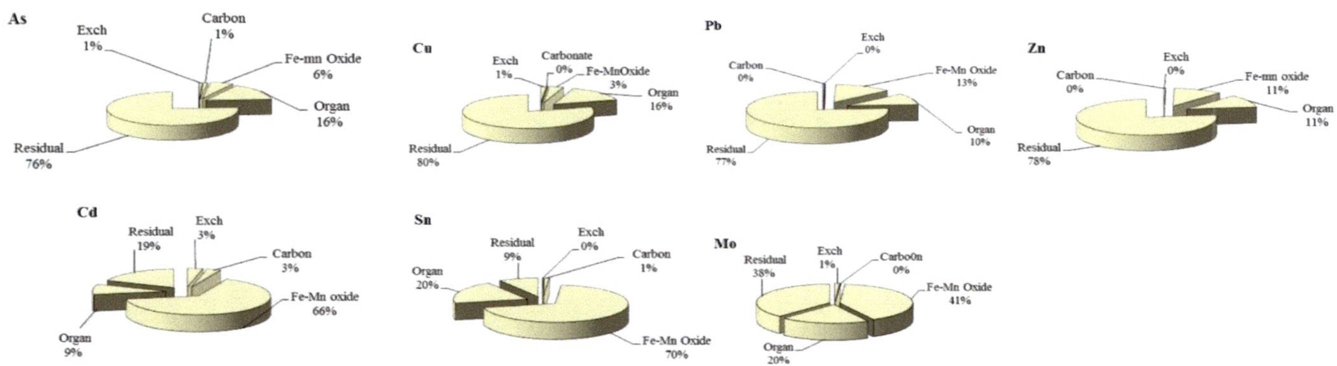

Fig. 2.7 Average concentration of the speciation of the heavy metal in the soil from the Hetao Area. Exch meant exchangeable species, Carbon meant carbonate, Fe–Mn oxide meant Fe–Mn oxides, Organ meant organic matter, and Residual meant residual. The percentages in the chart were speciation concentration, respectively, in their total concentration of the element, and they were from analysis for 10 samples from whole research area

Table 2.17 Average ratio in total concentration for speciation of heavy metals in the soil from the Hetao Area (%)

Element	Exch	Carbon	Fe–Mn Oxide	Organ	Residual
Cu	1.39	0.008	3.47	15.69	79.45
Zn	0.29	0.005	11.24	11.47	77.01
Mo	1.07	0.120	41.29	19.60	38.23
As	0.85	1.190	5.62	16.19	76.15
Cd	3.42	3.070	65.98	8.91	18.61
Sn	0.14	1.180	70.01	20.02	8.68
Pb	0.25	0.150	13.24	10.28	76.36

Note Exch meant exchangeable species, and Carbon meant carbonate, Fe–Mn oxide meant Fe–Mn oxides, Organ meant organic matter, Residual meant residual

Table 2.18 Ratio of efficient speciation in the total concentration of heavy metals in the soil from the Hetao Area (%)

Sam. No.	Cu	Zn	Mo	As	Cd	Sn	Pb
A1	30.27	20.32	56.52	27.32	68.18	99.98	31.82
A3	15.90	16.02	51.61	13.78	91.19	81.67	36.17
A5	14.29	18.11	59.46	19.34	89.07	71.97	34.07
A7	17.33	15.13	77.39	21.13	86.12	98.50	23.66
A9	22.40	29.29	57.69	35.89	81.42	91.32	21.05
A11	22.77	29.63	57.89	29.38	81.52	91.04	15.25
A14	22.01	24.16	65.00	6.67	76.35	93.76	16.67
C1	21.85	22.25	48.98	24.94	79.56	87.40	16.67
C5	19.22	26.94	76.47	27.15	84.56	97.59	21.05
C11	19.41	28.06	66.67	32.91	75.92	99.99	20.00
Average	20.55	22.99	61.77	23.85	81.39	91.32	23.64

Note Exch meant exchangeable species, and Carbon meant carbonate, Fe–Mn oxide meant Fe–Mn oxides, Organ meant organic matter, Residual meant residual

area was shown in Table 2.18. It was indicated that the concentrations of different elements in efficient fractions had relatively wide variation range. The highest concentrations were Sn, Cd, and Mo, with the average concentration of 61.77–91.32% and variation range of 48.98–99.99%. However, the concentrations of Cu, Zn, As, and Pb in efficient fraction were about 20% of the total concentrations (<24%). The spatial distribution characteristics of concentrations in efficient fraction are shown in Fig. 2.8.

These results indicated that the concentrations of Cu, As, and Pb in the efficient fraction appeared a slowly decreasing trend along the flow direction of the water (145° azimuth). Other elements had no obvious variation tendency at the direction of flow. The concentration in the efficient fraction is the mark for the active part that may

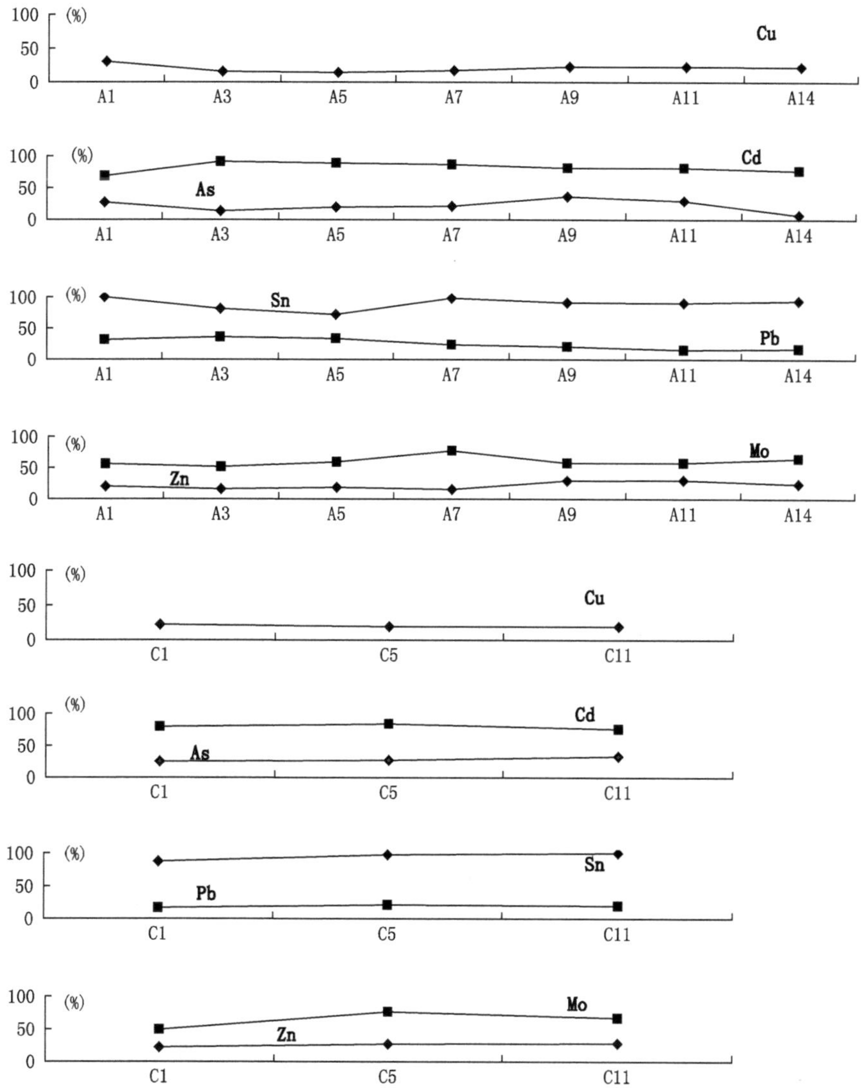

Fig. 2.8 Distribution of the concentration of efficient speciation of the heavy metals in soil from the Hetao Area. The numerals on the left of vertical axes of charts meant the percentage of the efficient speciation in the total content of the heavy metal element, and the symbols below horizontal axes meant the sample number and the working line. The arrowhead below the charts showed the direction of the working line

move with water under natural conditions. The spatial distribution of concentrations in the efficient fraction of the metals above indicated that the concentrations of Cu, As, and Pb in soil of the Hetao Area had associated with the migration of elements from upstream to downstream of the groundwater of the Hetao Area. This is because those parts of the elements can transform along with the water, especially in the form of solute, which should be in the efficient fraction. This is an important indicator, which means indicating significance for the trace metal pollution in the research area.

2.3 The Distribution of Trace Metals in the Groundwater

Hetao Area was identified as the arsenic poisoning area by the Department of Endemic Disease Prevention and Control, Ministry of Health, China, as early as 1992. At that time, it was mainly according to the skin keratosis, coarse hair, and other symptoms of arsenic poisoning disease in the local residents, and the concentration of arsenic in drinking water (well water) was generally exceeded 0.05 mg/L recommended by World Health Organization in 1984. The standard arsenic concentration of drinking water was below 0.05 mg/L for China in 1985, and it was changed to 0.01 mg/L in July 2007. According to the survey data of the drinking water of 9631 wells in Hetao Area by the Health Department of Inner Mongolia Autonomous, 24% of the investigated wells contained excessive levels of arsenic in drinking water [13], and the concentration of arsenic in drinking water showed a positive correlation with the incidence rate in the local residents.

2.3.1 The Distribution of Arsenic

This work mainly investigated the concentration, valence, and distribution of trace metal As, etc., in the drinking water of the study area, and the depth of sampling wells was 18–25 m. The elements were analyzed for samples after field site acidification. The analysis elements included As, Cu, Zn, Cd, Sb, and Pb; Cu, Zn, Sb, and Pb were measured by inductively coupled plasma atomic spectrometry (TJA-1100 ICP-AES, USA), the detection limits (mg/L) are 2×10^{-3} (Cu), 4×10^{-3} (Zn), 5×10^{-2} (Sb), and 2.5×10^{-2} (Pb). As and Cd were analyzed by AF-610 atomic fluorescence spectrometer (Beijing Rayleigh Analytical Instrument Co. Ltd., China), the detection limits (mg/L) were 8×10^{-5} (As) and 8×10^{-5} (Cd).

The average concentration of arsenic in the well water (97 samples) was 0.201 mg/L, and the highest was 0.97 mg/L. The heavy disease district had an obvious positive correlation with the arsenic concentration in the well water, as shown in Table 2.19. The distribution of main arsenic poisoning district in Hetao Area is shown in Fig. 2.9.

These results show that the districts with high concentration of arsenic in groundwater visible in the downstream of the large deposits at the foot of the Yin Mountains,

Table 2.19 Concentration of arsenic in drinking well water (phreatic water) in the Hetao Area (unit: μg/ml)

	Total As concentration		The ratio of As^{3+} in the total As (%)	
	Average	Range	Average	Range
Heavy poisoning area* (30 samples)	0.483	0.312–0.969	27	9–90
Secondary poisoning area (7 samples)	0.248	0.218–0.284	22	6–62
Light poisoning area (28 samples)	0.107	0.006–0.193	17	4.6–93
No poisoning area (32 samples)	0.008	0.001–0.046	12	10–35

*Heavy poisoning area means where the residents are poisoned heavy with the morbidity > 30% from drinking well water containing As > 0.5 μg/ml, and Secondary poisoning area means where the morbidity is < 30% but > 10% with the drinking well water containing As > 0.2 μg/ml, Light poisoning area means where the morbidity is < 10% with the drinking well water containing As > 0.05 μg/m, No poisoning area means where the residents are no poisoned with the drinking well water containing As ≤ 0.2 μg/ml [68]

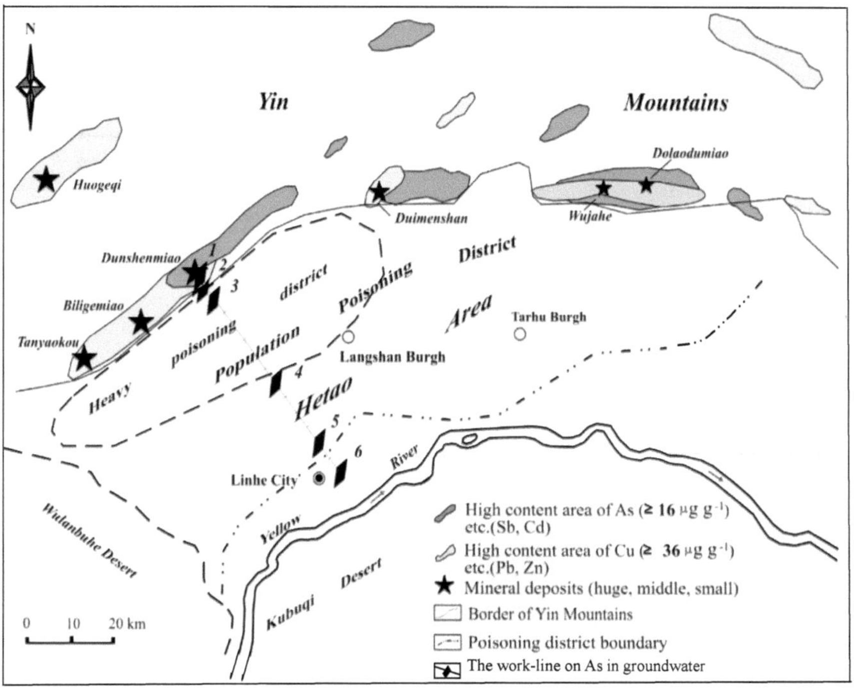

Fig. 2.9 Work-line for As in groundwater and the space relationship between the arseniasis and the background level of As, Cu, Pb, Zn, Cd, and Sb in medias of the Hetao Area

Table 2.20 Concentration of heavy metals in the groundwater from the Hetao Area (unit: μ g/ml)

Sample number on the work-line	The distance from deposit (km)	The sampling depth	Heavy metal elements					
			As	Sb	Cd	Cu	Pb	Zn
1	0	0	0.251	0.0064	0.005	0.053	0.047	0.069
2	0.5	21	0.236	0.0052	0.004	0.021	0.026	0.049
3	11.0	18	0.230	0.0050	0.004	0.030	0.022	0.053
4	29.6	15	0.004	0.003	0.002	0.015	0.007	0.038
5	44.0	15	0.005	0.003	0.002	0.012	0.006	0.042
6	The Yellow River water	0	0.003	0.002	0.002	0.014	0.007	0.037

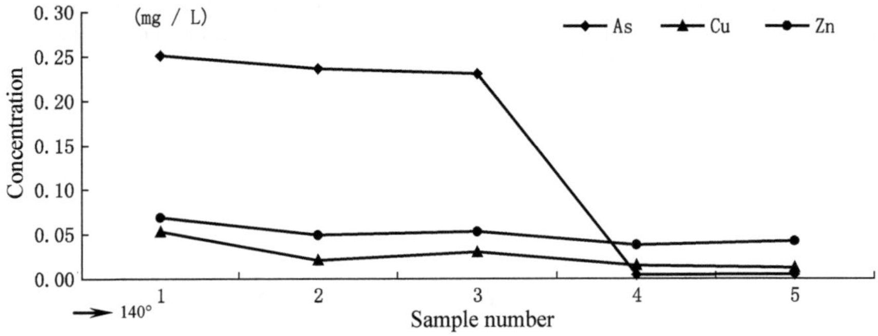

Fig. 2.10 Distribution of the concentrations of Cu, Zn, and As in the phreatic water from the Hetao Area

and people arsenic poisoning areas are obviously located in the downstream area near to the deposit area, especially concentrated on the downstream of large deposits (Dongshengmiao Pb, Zn, and S deposit), large marble deposits named Biligaimiao, and large Cu, Zn, and S deposits named Tanyaokou, which all of them have long mining history (>80 years). In the eastern part of the area, As was with relatively lower concentration in groundwater (<0.5 mg/L). Several small- and medium-sized polymetallic deposits (Cu and Pb) were located at the foothill of the mountains corresponding to the cases mentioned above with lower As in well water, which have not been exploited yet.

2.3.2 *The Distribution of Cu, Zn, Cd, Sb, and Pb*

The concentrations of trace metals (Cu, Zn, Cd, Sb, and Pb) in the groundwater (the drinking water) of the Hetao Area are shown in Table 2.20. The distribution of the concentration is shown in Figs. 2.9, 2.10, and 2.11, respectively.

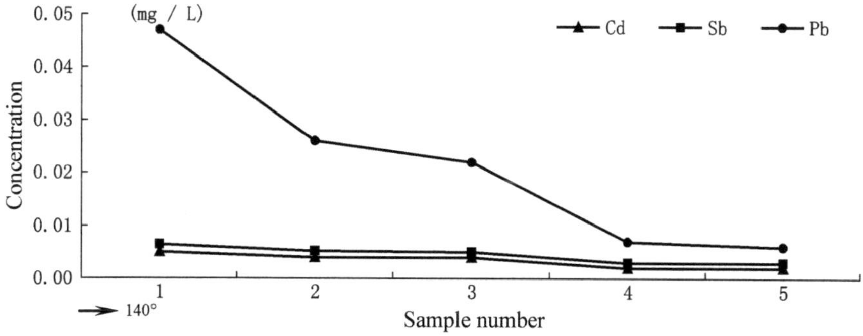

Fig. 2.11 Distribution of the concentrations of Cd, Sb, and Pb in the phreatic water from the Hetao Area

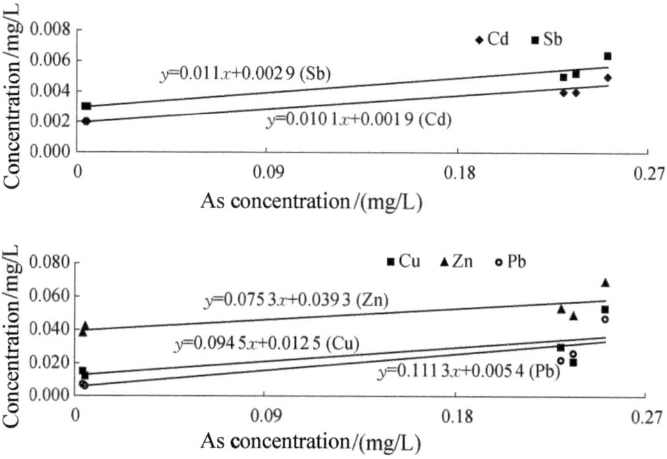

Fig. 2.12 Correlation between the concentrations of As and Cu, Cd, Sb, Zn, and Pb in the phreatic water in the Hetao Area

These results indicated that the concentrations of Cu, Zn, Cd, Sb, and Pb in Hetao Area could meet the standards of National Standard for Drinking Water of China. Only few samples (one sample) for few elements approached (Pb) or reached (Cd) this critical value. The concentration of these elements (Cu, Zn, Cd, Sb, and Pb) gradually decreased from the foothill of the Yin Mountains to the downstream, which were significantly correlated with the concentration of As, as shown in Fig. 2.12.

According to the distribution and variation features of trace metals in the groundwater of the Hetao Area, the results can be interpreted below:

A. With the exception of As, the other elements were satisfied to the national standard of drinking water of China.
B. The concentrations of As, Cu, Pb, Zn, Sb, and Cd in the groundwater of Hetao Area decreased from the deposits at the front of Yin Mountains to the down-

stream, and there was obvious correlationship between the concentrations of the elements.

C. In the Hetao Area, As in the drinking water was highly concentrated mainly in the downstream of the large deposits, which were with longer exploitation time.

2.4 The Distribution of Trace Metals in the Resident Hair of the Area

2.4.1 The Progress of the Research on Trace Metals in Human Hair as an Environmental Indicator

The most obvious environmental problem in this study area is the arsenic poisoning due to the high concentration of arsenic (As) in groundwater, known as the high-arsenic water environmental problem. The concentration of As in human hair as a potential indicator of arsenic exposure has received more attention and successfully applied in the research on the arsenic poisoning in related regions in recent years [14–17]. It was based on the fact that As can be delivered and accumulated in human hair through blood. The concentration of As in resident hair in the Hetao Area decreased along the direction of flow with the increasing distance from the foothill of the Yin Mountains with high concentration of As, which was in accordance with the variation rules of As in water and soil. It indicated that As in groundwater of Hetao Area was related to the long-term migration of As from the groundwater upstream. Also, the relationship showed that the concentration of As in human hair was inherently related with the environmental medium of the long-term life of people. As in the upstream of the groundwater was transported to the downstream with water, and As accumulated in the hair of local residents due to the long-term drinking water with the high concentration of As. Therefore, the concentration of As in the hair of people was gradually reduced away along the distance from the As source. Although the indication and significance of the concentration of As in human body to the arsenic poisoning needs to be further confirmed and the mechanism also needs to be further researched, the fact in Hetao Area is a very meaningful phenomenon.

The research on the biological indicator of hair reported as early as late twentieth century has been further carried out in recent decades and effectively used to solve practical problems. A unique attribute of the hair is that it can combine the elements absorbed during growing period, which makes it possible to indicate the history of contaminant exposure and the evolution of the biological growth process as a biomarker [15]. Thus, hair has been applied in the research on the measurement and evaluation of the consumption, metabolism, and pollutant exposure during, respectively, its occurrence period for organic and inorganic substances that come from the environmental medium [18–20]. The problem of trace element concentration in the hair of Napoleon, an famous suspense, was an typical instance to investigate the As concentration in the hair by using synchronous induced X-ray fluorescence method, and in this study, the related elements were found in 1–2 mm long hair [21]. So, some people think that the hair as a biological indicator can also be applied in the indication for Cu and Zn, even for other pollutants [21, 22].

Based on the new analysis and experiment technologies, some researches on the chemical structure of trace metals in the hair, the morphology of As, and the distinction of related compounds have made great progress in recent years [16, 23]. On the growth rate of human hair and the distribution characteristics of pollutants in different parts of hair, the concentration of pollutants in the hair was investigated to describe the change of pollutants in the environmental medium during the occurrence period [24, 25]. Results indicated that hair could precisely reflect the concentration change of As in the environmental medium. Based on synchronous induced X-ray fluorescence (SXRF) technology, the research on the distribution of trace elements in hair can be regarded as the significant monitoring indicator for the biological environment [15, 26, 27].

2.4.2 *The Distribution of Trace Metals in the Resident Hair of the Hetao Area*

The zone of resident arsenic poisoning area had positive correlation with the concentrations of trace metals (total concentration and efficient speciation concentration) such as Cu, Zn, As, Cd, Sb, and Pb as the concentration change in soil and groundwater in the Hetao Area. In contrast, Bi, Mo, and Sn which were in normal concentration in the flow upstream of the Hetao Area did not appear obvious change in all the Hetao Area where is only with natural fluctuation, as shown in Fig. 2.3. High concentration area of arsenic in the groundwater is also the heavy poisoning zone concentrated in the downstream of the large deposits at the front of Yin Mountains. Moreover, corresponding to the different geochemical background and mining history in the upstream of the area, there was difference in the concentrations of related elements in soil and groundwater. In the west of the research area near the Tanyaokou deposit of Cu, Zn, and S which has been in mining for more than 80 years, and Dongshengmiao deposit of Pb, Zn, and S for more than 70 years. The Duimenshan deposit of Cu, Zn, and S and other eastern smaller deposits, at central of the front Yin Mountains, which have relatively short mining history correspondingly with high concentration of As in groundwater too as shown in Figs. 2.1 and 2.9. The slopes of the concentration variation at the work-lines are obviously more intense at the downstream of Tanyaokou deposit, which was with longer mining history than that of with the short mining history deposit, Duimenshan deposit. It was distinctly shown that the slope of concentration change trend of trace metals in soil at the work-lines in the flow downstream of deposits was obviously intense, and the slope of BB' profile was smaller than that at the other two work-lines (Figs. 2.1 and 2.2). Correspondingly, the As concentration in the well water of resident poisoning area was significantly higher than that of the non-poisoning area, and As in well water of the serious poi-

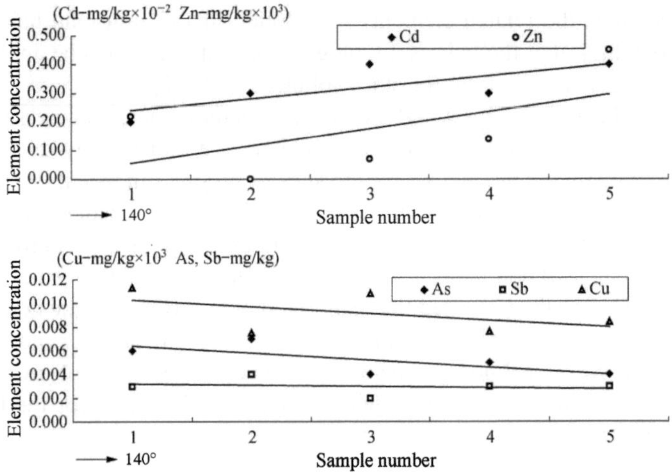

Fig. 2.13 Variation of the concentrations of Cu, As, Cd, Sb, and Pb in the hair from local inhabitants in the Hetao Area. The arrowheads are the direction of the sampling transacted line

soning area was significantly higher than that of the slight poisoning area relatively (Figs. 2.1 and 2.9).

In this study, it was carried out to investigate the hair of adult residents (35–60 years old) according to the work-lines of samples from soil and groundwater. The hair samples were obtained by hair cutting directly from the residents, then were analyzed after washing with deionized water, drying naturally, low-temperature ashing (<100 °C, LTA-154 low-temperature plasma ashing apparatus), and dissolution with acid solution. TJA-1100 ICP-AES (USA) was used to analyze the concentrations of Cu, Zn, and Sb in the hair, and the detection limits were 2×10^{-3} (Cu), 4×10^{-3} (Zn), and 5×10^{-2} (Sb). The concentrations of As and Cd were analyzed by AF-610 atomic fluorescence spectrometer (Beijing Rayleigh Analytical Instrument Co., Ltd., China), and the detection limits were 8×10^{-5} (As) and 8×10^{-5} (Cd). The analyzed work for all elements were evaluated by qualification rate (the relative deviation <5% or >95%), and the qualification rate was 100%.

Results show that concentrations of all element except for Cd in the hair of local residents gradually decrease from the flow upstream to the downstream, Fig. 2.13. In addition, they have positive relationship with the concentrations of As and Cu in the groundwater and soil, as shown in Figs. 2.14 and 2.15.

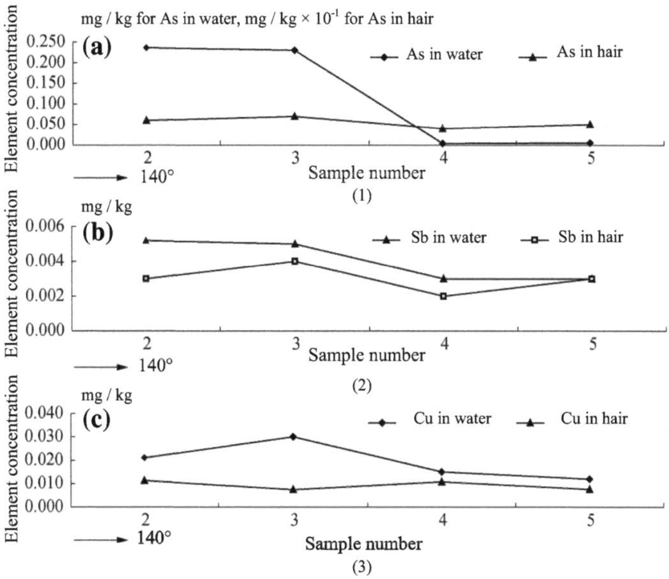

Fig. 2.14 Relationship between the concentrations of As (**a**), Sb (**b**), and Cu (**c**) in local resident hair and in the phreatic water from the Hetao Area. The numerals below *X* axes were the number of sampling site at the transacted line that was 44 km long, and the arrowheads below the chart showed the direction of the transacted line

2.4.3 The Relationship Between the Trace Metals in the Resident Hair and that in the Water and Soil of the Hetao Area

Based on above results, the variation of trace metals in the medium of the Hetao Area indicates the following facts: ① the high concentration of relevant elements in soil and groundwater of poisoning area has the relationship with the enrichment zone (deposits) of Cu, Pb, Zn, As, Sb, and Cd, and high background zone in the flow upstream of the research area (north side of the Hetao Area); ② the metal concentration in the soil from the foothill to the plain along the flow appears the slowly declining trend; ③ the zones with high concentration of arsenic in the groundwater are mainly the areas of the foothill, which are in agreement with the disease area, especially the serious disease areas. The results show also that the high concentration of trace metals in the environmental medium and the resident poisoning phenomenon seem related to the migration of the elements in the water from the flow upstream. Some studies have shown that the ore elements in the minerals and rocks can be released into the surface water and subsequently migrate and accumulate in soil and crops by the natural force under certain conditions (such as natural weathering, artificial mining, smelting, and processing), especially under the impact of human activities, and the corresponding influences to the environment should not be ignored

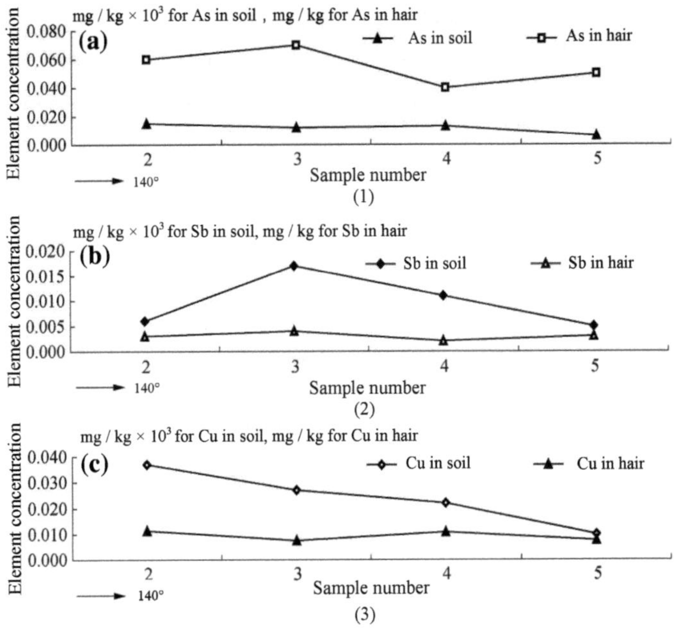

Fig. 2.15 Relationship between the concentrations of As (**a**), Sb (**b**), and Cu (**c**) in local resident hair and in the soil from the Hetao Area. The numerals below *X* axes were the number of sampling site at the transacted line that was 44 km long, and the arrowheads in the chart showed the direction of the transacted line

[28–31]. Accordingly, the elements fixed in soil can also transfer into the water and then into the crops or groundwater under certain conditions. The concentration of trace metals in the soil and groundwater is the action results of this cyclical migration [32]. The exploration and mining activities at the front of the Yin Mountains in the flow upstream of the area played an important role in the migration of the harmful elements. The mineral and rocks buried in the ground, which are rich in Cu, Pb, Zn, As, Sb, Cd, and other elements, are exposed to the surface due to the human activities, and dissolution of sulfide minerals and other acidic compounds in an oxidizing environment results in reducing the pH value of the environmental medium, leading to enhance the migration ability of metal elements with the actions of surface water and rainfall. Also, metal elements would migrate along with the direction of the flow in surface water and groundwater, even with long-distance migration [28, 33–36]. Hetao Area is the typical arid and semiarid ecological landscape, and the local water table is relative higher due to the impact of Yellow River. Under arid and semiarid natural conditions, the concentration of the elements in groundwater and soil in the downstream of the flow would be increasing due to the perennial groundwater circulation and surface evaporation, leading to accumulate in the human body through crops, food, and drinking water to harm the health of people. It is an ecological phenomenon due to the process of the development and evolution in the direction of substance homogenization, which is caused by the changes in the conditions of the distribution of elements in the geochemical cycle (such as the underground substance being exposed to the surface, and the chemical speciation of elements in surface substance being changed by the natural weathering, etc.) [37–40]; particularly, the human activities in the area (mainly mining) are the important factor to promote the process of migration and accumulation of some elements.

The distribution feature of trace metal concentration in soil and groundwater of Hetao Area indicates that:

A. The concentration of As, Sb, Cu, Pb, Zn, and Cd in the soil of Hetao Area is far greater than the background value of the region, and the concentration of Cu, Zn, As, Cd, Hg, Co, Ni, Cr, and V has reached the level of weak pollution and the concentration of Sb has reached the level of medium pollution, measured by the standard of geo-accumulation indexes recommended by Muller [3].
B. The As concentration in the groundwater, especially in the groundwater of poisoning areas, is much higher than that of the drinking water standard (0.05 mg/L, WHO1984; 0.01 mg/L, China GB5749-2006), and for some samples, the concentration of Cd had reached the critical value of drinking water (0.005 mg/L, China ZGB5749, 2006), and the concentration of Pb has exceeded the critical value of drinking water (0.01 mg/L, China GB5749-2006).
C. The concentration of trace metals in soil and groundwater of the Hetao Area has a positive relationship with the pathological proportion of the local residents, with the trend of slow decline along the flow direction from the front of Yin Mountain, northern side of the Hetao Area.
D. The distribution variation of Cu, As, and other trace metal elements in soil and groundwater has a positive correlation with the variation of As in the hair of local residents.
E. The elements with high background concentration (Cu, Zn, As, Cd, Sb, Pb) in the flow upstream of the Hetao Area have migrated to the soil and groundwater of the flow downstream, and the concentration of As, Cd, and Sb in the soil of the area of the flow downstream is three times of the background value; the highest concentration of As in the groundwater (0.97 mg/L) is near to 100 times of the China national drinking water standard, while the concentrations of Cu, Zn, and Pb concentrations in soil are less than 2 times of the background values (Table 2.1).
F. The deposits mainly are at the polymetallic ore (Cu, Zn, and Pb)-rich area at the area of the flow upstream of the Hetao Area, and the elements (metals) migrated to the flow downstream do not the elements that most rich in the media of the deposit's area. This suggests that the phenomenon might be induced by the different migration natures between different elements under the conditions of the area. These differences may be a theoretical issue in the study on the elemental geochemistry.

2.5 Study on the Isotope Tracing for Trace Metals in the Soil and Groundwater

The study on the isotope tracing of the environment pollutants in the field of environmental science and technology is one of the advanced methods arisen in the geochemical research in recent years. In the isotope tracing research, it is based on the principle of the isotope geochemistry method and premised on the intrinsic relationship of the isotopic composition and their ratios between the source substance and the fractionation substance to reveal the intrinsic relationship between target substances and source substances by analyzing changes of the isotope fractions of the relevant element. It is the expansion of the application of isotope geochemistry on the investigation objects, and it is a new growing direction in the research of environment pollution, especially in the research on environment pollution with a certain time and space span. As an effective way, it has become a powerful tool in the probing for the origin of environment pollutant. In recent years (1997–2010), the research on waste burying field research by Siegel et al. [41] the groundwater pollution of the landfill was studied in the slaughterhouse (Staten, New York). Deboudt K. investigated the aerosol pollution in the Strait of Dover (UK) by lead isotope ratios [42]. Zhu B. et al. investigated the lead isotope composition in aeolian dust, aerosol, and soil in the Pearl River Delta (Guangdong, China) in order to reveal the source of Pb in the environment [43]. The quantitative estimation of lead source was carried out for the dust of domestic bedroom in Jersy City, New Jersy, USA, by John L. et al. using the lead isotope ratios [44]. In recent years, there are many typical researches on the isotope tracing technique such as the research of the evolution process of the groundwater composition, the groundwater transport channel, the element migration in the geological history, the bio-geochemical tracer studies on the trace metal pollution through tree-ring recording method, and the material mixing and fluid source tracing studies in the geological processes [45–53]. Nevertheless, the isotope tracing of the pollutants in environmental media is still in the stage of development. There are many problems need to be further explored although some researches were carried out some work using this method. For example, in the research of groundwater pollution of the landfill in the slaughterhouse (Staten Island, New York) by Siegel D. I. et al., they tried to use the isotope ratio of Sr and Pb change in the groundwater to indicate the seawater pollution and pollution level resulting from the leaching water of the slaughterhouse based on the correlation between isotope ratios of Sr and Pb and the concentrations of B, NH_4^+, and organic carbon, and the result let them believe that Pb isotope could not indicate to the differences because of no significant correlation with the related parameters ($^{87}Sr/^{86}Sr$, B, NH_4^+) and Pb isotope. However, $^{87}Sr/^{86}Sr$ is an effective tool for the pollutant to be indicated on the pollution of the groundwater and seawater polluted by leaching water of the slaughterhouse in this area. From the fact, it was revealed by their work that $^{87}Sr/^{86}Sr$ could be as a good geochemical tool to trace the pollutants in groundwater and seawater in this area. It can be seen that isotopic tracing theory and method as an extension and expansion of the application must have a broad using prospects, especially as the development of the arising of new instruments, the improvement of experiment accuracy, and the constant discovery and deepened understanding on isotope theoretical significance and application categories.

The prerequisite of the isotope tracing research of pollutants in environment is that the spatiotemporal process of the matter evolution from the source to the target region should be relatively stable in the inevitable correlation. As mentioned above, arsenic concentration in soil and groundwater of the Hetao Area exceeds the standard of drinking water recommended by the Government of China and is generally higher than the background concentration of the region. A large number of data showed that the change of As concentration was obviously dependent on the distance from the large deposits of the flow upstream and geochemical anomaly zone with high concentrations of related elements in the front of Yin Mountains. Besides, the poisoning degree and quantity of the local residents have a positive correlation with the concentration change of As. In this research, the trace metal pollution in soil and groundwater and the variation of Pb and Sr isotope in soil and groundwater were studied in order to reveal the pollutant source, the origin of the serious environmental problems in this area, as well as to explore the applicability of the isotope tracing method in solving the problem of environmental pollution sources and the pollutant migration.

There are many reports that are related to the migration of metals in the deposits and high background geochemical zone from the flow upstream to the downstream, leading to the increase of element concentrations in the downstream and even to poison in crowd. There have been many typical cases since 1996, such as the trace metal pollution in water and soil of the flow downstream caused by Devon Great Consls mining areas in Southwest of England [54], the changes of As, Pb, and Mn concentrations in the flow downstream areas in the southwest of Colorado in the USA [55], the higher As concentration in the downstream aquifer caused by San Antomio-EI Triunfo mining areas in Mexico [56], the higher concentration of harmful elements in environmental medium caused by waste residue from mining areas in central Germany [57], and the increased concentration of As in the groundwater and the poisoning in human caused by the long-term migration of materials in the high-arsenic background areas from the flow upstream areas of the alluvial plain of Ganges River, Bangladesh [58]. Based on the results from recent studies [59–61], these researches suggested, almost without any exception, that the causes of the formation of the pollution were related to the migration and diffusion of the pollutants from the flow upper reaches of the locality, respectively. However, these work's conclusions were only based on the relationships among various appearances (such as the concentration, the elements and the relationship of them, the correlation with the variations of physiochemical conditions of environment). On the explanation for causes of the formation of the pollution, they only were in the extrapolated stage in discussion simply according to the evolution of regional geological history, which were lack of data about matter movement records. The most representative work in these cases was the research on the arsenic pollution in the large-scale groundwater in Bangladesh, published in the journal Nature in 1998 by Ross Nickson et al., Department of Earth Sciences, University of London [62].

In summary, the following facts can be obtained from above discussion:

A. The isotope tracing research of trace metal pollution in soil and groundwater has been in more attention in the field of environmental science today, and there is no systematical work on the environmental problems similar to the features of the pollution of the Hetao Area.
B. There are neither research data about isotope researches in the soil and groundwater in the regional scale of the area nor the data about isotope researches related to environmental medium in the Hetao Area.

Both of the two points above are disadvantageous facts obviously to this research. In particular, the data about regional isotope background in environmental media are very important to isotope tracing research, which may affect the interpretation and conclusion of the work. Therefore, the isotope tracing research in this book is an exploratory attempt based on a lot of previous work. These works associated with a series of researches on the pollutant total concentrations, the concentrations with different speciation, the distribution characteristics of the pollutants in soil and groundwater, the correlation between these characteristics and concentrations of the source matters through erosion and weathering in the flow upper reaches of the area, and the correlation between these characteristics and population-based disease, poisoning, in the Hetao Area.

2.5.1 Samples

The samples for Pb and Sr isotope analysis in this research were taken from the soil and groundwater at the work-line AA in the Hetao Area. Samples of the two kinds of medium, soil and groundwater (well water), were systematically taken from the mining area of Dongshengmiao deposit and along the work-line AA (145°, the flow direction of the groundwater), respectively. In addition, except the well water samples, the samples of surface water minefield and the Yellow River water also were taken to be analyzed. The sampling sites are shown in Fig. 2.9.

The soil samples were the mixed grain soil under surface for 20–30 cm. The sampling sites were selected in the more stable area, and the human disturbance was relatively slight, near the design sampling points such as lawn, wood field, or fixed boundary of the farmland. Samples after taking were packaged with white cloth bags and labeled, and then transferred to the laboratory for further processing. Water samples were taken from drinking wells of the local residents, and the sampling well depth is within 18–25 m. To prevent contamination by the metals from well head and well tube, every water sample was collected after the water from the well continuously drawing for 10 min and was labeled and packaged with PTFE bucket, which was washing 5 times with the well water, and then acidified to pH 2–3 with nitric acid before sealing. The samples were stored in the refrigerator (4 °C) timely for further processing. The surface water of the minefield was taken from the plash in the mining field, which was pumped from mining well. Water samples of the Yellow River were directly collected with the sampling buckets from the Yellow River. The sampling buckets to be used and the procedures of sampling, acidification, and preservation for the samples from minefield and the Yellow River were same as that of the well water.

For different samples of different media (water and soil) and different isotope elements (Pb and Sr), the treatment processes were different, respectively, as follows:

A. Water sample pretreatment for Pb isotope: 200 ml original sample was dried up, and then the obtained solid matter was dissolved in 0.5 mol/L HBr (quantitative). The solution was separated with anion-exchange resin chromatography column to wash out the ion Fe, etc. Then, the resin column was leached with 5 mL 2 mol/L HCl. Finally, the sample in the resin was dissolved out with 6 mol/L HCL, dried, and measured with the mass spectrometer (MAT261).
B. Soil sample pretreatment for Pb isotope analysis: The original samples were ground in agate bowls until no sand feeling with the hand touching. Dissolve the samples (quantitative) with HNO_3 and HF, and then evaporate it to dryness. The dry residue was dissolved with HBr (quantitative) and then separated with anion-exchange resin chromatography column. The following processes were the same as the process of water samples presented above.
C. Water sample pretreatment for Sr isotope analysis: 50 mL original sample was dried up, and then the obtained solid matter was dissolved in 0.3 mol/L HCl (quantitative) for the separation with anion-exchange resin chromatography column. After leaching with 27 ml 1.75 mol/L, Sr in the resin was dissolved out with 8 mL 3 mol/L HCl and dried up. Finally, the obtained samples were measured with the mass spectrometer (MAT261).
D. Soil samples pretreatment for Sr isotope analysis: The original samples were ground in agate bowls until no sand feeling with the hand touching. Dissolve the samples (quantitative) with HNO_3 and HF, and then evaporate it to dryness. The obtained solid matter was dissolved with 0.3 mol/L HCl (quantitative) and separated with anion-exchange resin chromatography column. The following processes were the same as the pretreatment process of Sr isotope analysis for water samples.

All samples were pretreated in the super-clean laboratory, and the reagents and water were purified before being used in the analysis experiments. The detection limits of analytical instrument, thermal ionization spectrometer (MT261), for Sr and Pb were 1×10^{-7}, respectively. Qualified rate was 100% according to the relative deviations of the methods ($\geq$95% for the relative deviation limit <5% in the repeated analysis samples). The pretreatment and analysis for all samples were carried out in the Beijing Analysis Center of Nuclear Industry Geology Research Institute.

Table 2.21 Minerals detected by X-ray powder analysis for the soil from the Hetao Area

Sample site number on the work-line	Mineral components	
	Detrital minerals	Clay minerals
A-3	Quartz, calcite, feldspar, dolomite	Illite, kaolinite, chlorite, talcum
A-7	Quartz, calcite, feldspar, dolomite	Illite, kaolinite, chlorite, talcum
A-11	Quartz, feldspar, calcite, dolomite	Illite, kaolinite, talcum, chlorite
C-5	Quartz, feldspar, calcite, dolomite	Illite, kaolinite, talcum, chlorite,

Note The mineral order arrangement in the table implied the amount of the mineral from more to less

2.5.2 The Distribution of Pb and Sr Isotope in the Soil and Groundwater

Previous studies on the Hetao Area hydrogeological features by Li et al. in 1994 showed that the Hetao Area was situated in a basin of lake depositions in Middle or Lower Pleistocene of the Quaternary [1]. To the Late Pleistocene, the lake water began to leak out and the lake area gradually reduced and it was incorporated into the Yellow River, leading to the appearance of a lot of lacustrine sediments distribution on the area. This is the main reason that the geological background formed with the flat terrain and homogeneous soil composition in this area. As mentioned above, the soil of Hetao Area was classified as yellow fluvo-aquic soil in the group of silt loam [2], as shown in the book Chinese Soil authored by Nanjing Soil Research Institute, Chinese Academy of Sciences. In this area, results of X-ray diffraction analysis for soil phase (Table 2.21 and Fig. 2.16), and the relationship between the speciation and the total concentration (Fig. 2.6) indicated that the sedimentary circumstance and evolution history of the soil and its mother substances were generally the same for the whole area.

The data of Pb and Sr isotope as well as the concentrations of trace metals of samples from the soil and groundwater of the Hetao Area are all shown integratedly in Tables 2.22 and 2.23. Sample No. 1 in the table was the surface water of the field of Dongshengmiao deposit, Sample No. 6 was from Yellow River, and other samples were from the resident arsenic poisoning area along the work-line AA. The samples, No. 1–6, are listed in order from the flow upstream to the downstream of the Hetao Area, plain in the front area of the Yin Mountain, as shown in Fig. 2.9.

As shown in Table 2.22, the ratio of $^{87}Sr/^{86}Sr$ in mine water (Sample No.1) was 0.7196, which was significantly higher than that in groundwater of arsenic poisoning area. The ratio of $^{87}Sr/^{86}Sr$ in Yellow River (Sample No. 6) was 0.7168, which was higher than that in groundwater of arsenic poisoning area and lower than that in mine water. Moreover, from the flow upstream to the downstream of arsenic

Table 2.22 Characteristics of Pb, Sr isotope and Cu, As, etc., concentration in the soil and groundwater from the Hetao Area (unit: mg/kg)

Sam. No.	$^{206}Pb/^{204}Pb$	$^{207}Pb/^{204}Pb$	$^{207}Pb/^{208}Pb$	$^{206}Pb/^{207}Pb$	$^{87}Sr/^{86}Sr$	As	Sb	Cd	Cu	Pb	Zn
W1	19.1940	15.9594	0.4123	1.2028	0.7196	0.2510	0.0064	0.0050	0.0530	0.0470	0.0690
W2	19.1927	15.9578	0.4147	1.2027	0.7194	0.2360	0.0052	0.0040	0.0210	0.0260	0.0490
W3	19.1871	15.7725	0.4056	1.2165	0.7109	0.2300	0.0050	0.0040	0.0030	0.0220	0.0530
W4	18.5321	15.9522	0.4064	1.1617	0.7108	0.0040	0.0030	0.0020	0.0150	0.0070	0.0380
W5	18.3817	15.7581	0.4059	1.1665	0.7100	0.0050	0.0030	0.0020	0.0120	0.0060	0.0420
W6	18.3495	15.5969	0.4054	1.1765	0.7168	0.0030	0.0020	0.0020	0.0140	0.0070	0.0370
S2	18.2626	15.5369	0.4051	1.1754	0.7181	15.0000	0.6000	0.1800	37.0000	26.0000	75.0000
S3	18.6450	15.5873	0.4024	1.1962	0.7152	12.0000	1.7000	0.1000	27.0000	23.0000	63.0000
S4	18.6691	15.6595	0.4030	1.1922	0.7150	13.0000	1.1000	0.0900	22.0000	23.0000	61.0000
S5	18.5722	15.6640	0.4039	1.1857	0.7154	6.1000	0.5000	0.0600	10.0000	15.0000	41.0000

Note W1-W5 were the sample number of groundwater, and S2-S5 were the sample number of the soil (refer to Fig. 2.9)

Table 2.23 Measurement of definition for isotope content analysis in soil and groundwater from Hetao Area

	$^{206}Pb/^{204}Pb$	2 ό	$^{207}Pb/^{204}Pb$	2 ό	$^{208}Pb/^{204}Pb$	2 ό	$^{87}Sr/^{86}Sr$	2 ό
W1	19.1940	±0.0814	15.9594	±0.070	38.6990	±0.169	0.7196	±58
W2	19.1927	±0.080	15.9578	±0.079	35.1950	±0.187	0.7194	±88
W3	19.1871	±0.195	15.7725	±0.17	38.8910	±0.41	0.7109	±50
W4	18.5321	±0.221	15.9522	±0.19	39.2540	±0.472	0.7108	±45
W5	18.3817	±0.15	15.7581	±0.129	38.8250	±0.318	0.7100	±64
W6	18.3495	±0.071	15.5969	±0.061	38.4170	±0.15	0.7168	±78
S2	18.2626	±0.05	15.5369	±0.043	38.3520	±0.108	0.7181	±45
S3	18.6450	±0.035	15.5873	±0.029	38.7320	±0.073	0.7152	±28
S4	18.6691	±0.036	15.6595	±0.031	38.8550	±0.078	0.7150	±36
S5	18.5722	±0.078	15.6640	±0.067	38.7840	±0.165	0.7154	±33

Note W1—W5 meant the sample number of groundwater, and W6 meant the Yellow River water sample, and S2—S5 meant the soil sample number

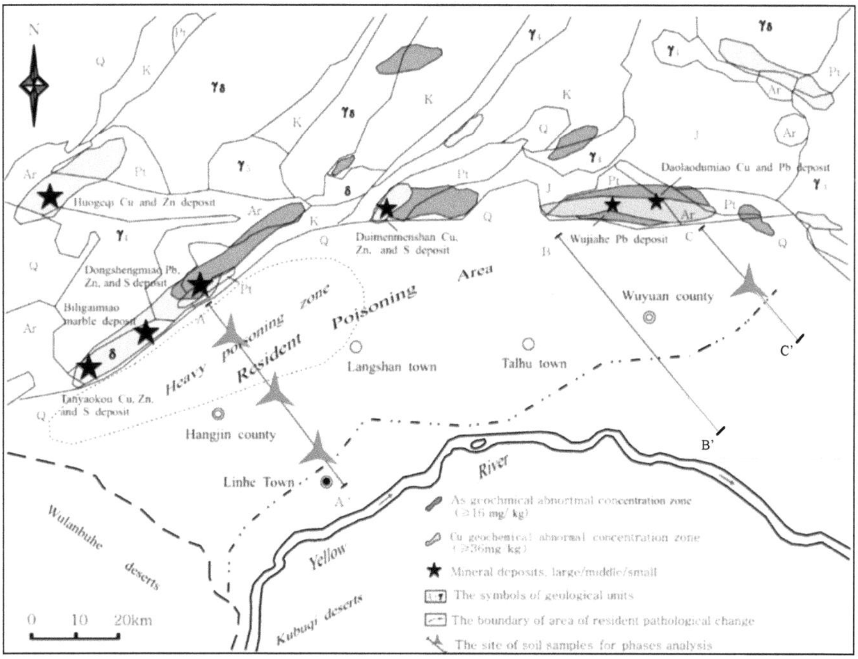

Fig. 2.16 Sampling sites for X-ray phases analysis in the soil and geochemical setting from the Hetao Area

poisoning area, the ratio of $^{87}Sr/^{86}Sr$ gradually decreased. The ratio of Pb isotopes ($^{206}Pb/^{204}Pb$, $^{207}Pb/^{208}Pb$) appeared with similar feature in this area. The ratio in mine water represented the isotope ratio in the water of the flow upstream (recharge area), which contained a lot of weathering materials in high background area for relevant elements. The isotope ratio in Yellow River water is considered more stable; thus, the change of isotope ratio in the groundwater of arsenic poisoning area is the result of the combined influence of the flow upstream supply water and the Yellow River water. The Yellow River affected the composition of groundwater from the infiltration of the river water and the infiltration of irrigation water because of the bed above the ground formed by long sedimentation from the river, as shown in Fig. 2.17. From Table 2.22 and Fig. 2.17, the facts that the change of Sr isotope was obvious, and the change of Pb isotope was not obvious may be due to some other factors (such as the participation of Pb in the atmospheric deposition), and this is a subject which needs to be further studied.

As shown in Table 2.22, the concentrations of As, Sb, Cd, Cu, Pb, and Zn in groundwater gradually decreased from the mining area to the lower reaches (Sample No. 1–5), which had obvious positive correlations with the change of $^{87}Sr/^{86}Sr$ and $^{206}Pb/^{204}Pb$, as shown in Figs. 2.18 and 2.19. The concentrations of As, Sb, Cd, Cu, Pb, and Zn of these samples in soil also had obvious correlation with the concentrations in groundwater, as shown in Fig. 2.20.

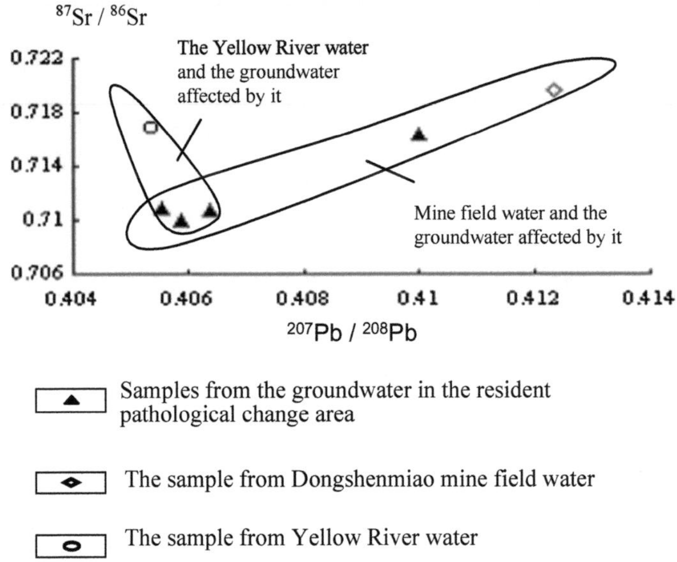

Fig. 2.17 Correlation of $^{87}Sr/^{86}Sr$ versus $^{207}Pb/^{208}Pb$ in groundwater from the Hetao Area

2.5.3 The Discussion of the Pb and Sr Isotope in the Soil and Groundwater

According to the research of Faure, et al., the ratio of $^{87}Sr/^{86}Sr$ bound to the mineral or pore water in sediment would change only when the geological characteristics of a catchment change (the material source change) [63, 64]. The research by Stille Peter et al. showed that the isotope composition of pore water in the sediment and the Sr concentration was resulted from the mixing of various substances of fluid phases generated from the weathering of previous mineral components [65]. Thus, the precondition, Sr isotope is used for the tracer of the flow upstream pollutants in soil and groundwater of Hetao Area, does exist. Firstly, the previous results showed that this area had not suddenly undergone geological events for the substances to be in overlying or commixture since the formation of the sedimentary basin, and the sedimentary environment and the evolution history were relatively stable. At these circumstances, the soil-forming process was the same and its composition was uniform generally in this area. So, the ratio of $^{87}Sr/^{86}Sr$ in the media would decrease over time if it was without the addition of exogenous Sr. And under the above geological background, the ratio of $^{87}Sr/^{86}Sr$ in pore water (groundwater) of sediments should be consistently stable in different sites in the Hetao Area if it was without the addition of exogenous Sr. Moreover, the hydrogeological characteristics of the Hetao Area are that the northern edge (flow upstream of the area) of the Yin Mountains is mainly the groundwater recharge area, the pediment plain is mainly the runoff area, and the southern part of the alluvial plain was mainly the discharging area. The southernmost

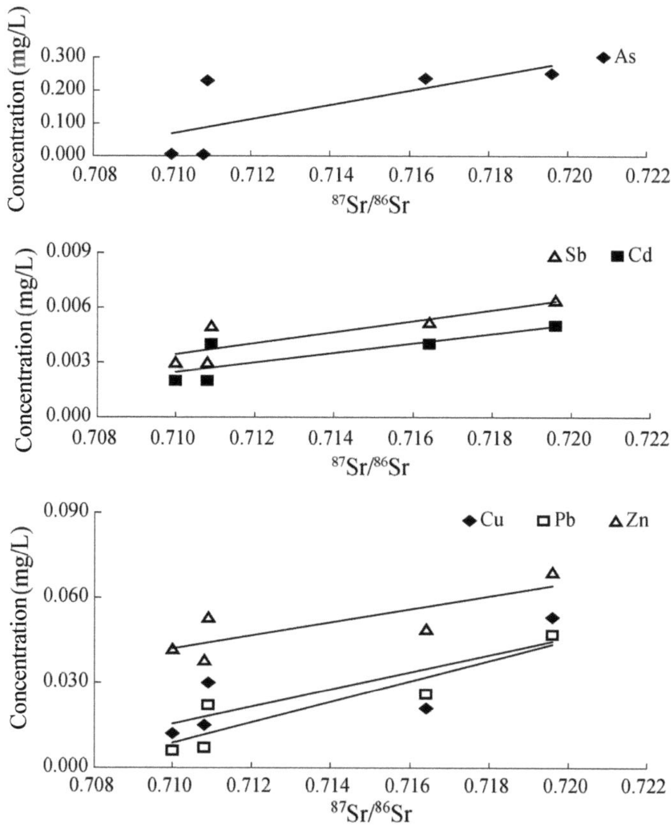

Fig. 2.18 Correlations of $^{87}Sr/^{86}Sr$ with the concentrations of As, Cd, Sb, Cu, Pb, and Zn in the groundwater

part of the area is the mainstream of the Yellow River, and due to the sedimentation over long time, the bed of the Yellow River here has been above the ground of its northern side. In the south of the Hetao Area, the Yellow River water and irrigation water in canals infiltrated or drained to the discharging area of the alluvial lacustrine plain [1]. Therefore, the groundwater discharging areas of the lacustrine plain of the southern side of the Hetao Area might be the convergence zone for groundwater from different sources and the carried substances by the water. Accordingly, the substances in the flow upstream of the area in the northern side migrated from there to the downstream with the groundwater must leave behind the migration traces, and the variation of the isotope Sr and Pb in soil and groundwater may be the record of the migration process leaved in natural media, soil and groundwater.

Based on the facts, revealed by the relationships among the isotopic data, the relationship between isotopic data and the concentration of As and other trace metals, and the relationship between the concentrations of As and other trace metals in groundwater and in soil, the items can be summarized as follows: (1) The groundwater

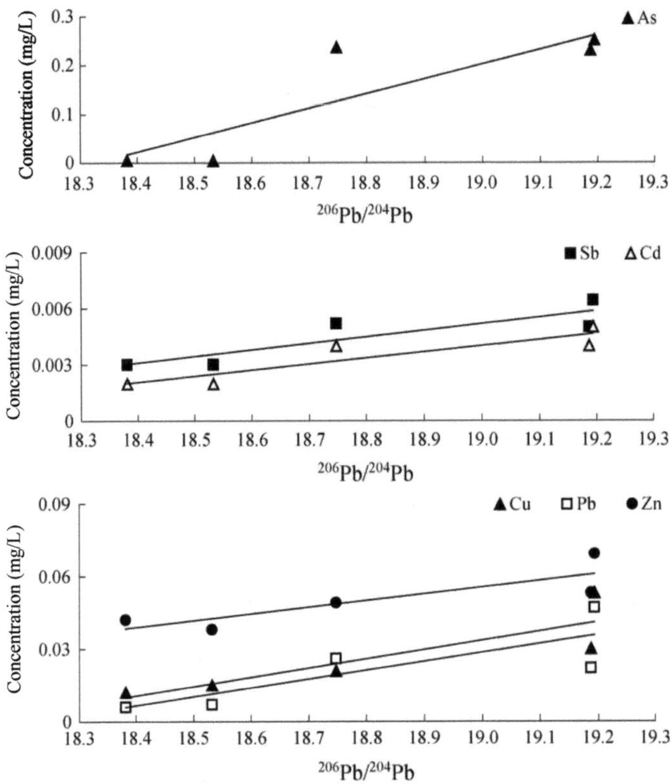

Fig. 2.19 Correlations of $^{206}Pb/^{204}Pb$ with the concentrations of As, Cd, Sb, Cu, Pb, and Zn in the groundwater from the Hetao Area

within the depth of 18–25 m in Hetao Area is contaminated by the surface water from the flow upstream of the area, and the degree of contamination decreases with the increasing distance away from the flow upper reaches. The weathering of the geologic body to be rich in trace metals in the upper reaches of the area and the natural runoff of the groundwater in the area are the possible mechanism of this contamination; (2) The concentration characteristics of As and other trace metals in groundwater and soil of the Hetao Area are the results of the migration of weathered substances in high background area in the flow upper reaches, and the objective evidences from substance records reveal that the concentration changes of related elements in soil and groundwater of the Hetao Area should be affected by the component of substances in the flow upstream of the area.

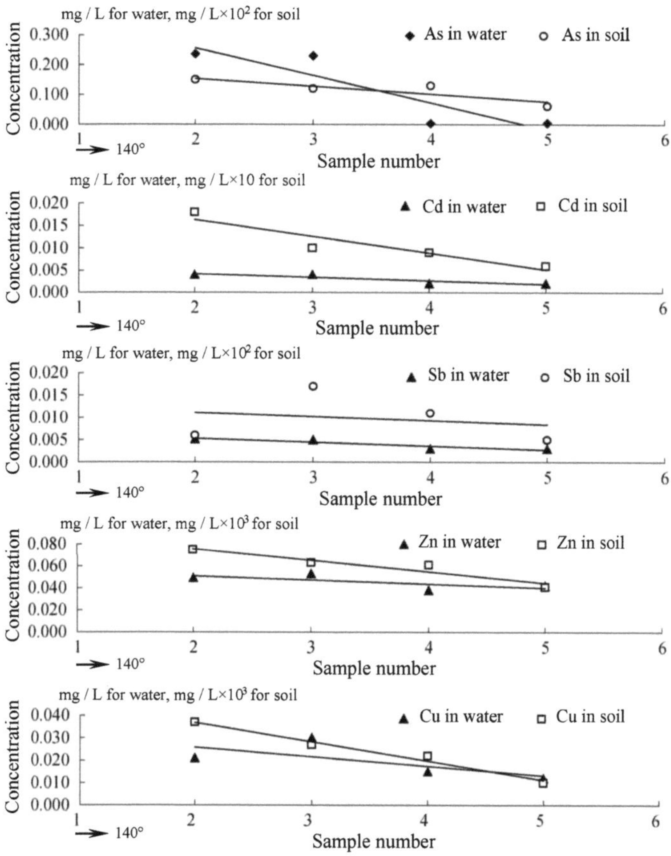

Fig. 2.20 Correlations of the concentrations of As, Cd, Sb, Zn, and Cu in groundwater and the concentrations of them in the soil from the Hetao Area

2.6 Discussion on the Distribution, Environmental Impacts, and Evolution Trends for Trace Metals in Soil and Groundwater

2.6.1 *The Correlation of the Concentration Distribution of Trace Metals Between Soil and Groundwater*

The concentration of As and other trace metal elements in soil and groundwater of the Hetao Area generally reflects a same change trend along the flow direction. As the exchange interface of surface living bodies, water, gas, and rock, the soil as an important medium involved in the actions. In the process of the geochemical cycle, or substance exchange, the elements in the media all have the opportunity with conditions to interact with the soil. Water is the basic transmission medium in the substance transport in the processes. There are the actions in the transformation and interactions such as the actions each other of the substances into the soil, the dissolving for the substances in the water, containing some substances in soil, and migration of the substance as solute with the water. If the substance composition in the soil is uniform at different sites of the research area, the changes of substances along with water flow in soil due to the movement of water should be related to the substances from other external places carried by the water. Particularly, there would be variation correlation within the substance changes along the moving path of the water. The above inference discussion was only an analysis based on a simplification for the process, while the practical situation in nature may be often much more complicated. The most important is that the soil is rarely uniform such as composition and physical and chemical conditions, and these factors were possible to cause the differences in the effects of the substances carried by water, leading to the irregular phenomenon of the distribution of migrating substances in water and soil. However, the rough regularity from this work could generally reflect the situation of the pollutant distribution and cause of formation to a certain degree. Facts revealed by a large number of case researches illustrated this point too.

2.6.2 The Difference and Correlation for the Distributions of Trace Metals Between the Soil and Groundwater

Based on the principle of geochemical cycle of substance, the concentration and its variation of trace metals in soil and water in the research area are the embodiment of the aspects in the problem, which is in the process of the substance mobilization. The exchangeable fraction is ionic species, which could easily migrate with water, and according to the feature of the speciation of trace metals of the Hetao Area, the exchangeable part of the metals in soil and groundwater may be in different level and variation because soil and water are in different phases with each attributes. The differences of the natural attributes of these two substances must cause the different and adaptable situation for the metals during the circulation, thus leading to different concentrations of the trace metal elements in soil and water to cause different environmental impact to the health of human in generally corresponding correlation.

2.6.3 The Evolution and Factors of the Trace Metal Pollution in Study Area

The facts above indicate that the high concentration of As and other trace metal elements in the groundwater of the Hetao Area caused the poisoning of the local residents, which was related to the high background concentration of the element in the flow upstream of the area and the migration with water. Some trace metals in the flow upper reaches of the area migrated and accumulated to the soil and groundwater in the flow lower reaches of the area along the water flow in long time, finally causing the poisoning of the local residents. This human poisoning problem of Hetao Area is related to the facts such as source of trace metals, interaction between the trace metal elements and the media, the physical and chemical conditions of the environment, the medium, and the rate of flow of the water. This situation of the substance migration will be continued under the action of natural force if the practical stratagem for prevent it was in the absence or not starting. There were a lot of researches on the trace metal pollution worldwide, especially in typical area contaminated by As, such as Bangladesh, India, Vietnam, Taiwan, Xinjiang, and other regions in China.

References

1. Li S, Li H. The geological characteristics and cause of arsenic poisoning in the Hetao Area, Innner Mongolia. Inner Mongolia Prophylaxis Treat. Endemia (In Chinese). 1994, 19:1–3.
2. The Institute of Soil Science of the Chinese Academy of Science. China Soil (In Chinese). Beijing: Science Press; 1980.
3. Muller G. Index of geoaccumulation in sediments of the Rhine Rive. Geo. J. 1969;2(3):108–18.
4. Tessier A, Campbell PGC, Bisson M. Sequential extraction procedure for the speciation of particulate trace metal. Anal Chem. 1979;51(7):844–51.
5. Tessier A, Fortin D, Belzile N, Devitre RR, Leppard GG. Metal sorption to diagenetic iron and manganese oxyhydroxides and associated organic matter: narrowing the gap between field and laboratory measurements. Geochim Cosmochim Acta. 1996;60(3):387–404.
6. Gleb P, Robert G, Jacques S, Alexandre Z, Jean-Claude H. Thermodynamic properties and stoichiometry of As, hydroxide complexes at hydrothermal conditions. Geochim Cosmochim Acta. 1996;60(5):737–49.
7. Ure AM, Quevauviller P, Muntau H, Griepink B. Speciation of heavy metals in soils and sediments. An account of the improvement and harmonization of extraction techniques undertaken under the auspices of the BCR of the commission of the European Communities. Int. J. Environ. Anal. Chem. 1993; 51:135–51.
8. Forstner U. Metal speciation-general concepts and applications. Int J Environ Anal Chem. 1993;51:5–23.
9. Lopez-Sanchez JF, Rubio R, Rauret G. Comparison of two sequential extraction procedures for trace metal partitioning in sediments. Int J Environ Anal Chem. 1993;51:113–21.
10. Salomons W. Adoption of common schemes for single and sequential extractions of trace metal in soils and sediments. Intern. J. Environ. Anal. Chem. 1993;51:3–4.
11. Gomez Ariza JL, Giraldez I, Sanchez-Rodas D, Morales E. Comparison of the feasibility of three extraction procedures for trace metal partitioning in sediments from south-west Spain. Sci Total Environ. 2000;246:271–83.
12. Dang Z, Liu C, Shang A. The progress on assessment for the heavy metal activity in the soil of orefield. Prog. Geosci. 2001;16(1):86–92 (In Chinese).
13. Wu K, Ma H, Yu G. Approach to the relationship between the poisoning and dosage of As intaking. Inner Mongolia Prophylaxis Treat. Endemia. 1994;19(1):52–4 (In Chinese).
14. Balakumar P, Kaur J. Arsenic exposure and cardiovascular disorders: an overview. Cardiovasc Toxicol. 2009;9:169–76.
15. Nicolis E, Curis P, Deschamps S, Benazeth. Arsenite medicinal use, metabolism, pharmacokinetics and monitoring in human hair. Biochimie. 2009; 91:1260–7.

16. Yáñez J, Fierro V, Mansilla H, Figueroa L, Cornejo L, Barnes RM. Arsenic speciation in human hair: a new perspective for epidemiological assessment in chronic arsenicism. J Environ Monit. 2005;7:1335–41.
17. Kempson IM, Henry DA. Determination of arsenic poisoning and metabolism in hair by synchrotron radiation: the case of phar lap. Angew Chem. 2010;49:4237–40.
18. Valkovic V. Human hair: trace-element levels. Boca Raton: CRC Press; 1988.
19. Tomic S, Lakatos J, Valkovic V. Analysis of trace elements in hair of pregnant women using XRF spectrometry. X-Ray Spectrom. 1989;18:73–6.
20. Toribara TY. Analysis of single hair by XRF discloses mercury intake. Human Exp. Toxicol. 2001;20:185–8.
21. Chevallier P, Ricordel I, Meyer G. Trace element determination in hair by synchrotron X-ray fluorescence analysis: application to the hair of Napoleon I. X-Ray Spectrom. 2006;35:125–30.
22. Kempson IM, Skinner WM, Kirkbride K. The occurrence and incorporation of copper and zinc in hair and their potential role as bioindicators: a review. J. Toxicol. Environ. Health Part B Crit. Rev. 2007;10:611–22.
23. Raab A, Feldmann J. Arsenic speciation in hair extracts. Anal Bioanal Chem. 2005;381:332–8.
24. Miyazawa N, Uematsu T. Analysis of ofloxacin in hair as a measure of hair growth and as a time marker for hair analysis. Ther Drug Monit. 1992;14:525–8.
25. Cookson JA, Pilling FD. Trace element distributions across the diameter of human hair. Phys Med Biol. 1975;20:1015–20.
26. Al-Delaimy WK. Hair as a biomarker for exposure to tobacco smoke. Tob. Control. 2002;11:176–82.
27. Hindmarsh JT. Caveats in hair analysis in chronic arsenic poisoning. Clin Biochem. 2002;35:1–11.
28. Yeongkyoo K, Cygan RT, Kirkpatrick RJ. 133Cs NMR and Xps investigation of cesium adsorbed on clay minerals and related phases. Geochim. Cosmochim. Acta. 1996: 60(6):1041–52.
29. Sikarwar SS, Guha S, Singh KN. Environmental pollution control through geo-botanical means—a case study of Manikpur block, Korba coalfield, M. P., India. Environ. Geol. 1999; 38(3): 229–32.
30. Thornton I. Applied environmental geochemistry. London: Academic Press; 1983.
31. Zhang H. The tutorialon on soil environtology. Shanghai: Shanghai Jiaotong University Press; 2009.
32. Zhang H. Soil environtology. Beijing: Chemical Industry Press; 2006.
33. Pugh CE, Hossner LR, Dixon JB. Oxidation rate of iron sulfides as affected by surface area, morphology, oxygen concentration and autotrophic bacteria. Soil Sci. 1984;137:309–14.
34. Regina NT, Lisa AS, Paul L, Jonathan P. Geochemical modeling approach to predicting arsenic concentration in a mine pit lake. Appl Geochem. 2000;15:475–92.
35. Zhang H, Ma D, Hu X. Arsenic pollution in groundwater from Hetao Area, China. Environ Geol. 2002;41(6):638–43.
36. Safiuddin M, Zhang H, et al. Aquatic arsenic toxicity and treatment. In Murphy T, Guo J, editors. Leiden, The Netherlands: Backhuys Publishers; 2003.
37. Valentine JL, He SY, Reisbord LS. Health response by questionnaire in arsenic exposed populations. J Clin Epidemiol. 1992;45(5):487–94.
38. Wu MM, Kuo TC, Hwang YH. Dose-response relation between arsenic concentration in well water and mortality from cancers and vascular diseases. Am J Epidemiol. 1989;130:1123–32.
39. Zhang H. Heavy-metal pollution and arseniasis in Hetao Region, China. Ambio. 2004;33(3):138–40.
40. Fendorf S, Michael HA, Geen AV. Spatial and temporal variations of groundwater arsenic in south and southeast Asia. Science. 2010;328:1123–7.
41. Siegel DI, Bickford ME, Orell SE. The use of strontium and lead isotopes to identify sources of water beneath the Fresh Kills landfill, Staten Island. New York, USA. Appl. Geochem. 2000; 15:493–500.

42. Deboudt K, Flament P, Weis D, Mennessier J-P, Maquinghen P. Assessment of pollution aerosols sources above the Straits of Dover using lead isotope geochemistry. Sci Total Environ. 1999;236:57–74.
43. Zhu BQ, Chen YW, Peng JH. Lead isotope geochemistry of the urban environment in the Pearl River Delta. Appl Geochem. 2001;16:409–17.
44. Adgate JL, Rhoads GG, Lioy PJ. The use of isotope ratios to apportion sources of lead in Jersey City, NJ, house dust wipe sample. Sci. Total Environ. 1998; 221:171–80.
45. Tommasini S, Gareth RD, Elliott T. Lead isotope composition of tree rings as bio-geochemical tracers of heavy metal pollution: a reconnaissance study from Firenze, Italy. Appl. Geochem. 2000; 15:891–900.
46. Iyer SS, Babinski M, Marinho MM, Barbosa JSF, Sato IM, Salvador VL. Lead isotope evidence for recent uranium mobility in geological formations of Brazil: implications for radioactive waste disposal. Appl Geochem. 1999;14:197–221.
47. Steven CA, Neil CS. Strontium isotopic evidence on the chemical evolution of pore waters in the Milk River Aquifer, Alberta, Canada. Appl Geochem. 1998;13(4):463–75.
48. Musgrove M, Banner JL. Regional ground-water mixing and the origin of saline fluids: Mid-continent, Unite States. Science. 1993; 259:1877–81.
49. Johnson TM, Donald JD. Interpretation of isotopic data in groundwater-rock systems: model development and application to Sr isotope data from Yucca Mountain. Water Resour. Res. 1994; 30(5):1571–87.
50. Thomas DJ, Waters SB, Styblo M. Elucidating the pathway for arsenic methylation. Toxicol Appl Pharmacol. 2004;198:319–26.
51. Phillips PJ, Busenberg E. The use of simulation and multiple environmental tracers to quantify groundwater flow in a shallow aquifer. Water Resour. Res. 1994; 30(2):421–33.
52. Solomon DK, Sudicky EA. Tritium and helium 3 isotope ratios for direct estimation of spatial variations in groundwater recharge. Water Resour Res. 1991;27(9):2309–19.
53. Thomas DB, David PK, Carol K. Kinetic and mineralogic controls on the evolution of groundwater chemistry and 87Sr/86Sr in a sandy silicate aquifer, northern Wisconsin, USA. Geochim Cosmochim Acta. 1996;60(10):1807–21.
54. Margaret EF, Thornton I, Kavanagh P. Exposure to high Arsenic contaminations in SW England. Book of the Abstracts of 30th International Geological Congress, Beijing. 1996; 3:50.
55. Apambire WB, Boyle DR, Michel FA. Geochemistry, genesis, and health implications of fluoriferous groundwaters in the upper regions of Ghana. Environ Geol. 1997;33:13–24.
56. Carrillo A, Drever JI. Adsorption of arsenic by natural aquifer material in the San Antomio-EI Triunfo mining area, Baja California, Mexico. Environ Geol. 1998;35:251–7.
57. Schreck P. Environmental impact of uncontrolled waste disposal in mining and industrial areas in Central Germany. Environ Geol. 1998;35:66–72.
58. Nickson RT, McArthur JM, Ravenscroft P, Burgess WG, Ahmed KM. Mechanism of arsenic release to groundwater, Bangladesh and West Bengal. Appl Geochem. 2000;15:403–13.
59. Zhang H. Arsenic movement and traces in the groundwater from the Hetao Area, Inner Mongolia. Environ Earth Sci. 2013;69:2127–8.
60. Zhang H. Erratum to: arsenic movement and traces in the groundwater from the Hetao Area, Inner Mongolia. Environ Earth Sci. 2013;69:2127–8.
61. Deng Y, Wang Y, Ma T. Isotope and minor element geochemistry of high arsenic groundwater from Hangjinhouqi, The Hetao Plain, Inner Mongolia. Appl Geochem. 2009;24:587–99.
62. Nickson R, John M, William B, Kazi MA, Peter R, Mizanur R. Arsenic poisoning of Bangladesh groundwater. Nature. 1998;395:338.
63. Guniter Faure. Principles and applications of Geochemistry. Upper saddle River, New Jersey: Prentice Hall; 1998.
64. Shen W. The tutorial on isotope geology: the geochemistry of Strontium and Rubidium. Beijing: Atomic Energy Press; 1997 (In Chinese).
65. Peter S, Shields G. Radiogenic isotope geochemistry of sedimentary and aquatic systems. Heidelberg: Springer; 1997.

66. Zhang H. The Specification of Geochemical Map K-48-24, K-48-30, K-49-19, K-49-25, K-49-31 (1/200000). Huhhot: Inner Mongolia Geological Bureau Press; 1991. p. 56–74 (In Chinese).
67. Bowen H. Environmental chemistry of the elements. London: Academic Press; 1979. p. 1–316.
68. The Institute of Endemic Prophylaxis and Treatment Inner Mongolia. The Criterion and types of the arsenic poisoning in Inner Mongolia. J. Endemiol. China (Album on Provincially Arsenic Poisoning, in Chinese). 1995; 58.

Chapter 3
The Metropolis Pollution of Trace Metals—Shanghai and Nanjing, China

3.1 The Pollution of the Trace Metals in Urban Rivers

As the case of the trace metal pollution of the urban river, the Suzhou Creek, the Huangpu River, and the Nanjing section of the Yangtze River are selected, respectively, to reveal the relationship and the mechanism between trace metals and modern metropolis development, Shanghai and Nanjing, China.

3.1.1 The Pollution of Trace Metals and Organic Matter (PCBs) from the Suzhou Creek, Shanghai

The Suzhou Creek is a time-honored urban river across whole town Shanghai, and this river has accompanied almost all the development process of Shanghai. The trace metal pollution always has been one of the important environmental issues of this creek.

3.1.1.1 Overview

The Suzhou Creek originates from the Gua Jing Estuary of the Taihu Lake. It is located in southeast Jiangsu province, Shanghai. Flowing through Kunshan city, Jiangsu province, this creek flows into Shanghai near Zhaotun, Qingpu District. Gliding through eight districts of Shanghai, including Jiading, Minhang, Putuo, Changning, Jingan, Zhabei, Hongkou, and Huangpu districts, respectively, it flows into the Huangpu River at the Waibaidu Estuary. The length of this creek is 125 km, including 53.1 km in Shanghai and 23.8 km in downtown region (following Huacao Harbor), and the average width is in the range of 70–80 m. The Suzhou Creek is the largest tributary of the Huangpu River. The whole terrain in the flow area is higher to lower from the north to south, and it is the main distributed tributary to

H. Zhang, *Behaviors of Trace Metals in Environment*,
https://doi.org/10.1007/978-981-13-3612-6_3

the Huangpu River with other creeks flowing in north Suzhou Creek. The main intersections of the creeks in north Suzhou Creek in Shanghai include Gupu Creek, Wenzaobang Creek, Yantie Creek, Fengbang Creek, Xincuopu Creek, Taopu Creek, Pengyuepu Creek, and so on, and those intersections in south Suzhou Creek include Xidaying Creek, Youdungang Creek, Jiedongdaying Creek, Xintongbotang Creek, Yancangpu Creek, Panlongtang Creek, and Xinjinggang Creek, and so on.

The whole basin of Suzhou Creek is low and flat, and the slope is generally small. The river twists and turns water flow slowly. The sediments and pollutants easily enter and precipitate in the creek by tidal backwater, resulting in the silted-up. The creek is difficult to leak out, and the lowland prone sometimes to be flooded. The maximum flow rate of ebb tide is 0.58 m/s. The creek takes about five or six days for the Beixinjing sewage flowing eastward to the Waibaidu Estuary of the Huangpu River. The creek water depth is about 2 m, and the flow velocity, sediment concentration, and water depth are all less than that of the Huangpu River. As a medium tide river flowing through the downtown of Shanghai, the Suzhou Creek tidal type belongs to shallow estuary and irregular semidiurnal tide, with average two high and two low tides during the of 24 h 50 min. In general, the daytime tide is greater than the diurnal tide in summer and vice versa in winter. At present, the Suzhou Creek has a variety of functions such as drainage, navigation, irrigation, and so on. The Suzhou Creek plays a very important role in the preventing flood and field waterlog. The watergates on the creek are scheduled according to the demand of flood discharge and therefore affect the hydraulic behavior of the Suzhou Creek. The Suzhou Creek is currently the main receiving creek for the water of the sewage from industry and people's life of the creek basin. It is about daily admission of 700,000 ton of industrial wastewater and 467,000 ton of domestic sewage. The pollution of the creek is mainly caused by it.

Since the founding of China in 1949, Shanghai has transformed from the original multi-functional city into an industrial center of China. Along Suzhou Creek and its tributaries, Beixinjing, Pengpu, Taopu, Anting, and other new industrial zones were built up in Shanghai, and the industrial plants along the Suzhou Creek are densely covered. By the end of the 1980s, residents on urban region of the Suzhou Creek increased to 300 million, and the population density was more than 15, 000 persons/km^2. By the end of 2000, the total population along the Suzhou Creek had reached to 5,578,600, including 93 streets (township or town). Those districts along the Suzhou Creek including Changning, Putuo, Jingan, Zhabei, Hongkou, and Huangpu are densely populated areas. The average population density is above 15,000 persons/km^2, and Huangpu district has the highest density, 53,326 persons/km^2. This high density of population distribution causes great pressure on the water environment of the Suzhou Creek. On the economic structure, the lower reaches of the Suzhou Creek are an important industrial concentrated area, including 7836 factories and enterprises in the entire region. Especially in Zhabei, Putuo, and Changning districts, the industry is mainly concentrated in the regions along the Suzhou Creek. Most of these enterprises are large-scale backbone enterprises in the economic development of Shanghai. At the same time, these enterprises are important sources of pollutions in the Suzhou Creek. In the outskirts of Jiading, Qingpu, Minhang district, in addition to municipal and regional belonged enterprises, there are many township enterprises and joint ventures. Otherwise, there are many service institutions along the Suzhou Creek, including 918 restaurants and hotels, 69 hospitals and health centers.

In recent years, accompanying with the adjustment of economic growth mode and industrial structure of Shanghai, the industrial components along the Suzhou Creek have also changed significantly. In particular, the downtown Huangpu and Jingan districts display outstanding performance. From the Tibet North Road to the estuary of the Huangpu River, a more important financial and trade development zone along the Suzhou Creek has formed. A commercial district has developed from Tibet North Road to TianMu West Road and Changshou Road. In these two regions, industrial enterprises (the original secondary industry) are migrated to other regions, and financial, trade, commerce, and other third industries are rapidly developed. However, from Tianmuxi Road to Xinjing north section of the coastal area of the Suzhou Creek, changes in the industrial structure are negligible, which can be attributed to the original industries are more concentrated in the Putuo District, Changning District, and Zhabei District.

The water quality of the Suzhou Creek was very clear, early days it was called as Wusong Creek, which had a wide variety of aquatic and fishery, and was regarded as a region of prosperity. From 1920 onward, with the rapid expansion of the size of the Shanghai, a large number of industrial wastewater and domestic sewage directly discharged into it without any treatment, resulting in the water quality deteriorating of the Suzhou Creek. After 1949, along the Wusong River and its tributaries, Beixinjing, Pengpu, Taopu, Anting, and other industrial districts had been built up. Residents in Shanghai downtown along the Suzhou Creek increased to 3 million. Residential and factory housing became density then. With the rise of the chemical industry, printing and dyeing, textile, paper, leather, food, and pharmaceutical industries, those amounts of domestic sewage, industrial wastewater, associated with the pesticide and fertilizer from the drainage basin of it entering into the Suzhou Creek, increased gradually. Before the implementation of sewage treatment projects, the quantity of sewage discharged into the creek accounted for about 25–30% of the total amount of the city's sewage of Shanghai. Due to pollution sources increased year by year, water pollution had becoming more and more serious; pollution scope gradually ascended and expanded. After 1978, the Suzhou Creek in the territory of Shanghai downtown was all suffering from the pollution. Water quality in urban regions was inferior to the national standard grade V of the surface water, with black and odorous in whole years. On spatial distribution, the pollution in the Suzhou Creek presented from upstream to downstream gradually worsened. Especially after entering the downtown of Shanghai, since the Huacao section, pollution became worse and worse, and was most prominent in Beixinjing section. Since 1996, in order to improve the water environmental quality in Shanghai and eliminate the black and odor in the Suzhou Creek, Shanghai Government continued to carry out the work in the comprehensive improvement of water environment of the Suzhou Creek. After a series of comprehensive treatment projects, the point pollution sources along the flow path of the Suzhou Creek have been effectively controlled, and water pollution in the downstream of the Suzhou Creek in the urban area has been significantly improved, the indexes of water quality have reached to the standard class IV of the national surface water of China.

3.1.1.2 Samples and Analyses

1. **Sampling and processes**

Six representative sampling sites were selected along the river way of the Suzhou Creek. They are Jiangnan Paper Mill, Shanghai secondary grease factory, Pengyuepu Creek, Zhonghuaxinlu Bridge, Wuzhenlu Bridge, and Waibaidu Bridge, as shown in Fig. 3.1.

The Jiangnan Paper Mill and grease factory, Shanghai, are located near Caojiadu Wuning Road. The Jiangnan Paper Mill was originally built in 1925, and it is the most main paper mill in the enterprises in Shanghai. Currently, it is mainly engaged in the production of coated paper. This plant was in a stop business for rectification when we took samples. Shanghai secondary grease factory was founded in 1902, which was the largest and leading suppliers of food grade grease in China. Therefore, this region was used to be the main industrial area in Shanghai, and it had concentrated on the large number of heavy industry factories. Therefore, its pollution can reflect the overall characteristics of the situation of the Suzhou Creek. Pengyuepu Creek and Zhonghuaxinlu Bridge are located in the junction of the Suzhou Creek and its tributaries of Pengyuepu Creek. The pollution from the tributaries of the Suzhou Creek has important influence on the Suzhou Creek, and Pengyuepu Creek is one with the most serious pollution in the tributaries into the creek. The two sampling

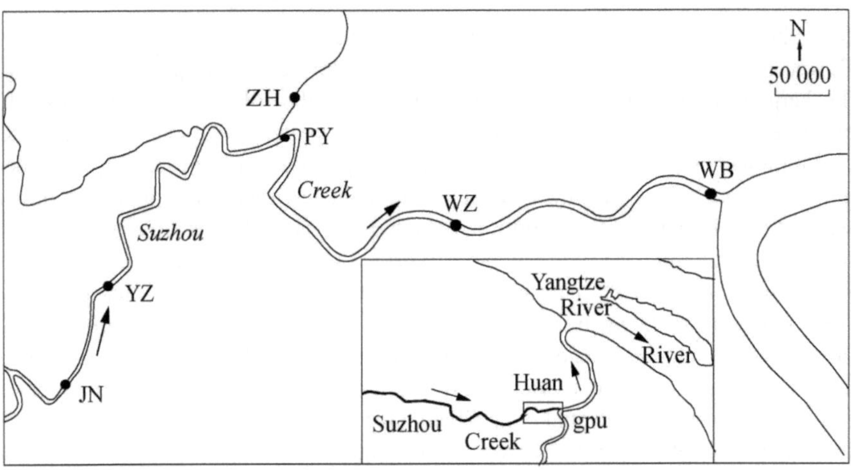

Fig. 3.1 Sampling sites in the Suzhou Creek. JN means Jiangnan Paper Mill site, YZ means Shanghai grease plant site, PY means Pengyuepu Creek site, ZH means Zhonghuaxinlu Bridge site, WZ means Wuzhenlu Bridge site, and WB means Waibaidu Bridge site

Table 3.1 Sampling record for sediment from the Suzhou Creek

Site	Sediment core length (cm)	Water depth (m)	Feature of the sample
Jiangnan Paper Mill	70	3.0	Black clay and fine silt with stink
Shanghai grease factory	65	2.5	Black clay and fine silt
Wuzhenlu Bridge	30	2.0	Black clay and fine silt
Zhonghuaxinlu Bridge	30	0	Black clay and fine silt with domestic waste and stink

points selected at the sites, respectively, above reflected the contribution of pollutions from the tributaries. Sampling site at Wuzhenlu Bridge is located in the commercial storage area in the downstream. It is a business and traffic concentrating section and can reflect the pollution from the service sector. The sampling site at Waibaidu Bridge is close to the estuary of the Suzhou Creek, which can be compared with the sampling points in the upper reaches, and reflects the overall pollutant contribution of the Suzhou Creek to the Huangpu River [1].

Water sample was taken in the flow centerline of the river below 50 cm of water surface when was at low tide estimated according to the Wusong tidal time announced by Shanghai hydrologic station. The sample was put in polyethylene barrels, pH value was determined on-site, and then it was acidified with HCl to pH = 2, transported to the laboratory, and stored at 4 °C in an icebox. Sediment samples were taken out by putting the sampling tube (PVC tube) into the sediment with a boat in the inside part of the creek convex bend. Then the mud section was removed out, and two ends were sealed and taken back to the laboratory for further processing. Sampling tools and containers were all immersed in a 10% nitric acid for 12 h. The records of the samples are shown in Tables 3.1 and 3.2.

All pretreatments, experiments, trace metal analyses were completed in the Lab of Pollutant Behavior Chemistry in School of Environmental Science and Engineering, Shanghai Jiaotong University, and PCB measurements were completed in the East China Sea Environmental Monitoring Center, the State Oceanic Administration.

Water samples for composition analyses were taken directly from the original water or those filtrated water according to the different composition characteristics.

The processing procedure for the measurement of suspended matter in the water sample is as follows: 20 L raw water sample was filtered through a 0.45 μm membrane. The 500 ml acidified filtrate was stored for the relative item measurement. Particles on the filter were collected by a rhino horn spoon scrape, natural air-dried (drying in an oven would change the solid distribution of trace metals), ground and sieving by a 200-mesh sieve. Then these particles under the sieve were preserved in a kraft bag in a cool and dry place.

The method for measuring the content of particulate matter in the creek water was as follows: After shaking the water sample, 250 ml water was taken and filtered

Table 3.2 Sampling data recorded for water from the Suzhou Creek

Sites	Sampling time	Width of the creek/m	Depth of the water/m	pH	Flow velocity (m/s)
Jiangnan Paper Mill	2006.3.11 10:15–11:40	30	1.2	7.2	1.2
Shanghai grease factory	2006.3.11 11:10–11:30	30	1.5	7.1	1.5
Wuzhenlu Bridge	2006.3.18 10:00–10:20	55	1	6.75	0.22
Outer Baidu Bridge	2006.3.18 11:30–11:45	60	1.3	6.82	0.33
Pengyuepu Creek	2006.3.29 18:20–18:45	80	2	7.98	0.5
Zhonghuaxinlu Bridge	2006.3.29 17:20–17:50	18	2–3s	7.91	0.3

by 0.45 μm microporous membranes, which were dried before and after filtration at low temperature (to the weight of the membrane remaining constant), and then the weight of suspended particulates could be obtained from the difference between the two values of the weight before and after filtration. The average value of ten time duplications was calculated as the particle content in the creek. Handling procedures for sediment were as follows: The collected column sediments were divided into sections in every 10-cm interval; after dried in natural air, they were ground and passed over 200-mesh sieve, and then put into kraft paper bags and stored in a dry and cool place.

Digestion procedure for arsenic and mercury: Put 0.5 g powder in a colorimetric tube (25 ml) with 10 ml aqua regia (HNO_3: HCl = 1:3), then the tube was covered and put in a boiling beaker for 1H, and then the water was added into the tube to maintain the final volume 25 ml for further measurement. Digestion procedure for other trace metals: 0.5 g powder was placed in a 50-ml Teflon beaker with 50 ml HNO_3 and heated to nearly dryness on an electric heating plate, then 50 ml HF was added and heated to dryness (the residue on the tube wall should be dissolved by intermittently shaken), and then HF was added to over the residue with 2–3 drops of $HClO_4$, and then heating it till no white smoke, finally the residue was dissolved by 1:10 HCl and fixed volume to 50 ml for further measurements.

Samples used for PCB measurement: Weighing 10 g sample and placing it in a 200-ml conical flask, adding 60 ml 1:1 hexane/acetone solution and 1 g high pure copper powder (to remove sulfur impurity in the sample) and staying in an ultrasonic oscillating water bath for 1 h, then soaking in water overnight. Collecting the extract, and then adding 40 ml 1:1 hexane/acetone solution, and collecting the extract after shaking for 30 min; and then adding 20 ml 1:1 hexane/acetone solution and shaking for 30 min. All extracted solution was collected into a pear-separating funnel, and 100 ml 5% Na_2SO_4 solution was added to the liquid, and let the lower layer be removed with repeating for twice. Then concentrated sulfuric acid with 10 ml for

Table 3.3 Operational conditions and parameters of the instruments used in the work

Instrument	Element	Detecting limit (μg/g)	RSD (%)	Adding reference recovery (%)
Double-channel atomic fluorescence spectrophotometer (AFS-810, Beijing)	Pb	0.00025	3.2	96.9
	As	0.00018	5.7	95.4
	Cd	0.00021	4.6	101.2
Automatic mercury analyzer (Milestone-AMA254, Italy)	Hg	0.0001	2.5	100
Atomic absorption spectrophotometer (Perkin-Elmer 5100, USA)	Zn	0.0326	1.6	94.5
	Cu	0.0563	1.3	102.8
Gas chromatograph (Varian-CP-3800)	PCBs	0.00025	8.4	86.7

each time was used to wash the sample until the acid layer colorless. Then 10 g Na_2SO_4 was used to remove water, and the upper layer substance was collected and was concentrated to 0.5 ml by using the nitrogen injection instrument for further measurements.

2. **Analysis**

The instruments used for the measurement mainly were atomic absorption spectrophotometer (Perkin-Elmer, USA), double-channel atomic fluorescence spectrophotometer (Jida-Cygnet-AFS-81, Beijing), automatic mercury analyzer (Milestone-AMA254, Italy), gas chromatograph (Varian-CP-3800), as well as other auxiliary instruments such as high-speed refrigerated centrifuge (SIGMA-3-18 K, Germany), multipurpose vacuum pump (SHB-III, China), thermostatic shaking tank 9DKZ-2, China), stainless steel electric heating board (BD-3, China), and so on.

The ware glass and Teflon vessels used in the processes and analysis all were soaked in 10% HNO_3 solution over one night, washed with deionized water, and dried at lower temperature (∠50 °C) before use. The agents used in the processes all were pure analysis grade. The water used all are double time distilled water generated by automatic dual water distiller (SZ-93, Shanghai). The rate of recovery for the analysis was detected by the parallel determination times for every type of samples associated with the detection for the detecting limit and sensitivity of the used instruments.

The data on these features are integrated in below Table 3.3.

The sequential extraction method recommended by Tessier (1997) was referred for the analysis of trace metals from the sediments [2]. The sample used in experimental extraction for every sample was 1 g, and the steps and the reagents are shown in Table 3.4.

Table 3.4 Sequential extraction procedure and conditions for trace metal speciation analysis

Step number	Speciation	Extraction condition
1	Exchangeable	The soil was extracted with 8 ml of 1.0 M MgCl2 at pH 7.0 for 1 h, with continuous agitation at 18 ± 0.5 °C.
2	Carbonate	The residue from Step 1 was leached for 17.5 h with 8 ml of 1 M NaOAc (adjusted to pH 5.0 with HOAc) at 18 ± 0.5 °C. Continuous agitation was maintained during the extraction.
3	Fe–Ma oxide	The residue from Step 2 was extracted with 20 ml of 0.04 M $NH_2\cdot$ OH·HCl in 25% (v/v) HOAc for 3 h. The extraction was performed at 96 ± 2 °C with occasional agitation.
4	Organic matter and sulfate	The 3 ml of 0.02 M HNO_3 and 5 ml of 30%(v/v) H_2O_2 (adjusted to pH 2.0 with HNO_3) were added to the residue from Step 3. The sample was heated progressively to 83 ± 2 °C and maintained at this temperature for 1.5 h with occasional agitation. A second 3-ml aliquot of 30% (v/v) H_2O_2 (adjusted to pH 2.0 with HNO_3) was then added, and the mixture was heated again at 83 ± 2 °C for 1.5 h with intermittent agitation. After cooling to 18 ± 0.5 °C, 5 ml of 3.2 M NH_4·OAc in 20% (v/v) HNO_3 was added, followed by dilution to a final volume of 20 ml with deionized water. The tubes were then continuously agitated for 30 min.
5	Residual	The residue from Step 4 was digested with 4 ml concentrated HNO_3 (70%), 2 ml $HClO_4$ (60%), and 15 ml HF (40%) to dryness using the following heating regime: 90 °C for 6 h, 120 °C for 10 h, 190 °C for 6 h. The remaining material was then taken up in 5 ml of 4 M HCl at 70 °C for 1 h, and diluted to 25 ml with 0.3 M HCl.

After the processes mentioned above, the first four fractions of the speciation from the extraction were, respectively, centrifuged at rotating speed 12,000 r/min for 15 min, and the supernatant was, respectively, analyzed for the relative elements by atomic absorption spectrophotometer and double-channel atomic fluorescence spectrometer. The residual was, respectively, digested directly by acids to be analyzed.

(1) **Measurement for the trace metals in the water, suspended materials, and sediment**

Zinc and copper were measured, respectively, by atomic absorption spectrophotometer, and the working parameters are shown in Table 3.5. Cadmium, lead, and arsenic were measured, respectively, by atomic fluorescence, and the working parameters are shown in Table 3.6. Mercury is measured by automatic mercury analyzer.

The data, correlation coefficient, and correlation equation of standard curve of analysis for each element are shown in Table 3.7.

Table 3.5 Parameters of the atomic absorption spectrometer

Parameters of instrument	Zn	Cu
Wavelength (nm)	213.9	324.8
Lamp current (mA)	10	7.5
Ethyne flux (km/m^3)	0.10	0.15
Air flux (kg/cm^3)	1.6	1.6
Slit (mm)	1.3	1.3
Flame height (mm)	5.0	5.0

Table 3.6 Parameters of the atomic fluorescence spectrometer

Parameters of instrument	As	Hg	Cd	Pb
Photomultiplier negative high voltage (V)	270	270	270	270
Temperature of atomizer (°C)	200	200	200	200
Atomizer height (mm)	8	8	8	8
Lamp current (mA)	60	30	60	80
Carrier gas flux (mL/min)	400	400	400	400
Shielding gas flux (mL/min)	1000	1000	1000	1000

(2) **Measurement for PCBs in the sediment**

Total PCBs in surface sediments in different sections of the Suzhou Creek were determined referring to the US EPA method.

3.1.1.3 The Distribution and Environmental Implications of Trace Metals in the Water–Sediment System

Under certain conditions, most of trace metal ions in water eventually enter into sediment after the interactions including ion exchange, coprecipitation, adsorption, hydrolysis, complexation, flocculation, and other physicochemical actions. However, when conditions change, there are also a number of trace metals would discharge into water from the sediment due to the interactions including diffusion, desorption, dissolution, oxidation or reduction, complexation, biological, and other physical disturbance. In these processes, some of them would improve the water quality to a certain degree, and some of them would make the active concentration of trace metals in water increased to cause significant secondary pollution. These two processes almost take place at the interface of water and sediment, and only a part of trace metals in sediments involves these reactions. The amount of trace metals involving in the reactions, the modality, and extent of reactions depends much on trace metal species in water and sediments, water quality, the components of sediments as well as other environmental factors. Therefore, to determine the complicate chemical processes

Table 3.7 Parameters of standard curve on the analysis of metals

Sample number	Zn		Cu		Pb		As		Cd		Hg	
	$C^a/10^{-6}$	A^b	$C/10^{-6}$	A	$C/10^{-6}$	A	$C/10^{-6}$	A	$C/10^{-6}$	A	$C/10^{-9}$	A
1	0.000	0.000	0.000	0.000	0	0.0	0.00	0.00	0	0.00	0.0	0.00
2	0.500	0.131	1.000	0.011	1	2.2	1.00	2.60	1	0.40	0.1	12.30
3	1.000	0.218	2.000	0.020	5	7.3	5.00	6.80	5	1.20	0.2	25.40
4	3.000	0.684	3.000	0.031	10	13.2	10.00	12.90	10	4.70	0.4	48.10
5	4.000	0.931	4.000	0.039	20	26.1	15.00	18.80	20	7.60	0.6	76.30
6					40	51.7	20.00	24.80	40	15.70	1.0	112.50
FEqua[c]	$Y = 0.2306x + 0.0008$		$Y = 0.0098x + 0.0006$		$Y = 1.2788x + 0.5515$		$Y = 1.2077x + 0.7177$		$Y = 0.3941x - 0.0586$		$Y = 114.14x + 2.012$	
CCoe[d]	$R = 0.9995$		$R = 0.9987$		$R = 0.9998$		$R = 0.9989$		$R = 0.9966$		$R = 0.9970$	

[a]C means concentration
[b]A means absorbancy
[c]FEqua means fitted equation
[d]CCoe means correlation coefficient

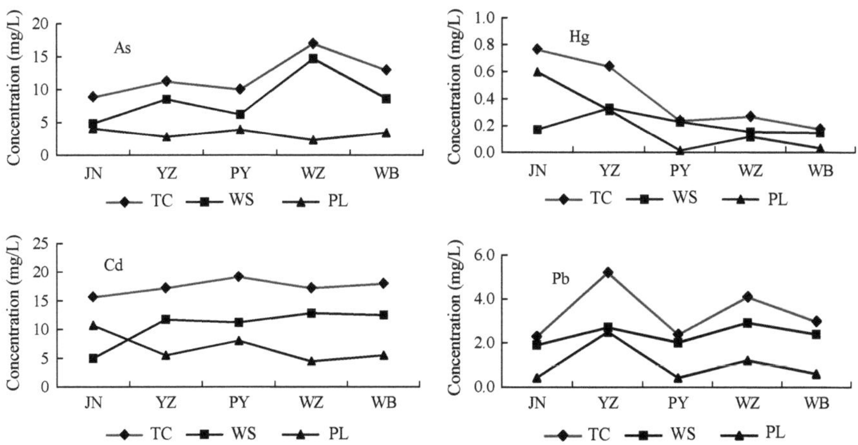

Fig. 3.2 Distribution of trace metals in the Suzhou Creek. JN means the Jiangnan Paper Mill sampling site, YZ means the Shanghai grease plant site, PY means the Pengyuepu Creek site, ZH means the Zhonghuaxinlu Bridge site, WZ means the Wuzhenlu Bridge site, and WB means the Waibaidu Bridge site. TC means the total concentration of the metals, WS means water soluble of the metals, and PL means the metals in the particulate, respectively

of trace metal contaminants in rivers and lakes, the water–sediment systems and speciation (species) of trace metals must be studied firstly.

1. **The distribution of trace metals in the sediment from the Suzhou Creek**

(1) **The distribution of trace metals in the creek water**

The total amounts and dissolved species of six trace metal elements, Zn, Cu, Pb, Cd, As, and Hg in the creek water were determined, and their concentration and the variation trend along the river are, respectively, shown in Fig. 3.2 (Cu and Zn were below the detecting limits). The total amount of trace metals in water samples was measured directly. The dissolved species of trace metals were obtained from the filtrated water that was filtered by 0.45 μm membrane. And the particle species of trace metals were obtained by calculating the difference between the total concentration and the dissolved species concentration.

The data shown in Fig. 3.2 suggest that the range of total amount of trace metals in the Suzhou Creek is as follows: As 8.9–17.0 μg/L, Hg 0.24–0.767 μg/L, Cd 15.7–19.2 μg/L, Pb 2.3–5.2 μg/L. According to the prospected function of the water body in Shanghai, the water quality of the Suzhou Creek should conform to the standard of surface water class III. Measuring with the national water quality standard of surface water class III, As, Hg, and Cd in the Suzhou Creek all have exceeded the limits, respectively. Among the elements, arsenic in the water at the Wuzhenlu Bridge section is the highest, Hg and Cd are also relatively higher, the highest value of Hg is found in the Jiangnan Paper Mill section, and the highest Cd appears in

Table 3.8 Comparison for trace metal concentrations between in the Suzhou Creek and other water systems (μg/L)

Water system name	As	Hg	Cd	Pb
Suzhou Creek	8.9–17.0	0.24–0.767	15.7–19.2	2.3–5.2
The headstream of the Yangtze River [3]	3.32	0.009	0.046	3.18
Black water lake [4]	1.27	0.08	0.12	18.3
Bohai Sea Gulf [5]	1.31–4.22		0.056–0.129	0.73–16.74

Pengyuepu Creek section. The cadmium is generally higher along the creek flow, indicating that Cd pollution in the Suzhou Creek should be paid special attention to.

Suzhou Creek is a typical urban river. To compare with that of other cities of China, trace metals in the water of the industrial areas are, respectively, shown in Table 3.8. It is shown that the concentrations of Cd, Hg, and As all were relatively higher, while Pb is lower in the water of the Suzhou Creek. In fact, Cd, Hg, and As are all highly toxic, and they should be paid special attention to the problem of the migration and diffusion of their pollutions in the Suzhou Creek.

The smelting process of Zinc is a major source of cadmium in environment, and the lead or copper smelting would also discharge cadmium. The use of element Cd in industry is very wide, as we know including electroplating, plasticizer, pigment production, Ni-Cd battery production, etc. The famous case is the solution from process in electroplating factory, in which the wastewater sometimes contains cadmium content as high as 2200 mg/L, and usually it was discharged directly into the surrounding water body. In addition, a small amount of cadmium is also contained in phosphate fertilizer, sludge, and fossil fuel. It is known that the harm of cadmium to human body is serious. In addition, when cadmium entering into the human body, kidney and liver would be greatly damaged, and then a serious deterioration on metabolism in the human body would happen. The most famous event on cadmium pollution incident had taken place in Japan, the bone pain disease events in the early 1900s. Moreover, due to Cd^{2+} radius is very close to Ca^{2+}, Cd will hinder the absorption of calcium and destroy the bone having formed. According to the investigation by Zu et al. (1998), the discharging of Cd is about 196.77 t/a from industrial and domestic wastewater to the Suzhou Creek, which is the main source of cadmium in Suzhou Creek, with a large number of trace metals every year. It is an important means to control the pollution by prohibiting the discharge [6].

The variation trend of the concentration of trace metals in Suzhou Creek shown in Fig. 3.2 suggests that there was no obvious rule in the distribution of Pb in the creek indicated by those samples from the sits, while those of As, Hg, and Cd were relatively stable. Especially the variation was obvious for those across Putuo, Jingan, Zhabei, and Huangpu district areas and downward from Shanghai oil plant. The social and economic structures in these regions are very complicated. It is the industrial concentration area in the range from Shanghai oil plant to Pengyuepu Creek, while the population and traffic in this zone were very density too from the Pengyuepu

Table 3.9 Total concentrations of the trace metals discharged into the Suzhou Creek per year [6]

Element	Pb	Cd	Cu	Hg	As	Zn
Discharging value (t/a)	2744	196.77	9451.91	148.57	638.14	70881.91

Creek to the Wuzhenlu Bridge. There was vast industrial wastewater and domestic sewage from this zone taking along serious pollution load to the Suzhou Creek with different features, respectively.

(2) **The distribution of trace metals in the sediments and suspended matter**

A. The distribution of trace metals in the suspended matter

The particulate matter in the river is one of the most important substance components in the river environment, which plays a significant role on the bioavailability of pollutants. Particulate matter is often as a strong agent to act with adsorption for the pollutants discharged into the water, especially for trace metals. These substances, particulate matter, are the main factor that controls the migration, transformation of trace metals, and river water self-purification. Suspended matter is the main carrier for the migration of pollutants in river, and sediments are the main storage sites of pollutants, in which the trace metals are often many times higher than the water, and sometimes can be higher than a few orders of magnitude. When pH, Eh, and other conditions of the water change, the adsorbed pollutants can be released from the sediments and cause secondary pollution. In addition, many benthic and filter feeding organisms can inhale a large amount of particulate matter through their digestive system, so that they may contain toxic substances. Moreover, the organisms in the sediment take organic debris as their basic food to accumulate toxic substances again. The organisms in water such as fish can also accumulate toxic substances directly from using the particulate matter as their food. Therefore, the monitoring of particulate matter is of great significance to reveal the pollution of water environment and the rule of water self-purification. So, at the certain extent the pollution of the water environment can be understood through the study on particulate matter in it.

As one most important surface river in Shanghai, the emission of main trace metals into the Suzhou Creek is shown in Table 3.9.

Trace metal pollutants discharged into the water body, usually the vast majority of them may be attached to the suspended matter, and move with the water flow or enter the sediment by the process of the flocculation or sedimentation. Subsequently, they may move toward the downstream at the bottom of the water or as the suspended matter in the water. Therefore, the adsorption of trace metals by the suspended sediment is one of the most important steps in the process of migration and transformation. Generally, suspended matter often shows greater adsorption capacity toward trace metals than the sediments, which is attributed to that suspended particles contain more clay and mica, while sediments contain the higher content of carbonate and lower content of clay. This has been reported in a lot of references [7]. The particulate matter concentration in the Suzhou Creek is shown in Table 3.10.

Table 3.10 Concentration of the suspended particles in the Suzhou Creek

Sampling site	Jiangnan Paper Mill	Shanghai oil and grease factory	Pengyuepu Creek	Wuzhenlu Bridge	Waibaidu Bridge	Zhonghuaxinlu Bridge
Concentration of particles (mg/L)	20.8	21.2	38.8	89.2	156	47.3

Table 3.11 Concentrations of the trace metals in the suspended matters of the Suzhou Creek (μg/g)

Sampling site	Pb	Cd	As	Hg	Zn
Jiangnan Paper Mill	0.6	16.39	5.99	0.943	2889.34
Shanghai grease factory	0.8	6.53	4.55	1.33	1375.5
Pengyuepu Creek	0.9	4.7	5.93	0.465	905
Wuzhenlu Bridge	0.45	3.19	5.04	0.694	753.49
Waibaidu Bridge	0.3	3.95	5	0.73	825
Zhonghuaxinlu Bridge	1.2	4.24	6.34	1.627	1165.84
Average value	0.71	6.5	5.48	0.965	1319.03

There is no relevant report on the trace metals in the suspended matter in the Suzhou Creek up to now. The total amount of trace metals in those was measured in our work, and the results are shown in Table 3.11.

The results show that the trace metals in the suspended matter of the Suzhou Creek indicate the characteristics as follows:

Overall, the concentrations of the trace metals in the Suzhou Creek were high except Pb. The concentrations of the elements were, respectively, Cd 3.19–16.39 μg/g, As 4.55–6.34 μg/g, Hg 0.465–1.627 μg/g, Zn 753.49–2889.34 μg/g. Among them, the highest values of both Zn and Cd were in the Jiangnan Paper Mill site, and gradually decreased toward the lower reaches; the highest value of Hg appeared in Shanghai oil plant site, and gradually decreased toward the downstream. This corresponded obviously to the feature of the industrial distribution along the Suzhou Creek.

The amount of trace metals in suspended matter in Suzhou Creek is relatively higher, and the suspended matter is easy to migrate under the action of water flow because of its smaller particle size, which is with destructed impact to the water quality of the Huangpu River. It is necessary strictly controlling for the discharge of all kinds of polluted wastewater to the creek in order to prevent the adsorption and re-release of trace metals in the river water.

B. The amount and distribution of trace metals in the sediments of the creek

Trace metals are accumulated easily in river sediments and have obvious distribution rule or features [8]. Most trace metals have a great variety of element valence states. After hydrolyzing, trace metal salts are easy to produce hydroxide, sulfide, carbonate,

Table 3.12 Concetrations of the trace metals in the sediment of the Suzhou Creek (μg/g)

Sampling site	Pb	Cd	Cu	Hg	As	Zn
Jiangnan Paper Mill	47.53	1.32	97.88	0.069	14.68	852.63
Shanghai grease factory	64.36	2.53	137.38	0.33	17.95	1186.07
Zhonghuaxinlu Bridge	103.54	0.82	190.17	1.986	43.24	1269.79
Wuzhenlu Bridge	48.49	0.15	147.61	0.171	19.56	796.05
Average value	65.98	1.21	143.26	0.639	23.86	1026.14
Background value of soil in Shanghai [10]	19.85	0.148	23.85	0.152	9.07	79.22
The criterion[a] [11]	1000	20	500	15	20	1000

[a]The state standard for sludge to be used for agriculture

etc., which are easily formed precipitation. The reaction between trace metals and anions can produce clathrate, which can be adsorbed by minerals, organic matter, and accumulate in sediment [9]. Trace metals in sediments may be re-released, because of the change of salinity, the change of redox conditions, the decrease of pH, the use of natural or the synthetic complexing agent, the biological effect, etc., which can lead to secondary pollution.

The trace metals in the sediments of the Suzhou Creek have been reported in the relevant reports. In this work, samples were taken at four sampling sites, the amount of the concentration of trace metals in the water was compared, and the speciation distribution of the samples was studied. The results of the determination of trace metals in the sediments of the Suzhou Creek are shown in Table 3.12.

The results showed that the amount of trace metals in the sediments of the Suzhou Creek was relatively higher. The amount of the trace metals were as follows: Pb 47.53–103.54 μg/g, Cd 0.15–2.53 μg/g, Cu 97.88–190.17 μg/g, Hg 0.069–1.986 μg/g, As 14.68–43.24 μg/g, and Zn 796.05–1269.79 μg/g. The pollutions of Cd, Zn, and Hg were relatively serious in the part of the creek. The amount was ten times or higher than the soil background value of the region. The amount of Zn had exceeded the control standard of the sludge used for agriculture, in some part of the creek.

Research has shown that the more developed the industry, the more serious the pollution is. As shown in Table 3.13, the amount of Zn, Cu, and Hg in the sediments of the Suzhou Creek were significantly higher than the urban rivers in developed countries, the Yangtze River, and the Yellow River of China.

According to the statistical data from 2000, wastewater discharged by the top three enterprises of the sewage discharge around the Suzhou Creek (Qingpu Chemical Plant, Jiangnan Paper Mill, Shanghai No.8 Dyestuff Chemical Plant) contained one or more kinds of metals in Cd, Hg, Zn, and Cu [21], and the development of cleaning production and wastewater treatment for these enterprises is an important part of the prevention of trace metal pollution in the sediments of the creek.

Table 3.13 Trace metals in the sediments from city water systems (μg/g)

River name	Pb	Cd	Cu	Hg	As	Zn
Suzhou Creek	47.53–103.54	1.02–2.92	97.88–190.17	0.085–1.986	14.68–43.24	796.05–1186.07
The Yangtze River [12]	23.7	0.215	26.7	0.044	8.8	80.1
The main sections of the Yangtze Rive of passing cities [7]	53.41	0.176	35.35	–	–	90.29
The Baotou section of the Yellow River [13]	22–31	1.7–3.2	16–20	–	–	43–80
The Hangzhou section of the Grand Canal [14]	81.41	1.4	127.3	–	–	659
The Delft Canal of the River Rhine [15]	109	2.3	83	–	–	515
The Shenzhen River of Hong Kong [16]	13.9–407.89	0.62–3.95	8.29–768.05	–	–	63.26–1095.56
The Kanpur section of the Ganges River [17]	32	0.31	– [a]	–	–	204
The Aswan section of the River Nile [18, 19]	110	13	86	–	–	201
The River Nile [20]	59.9–333	9.0–18.0	22–535	–	–	78–449

[a]The symbol – means no data for this element

C. Relationship between the trace metals in suspended matter and that in sediment

The water suspended matter and the river sediment can transform each other, so that they have a certain similarity in composition. As shown in Table 3.13, the trace metals in the sediments and suspended matter of the Suzhou Creek have the following characteristics:

a. The amount of Zn and Cd in suspended matter are relatively higher in the section of Jiangnan Paper Mill. Other than that, the change trend is basically the same as that of the sediment and shows a good correlation (the correlation coefficient of Pb, Cd, As, Hg, and Zn were 0.97, 0.86, 0.98, 0.82, and 0.88, respectively) as shown in Fig. 3.3. There are an internal dependence relationship between the migration and transformation of trace metals in suspended matter and sediment under hydrodynamic and environmental conditions in the Suzhou Creek.
b. The amounts of other trace metals in the suspended matter are higher than that in the sediment, except Pb and As. It shows that the water quality evolution of the Suzhou Creek has been influenced by the important sources of trace metals. The part of these metals flows into the Huangpu River along with the creek, which leads to the pollution in downstream water. Under certain conditions, another part of these metals would be settled into the sediment, which led to secondary pollution.

In summary, the trace metal pollution in the Suzhou Creek is relatively serious, especially the pollution of Zn, Hg, and Cd, a large number of trace metals are enriched in suspended solids and sediments, which is one of the important hidden dangers of trace metal pollution in the Suzhou Creek. According to the natural conditions and trace metal distribution characteristics of the Suzhou Creek, it should be took measures to repair its governance and eliminate hidden dangers as soon as possible.

2. **Discussion on the migration and accumulation of trace metals in the water and sediment system of the Suzhou Creek**

The material exchange cycle in the phases of water, suspended matter, and sediment leads to the release or accumulation of metals. So interaction of trace metals in the three phases is a key to understanding the behavior and end result of trace metals in the Suzhou Creek. By analyzing the amount, distribution rule, and chemical speciation distribution of Hg, Cd, Pb, and As in river water and sediment, the characteristics of their migration and accumulation were studied in this work to explain the rule of distribution and transformation of trace metals in the Suzhou Creek.

(1) **The accumulation of trace metals in the sediment system of the Suzhou Creek**

The percentage content of the water-soluble Pb, Hg, Cd, As and particulates Pb, Hg, Cd, As in the Suzhou Creek were listed in Table 3.14.

The soluble trace metals in water can directly flow into the Huangpu River along the creek, its rate of migration might be the fastest, its distance of migration might be furthest, and its influence on water quality of downstream and Huangpu River

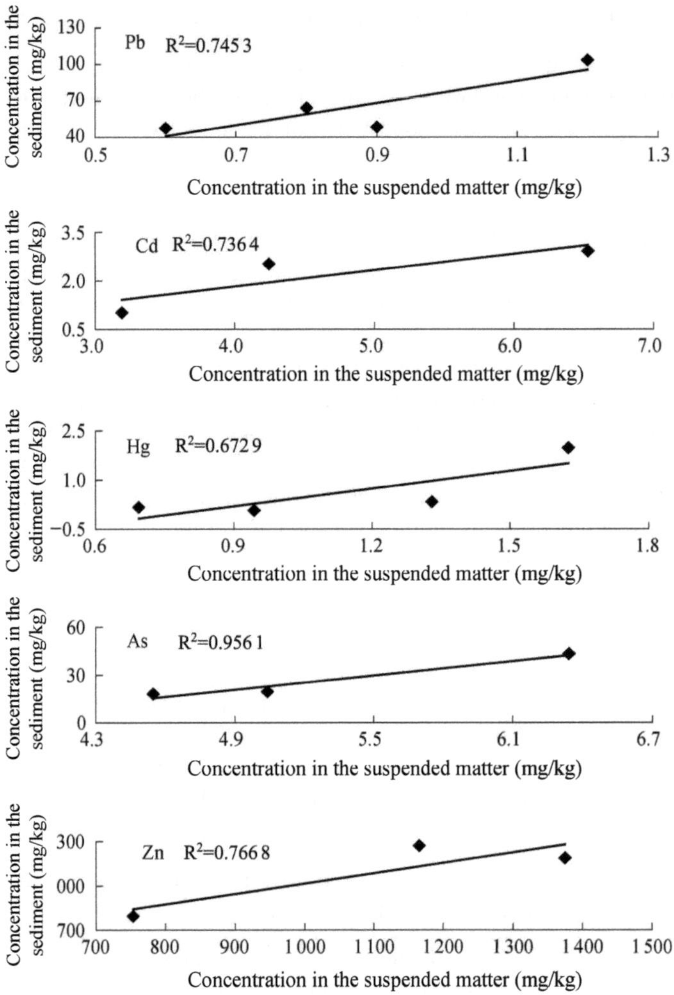

Fig. 3.3 Relationship between concentrations of trace metals in the sediment and suspended matter of the Suzhou Creek

might be the largest too. Particulate trace metals can enter the sediment by migration or flocculation deposition. The watercourse of the Suzhou Creek is sinuous and the flow rate of the water is slow, which leads to a large number of particles settle down. As shown in Fig. 3.3, the correlation of trace metals in sediment and that in suspended matter is good, which explains the migration and transformation of trace metal elements are relation on intrinsic dependence, under hydrodynamic and environmental conditions in the Suzhou Creek. Therefore, this part of the metals could be regarded as trace metal deposition amount.

Table 3.14 Distribution of trace metals in the water from the Suzhou Creek (%)

Element	Media	Jiangnan Paper Mill	Shanghai grease plant	Pengyuepu Creek	Wuzhenlu Bridge	Waibaidu Bridge	Average value
As	In water	53.9	75.2	61.4	86.4	66.2	68.6
	In particle	46.1	24.8	38.6	13.6	33.8	31.4
Hg	In water	21.9	51.3	94.6	57.1	82.8	61.5
	In particle	78.1	48.7	5.4	42.9	17.2	38.5
Cd	In water	73.1	68	58.3	74.4	69.4	73.1
	In particle	26.9	32	41.7	25.6	30.6	26.9
Pb	In water	82.6	51.9	83.3	70.7	80	72.1
	In particle	17.4	48.1	16.7	29.3	20	27.9

Table 3.14 data showed that As, Hg, Cd, and Pb in the Suzhou Creek were mainly soluble form in water. The order of the average values of them was as follows: Cd > Pb > As > Hg. This fact suggested that the mobility of Cd is the strongest, which led to a wide range of toxic effects and the mobility of Hg was the weakest. The above situation showed that trace metals in the water system of the Suzhou Creek flowed into the Huangpu River with the creek, the cumulant of Cd in the downstream, and the East China Sea could be largest accumulated.

The Waibaidu Bridge is located at the entrance of the Suzhou Creek to the Huangpu River. The amount of trace metals in this water site can be estimated to understand the trace metal migration and the contribution for the Suzhou Creek to the Huangpu River in the trace metal pollution. The formula is calculated as follows [22]:

The migration volume of trace metals = The migration volume in water-soluble trace metals + The migration volume of in particulate trace metals;

The migration volume in water-soluble trace metals = Trace metal concentration in water × The volume of runoff;

The migration volume of in particulate trace metals = Trace metal concentration in the suspended matter × The sediment runoff.

The results of calculation are shown in Table 3.15, trace metals seriously affect the water quality of the Huangpu River, and trace metals in the Suzhou Creek are an important source of that in the Huangpu River, which is a problem that needs to be highly valued.

Enrichment factors (K_S and K_D) of trace metals [23] in suspended particulates and sediments should be calculated to understand the concentration of As, Hg, and Cd in suspended matter in the Suzhou Creek. The calculation formula was as follows. The calculation results are shown in Table 3.16.

Table 3.15 Amount of the trace metals transferred to the Yangtze River through the estuary of the Suzhou Creek (t/a)

Media	Pb	Cd	As	Hg	Zn
In water	0.448	0	1.604	0.027	0
In suspended substances	0.009	0.115	0.145	0.021	24
Total amount	0.457	0.115	1.749	0.048	24

Table 3.16 Accumulation indexes of the trace metals in the sediment of the Suzhou Creek

Indexes	Element	Jiangnan Paper Mill	Shanghai grease factory	Pengyuepu Creek	Wuzhenlu Bridge	Average value
K_S	Hg	1229	4043	15,644	4536	6363
	Cd	423	555	446	249	418
	As	951	433	556	302	561
K_D	Hg	90	1003	19,096	1118	5327
	Cd	264	250	266	80	215
	As	2330	1709	3793	1171	2251

$$K_D = \frac{\text{concentration of trace metals in sediment } (\mu g/g)}{\text{concnetration of water soluble trace metals } (\mu g/g)}$$

Table 3.16 data show that the accumulation of Hg in suspended matter and sediment are the most, followed by As and Cd which are least. When the Hg concentration in the creek water is lower or other environmental conditions are changed, the Hg in the sediment is likely to be released and cause secondary pollution, which is one of the important hidden dangers of trace metal pollution in the Suzhou Creek.

(2) **The chemical speciation of trace metals in the water–sediment system of the Suzhou Creek**

It is known that the total amount of trace metals has not been very good to indicate the bioaccessibility and toxicity for the trace metals. In water–sediment system, the potential harm impact of trace metals depends on the amount of bioavailable trace metals [20, 24]. Trace metals consisting in various liquid- or solid-state materials exist in the different forms in the environment affected by the physical and chemical actions in the environment [23, 24]. Different forms of trace metals have different environmental behavior and biological effects. Generally, only a certain form or a few forms of the metals may be absorbed by the biological, and cause harm to the biological. Some scholars have used multiple correlation analysis to study the relationship between trace metals in sediments of geochemical facies and the amount of metal accumulation in benthic invertebrates and found that the copper accumulation in living organisms had a good correlation with the amount of copper in exchangeable fraction and reduction fraction, instead of the total amount of the copper in sediment, which showed that copper in the these two fractions was easily absorbed

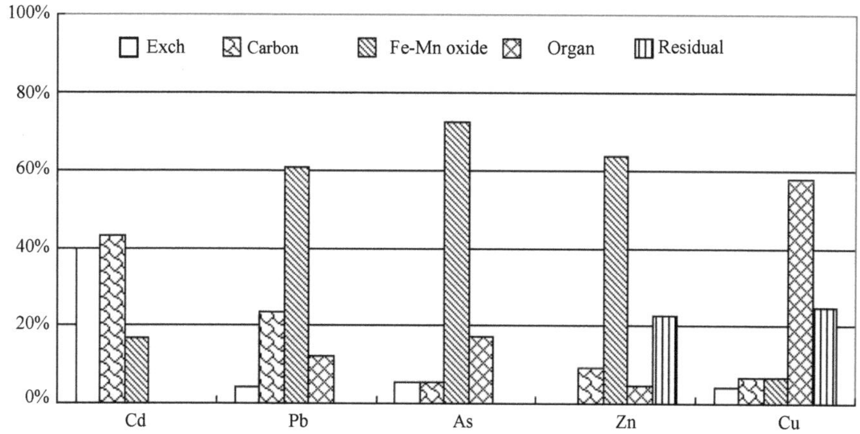

Fig. 3.4 Distribution of the speciation of trace metals in the sediment from the Jiangnan Paper Mill site of the Suzhou Creek. Exch means exchangeable speciation, and Carbon means carbonate, Fe–Mn oxide means Fe–Mn oxides, Organ means organic matter, Residual means residual

and retained by the organism [25]. It can be seen that the speciation of trace metals is an important parameter for the bioactivity of trace metals in sediments and main factor to determine the environmental behavior and biological effects of the trace metals in environment [25]. The speciation is significant for understanding the ecological effects of the trace metals [26].

The five-step sequential extraction procedure for trace metal speciation analysis recommended by Tessier et al. was used in this study [2]. The forms of trace metals are divided into exchangeable, carbonate, Fe–Mn oxide, organic sulfides, and residual fraction. Generally, ion-exchange species is easy to be absorbed by organism. The carbonate can be used by organism when the pH changes. Under the condition of reduction, iron–manganese oxide hydrate can be released and easy to be absorbed by organism, and only under the condition of strong oxidation, organic sulfide species can be released. The species of residual of trace metals could not be used by organisms at common conditions of the environment and only under special conditions may be absorbed by organism for the species of residual of trace metals, because most of the metals combined with residual species are bounded in the mineral lattice, affected by mineral composition and soil or sediment erosion.

A large number of studies show that the biology ability to absorb trace metals from sediment first depends on the speciation of trace metals in the sediments. Distribution characteristics of various fractions of trace metals in the sediments of the Suzhou Creek are shown in Figs. 3.4, 3.5, 3.6 and 3.7 respectively.

The results show that the distribution of trace metals in the sediments of the Suzhou Creek has the following characteristics:

A. The exchangeable fraction Cd is accounted for its total concentration amount of 24.8–40.0%, this proportion of exchangeable fraction of Pb and Cu is less than

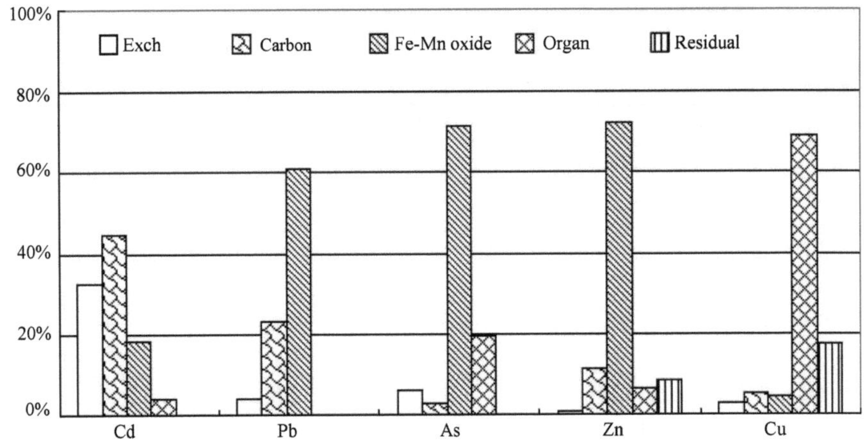

Fig. 3.5 Distribution of the speciation of trace metals in the sediment from the Shanghai oil and grease plant site of the Suzhou Creek. Exch means exchangeable species, Carbon means carbonate, Fe–Mn oxide means Fe–Mn oxides, Organ means organic matter, Residual means residual

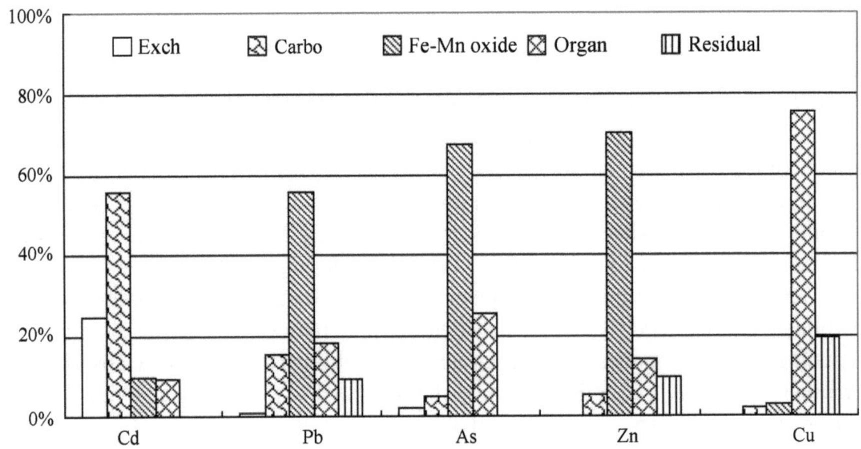

Fig. 3.6 Distribution of the speciation of trace metals in the sediment from the Zhonghuaxinlu Bridge site of the Suzhou Creek. Exch means exchangeable species, Carbon means carbonate, Fe–Mn oxide means Fe–Mn oxides, Organ means organic matter, Residual means residual

5%, the proportion of exchangeable Zn was less than 1.4%, and exchangeable As accounted for the total amount of 2.1–6.1%. The order of the amount of exchangeable trace metals is as below: Cd > As > Pb > Cu > Zn. The high amount of exchangeable Cd is related to itself adsorbed by sediments. Some studies have showed that the element Cd is only adsorbed on the exchange sites of the adsorbate, rather than the fixed internal position of the crystal lattice [27], so it is difficult to combine with clay minerals, metal oxides, organics, and so on.

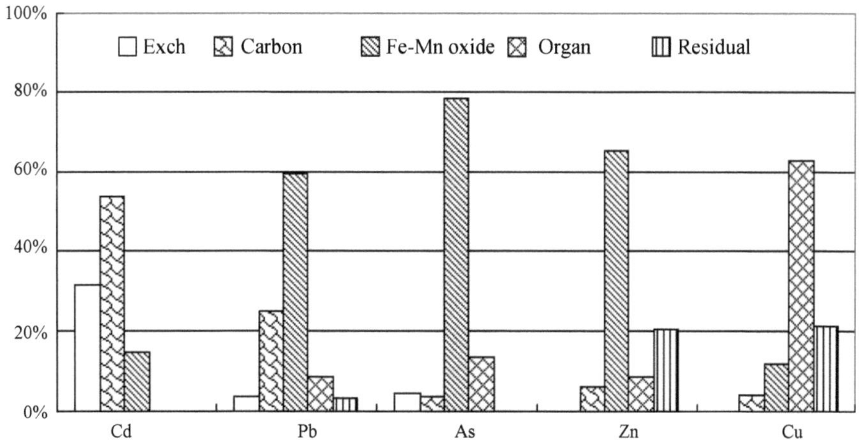

Fig. 3.7 Distribution of the speciation of trace metals in the sediment from the Wuzhenlu Bridge site of the Suzhou Creek. Exch means exchangeable species, Carbon means carbonate, Fe–Mn oxide means Fe–Mn oxides, Organ means organic matter, Residual means residual

In addition, Cd is relatively active in environment, when pH is less than 8 there are only Cd^{2+} exist in the environment, and when pH reaches 9, the $Cd(OH)_2$ is generated, that is why the higher amount of Cd existed in this species.

B. There are various metal carbonates in the sediments. The ratio of Cd carbonate is 43.2–55.7% which is the highest, the ratio of Pb carbonate is 15.6–24.9%, the ratio of Zn carbonate is 5.6–11.6%, the ratio of Cu carbonate is 2.3–6.8%, the ratio of Zn carbonate is 5.6–11.6%, and the ratio of As carbonate is 2.9–5.3%. The different metal carbonate amount order is as follows: Cd > Pb > Zn > Cu > As. Zhao Jian et al. reported that soil in Shanghai area is alkalescent and the pH value is within 7.0–8.0 [28], which may be one of the reasons for the high carbonate meatal amount in the sediments of the Suzhou Creek, because in alkaline conditions the metal is not easy to release from carrier to the system. However, according to the bulletin of the environmental situation of Shanghai in 2005, the average pH value of precipitation in Shanghai in 2005 was 4.93 and acid rain frequency is 40%. Compared to 2004 year, the frequency of acid rain increases 7.3%. Under such environmental conditions, acid rain might lead to the release of metals in the carbonate in the sediments of the Suzhou Creek, which may lead to a more serious secondary pollution [29].

C. When the redox potential of the water is decreased or the water body is anoxic, Fe–Mn oxide metal is easy to be reduced, which leads to secondary pollution. The chemical fractions of Pb, Zn, and As are mainly Fe–Mn oxides in general environment. The Pb, Zn, and As in the sediment of the Suzhou Creek are contained mainly in the fraction of Fe–Mn oxide speciation. Their respective proportions in their total concentration were as follows: Pb 55.7–75.9%, Zn 63.5–72%, As 67.5–78.5%. The amount of Cd and Cu for Fe–Mn oxide fraction is relatively lower, Cd was 14.7–20.9%, and Cu was 2.8–11.8%. The Fe–Mn oxide fraction ratio in the total concentration of metals in the sediment of the Suzhou Creek is as the order as follows: As > Zn > Pb > Cd > Cu. The redox potential of the sediment in the Suzhou Creek is relatively lower, because of the low dissolved oxygen and high organic matter content in the creek, some organic ligands in organic pollutants are easy to form stable complex with Fe–Mn and so on, which releases Fe–Mn ion from the solid phase to water. The obvious decrease of the trace metal amount in Fe and Mn oxide fraction in the suspended matter of the creek may be related to that.

D. The Cu amount of organic sulfur compounds in the sediments of the Suzhou Creek is highest in the researched metals, 57.7–75.5% in its total concentration. The reason for this phenomenon may be that copper has high affinity for organic compounds and sulfides. The amount in the sediment of the Suzhou Creek for other metals is as follows: Cd 0–9.5%, Pb 8.5–18.2%, As 13.6–25.4%, Zn 4.7–14.3%. The Cu in the sediment is more difficult to release, and is less threatening to the environment, because organic compounds and sulfides can be decomposed only under strong oxidizing conditions, while the Suzhou Creek is in the condition of anoxic environment for a long time.
E. The metal in residual can be released only in the weathering process because it is in mineral lattice, so it has a strong stability in the sediment and basically cannot be used by organisms. In this work, Cd and As are not detected out in the residual species, and Pb and Zn are not detected out in some samples. In the samples, the ratio of Pb in residue fraction is 8.6–22.6% and Cu was 17.8–24.7%, which could proved that the amount of trace metals in the sediments of the Suzhou Creek is not very high, and most metals exist in the effective fraction. It indicates that the potential ecological harm from the trace metals in the sediments of the Suzhou Creek is higher, and the possibility of the secondary pollution of the water body is larger.

The distribution of the effective fraction amount of trace metals in the sediments of the Suzhou Creek is as follows:

A. Exchangeable species: In the five metals studied, the amount of the exchangeable species is highest in Shanghai oil plant samples, which suggests that the mobility and bioactivity of trace metals in the sediments here is the largest.
B. Carbonate species: The highest amount of Cd, Pb, and As in this fraction also appears in Shanghai grease plant, the highest amount of Cu and Zn in this species appears in Zhonghuaxinlu Bridge site.
C. Fe–Mn oxide species: The highest amount of Pb, Zn, and As in this fraction appears in Zhonghuaxinlu Bridge, the highest amount of Cd in the fraction appears in Shanghai grease plant too, and the highest amount of Cu in the fraction appears in Wuzhenlu Bridge site.
D. Oxidation species: The amount of the five metals in the organic fraction is highest in Zhonghuaxinlu Bridge site, which may be related to the serious organic pollution in this creek.

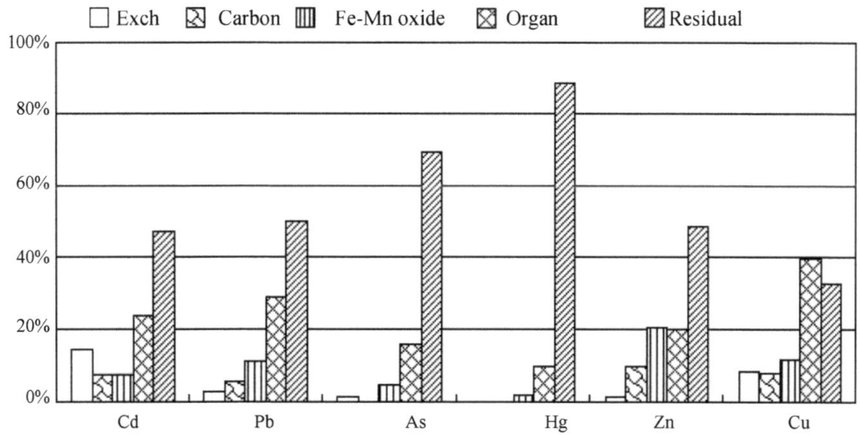

Fig. 3.8 Distribution of the speciation of trace metals in the suspended matter from the Wuzhenlu Bridge site of the Suzhou Creek. Exch means exchangeable species, Carbon means carbonate, Fe–Mn oxide means Fe–Mn oxides, Organ means organic matter, Residual means residual

To sum up, the proportion of residual species trace metals in the sediments of the Suzhou Creek is relatively lower, and the effective species is relatively higher. The exchangeable and carbonate species of Cd is relatively higher, which is more harmful to the biont in the environment. The Fe–Mn oxide species of Pb, Zn, and As is relatively higher; these metals are easy to cause secondary pollution when the redox potential of the water system is low. Cu was mainly in the organic species. At each sampling point, the exchangeable and carbonate of trace metals in the sediments of Shanghai grease plant site are the highest, and the organic fraction of trace metals in the sediment of Zhonghuaxinlu Bridge site is the highest. Generally, the potential ecological hazards of trace metals in the sediments of the Suzhou Creek should be paid highly attention to.

Compared with the study about the speciation of trace metals in sediments, the study about the speciation of trace metals in suspended matters is less, the reason of that it might be difficult to collect a large number of suspended matter samples, and the analysis work is so heavy. Suspended matter is one of the direct carrier media of substances in water, which is influenced by the environmental conditions of the water body (pH, Eh, etc.), and the trace metal carried by suspended matter has a strong biological activity. In this study, the species of the trace metals in the suspended matter in Wuzhenlu Bridge site and Waibaidu Bridge site were analyzed, respectively. Results are shown in Figs. 3.8 and 3.9.

The results show that the trace metal speciation distribution in the suspended matter of the Suzhou Creek has the following characteristics:

The order of the amount of exchangeable fraction is as follows: Cd > Cu > Pb > As > Zn > Hg. The ratio of exchangeable Cd was 11.1–14.2%, the ratio of Cu is 5.1–8.3%, the ratio of Pb, Zn, and As is all less than 4%, and Hg was not detected. Commonly, the exchangeable metal is combined with the surface of particles by electrostatic

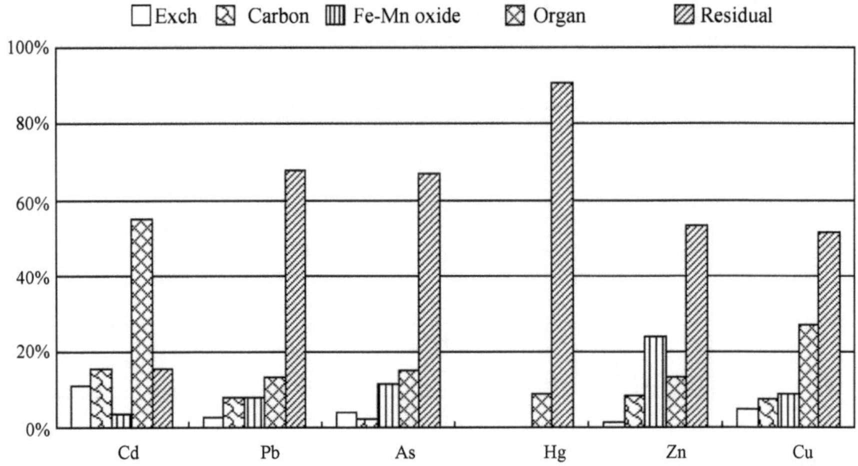

Fig. 3.9 Distribution of the speciation of trace metals in the suspended matter from the Waibaidu Bridge site of the Suzhou Creek. Exch means exchangeable species, Carbon means carbonate, Fe–Mn oxide means Fe–Mn oxides, Organ means organic matter, Residual means residual

interaction, exchange, and adsorption. This amount of the metals is controlled by the concentration of trace metals in water and the distribution coefficient between the ion in water and the particle surface, which has important influence on water trace metal pollution.

The ratio of carbonate in its total concentration for Cd was 7.3–15.3%, for Zn is 8.5–9.8%, for Cu is 7.4–7.8%, for Pb is 5.4–8.1%, for As less than 3%, and Hg is not detected out. The order of accumulation ability of each element is as follows: Cd > Zn > Cu > Pb > As > Hg. Carbonate species is affected by the pH value of the system, and it is easy to release when the river is in the acidic.

In Fe–Mn oxide fraction, the ratio of Zn was 20.5–23.8%, Cu was 9–11.5%, Cd is 3.5–7.3%, Pb is 8.1–10.8%, As was 4.7–11.6%, and Hg less than 2%. The atomic radius of Zn^{2+}(83 pm), Fe^{2+}(83 pm), and Mn^{2+}(92 pm) are similar, and the silicate minerals and iron oxide minerals of Fe^{2+} and Mg^{2+} often contain zinc, which may be the reason for the high amount of zinc in Fe–Mn oxide species. The order of the accumulation ability of trace metals in Fe–Mn oxide fraction in suspended matter of the Suzhou Creek is as follows: Zn > Cu > Pb ~ As > Cd > Hg.

In the suspended matter of the Suzhou Creek, the proportion of organic sulfide trace metals is the highest except the residual fraction. The ratio of Cd and Cu in this form is relatively higher, Cd in this fraction is 23.8–54.9%, Cu is 27.3–39.36%, Pb is 13.5–28.9%, Zn is 13.4–19.8%, and Hg is 9.1–9.8%, which are easy to release under the oxidation condition. About thirty percent of the Suzhou Creek has come from sanitary sewage and industrial wastewater with high organic content. Thus, a considerable amount of trace metals may exist in the fraction of organic matter in the Suzhou Creek. There are two kinds of copper species compounds (Cu^{+} or Cu^{2+}) in nature systems, which are mainly in sulfide or sulfosalt. Copper can exist as an independent mineral and can also replace some of the elements in the sulfide. In addition, Cu^{2+} and Cu^{+} are also easily react with inorganic, organic active group to form complex or chelate [30, 31]. In natural water, the accumulation coefficient of humus on Cd is far greater than that of silica and Kaolinite on Cd, and humus is the main adsorbent of Cd ions in river water, which may be the reason for the high fraction concentration of sulfide organic Cu and Cd.

The residue species is the most species of trace metals in suspended matter of the river, and commonly exists in crystalline minerals, which is considered inert in environment. The ratio of residue in total concentration for Hg was 89.7%, which is highest in the metals researched. As and Zn are little above 50%, Cu and Cd are little greater than 30%. The data show that the trace metals in suspended matter of the Suzhou Creek are relatively stable, and the possibility of their release to the water may be relatively small.

In summary, the residual fraction is the main form of all kinds of metals in the Suzhou Creek. The amount of carbonate metal is relatively low. In the Fe–Mn oxide fraction, the amount of zinc is the higher in the researched metals, the organic fraction is the highest except the residual species for all metals researched, and in the organic fraction Cd and Cu are the highest in metals researched. In exchangeable and carbonate fractions, Cd has the highest accumulation capacity so that it would be with strong mobility and relatively bioavailability.

Speciation distribution of trace metals in suspended matters and that in sediments of the Suzhou Creek is different. Compared with the sediment, most of exchangeable fraction, carbonates, and Fe–Mn oxide fraction of metals are all relatively less. In particular, Fe–Mn oxide fraction is reduced, and the residue fraction is increased. The reason of the change may be caused by the change of the complex geochemical processes in the long-term solid–liquid interaction and deposition, and Fe–Mn oxide is a very good adsorbent for trace metals in particulate matter, under certain conditions, other metals may turn into Fe–Mn oxide during the deposition process. On the other hand, with the change of the system conditions, it is possible for suspended matter to release the metal element to the water, which may be the reason of the difference between both suspended matter and sediment. After point-source pollution has been effectively controlled, the sediment has become the main source of pollution in river water. Exchangeable metals are considered to be directly used by the biological, under acidic conditions, the metal ions in carbonate species will be released again to the environment. These two species of the metals are not stable, as shown in Fig. 3.10. Figure 3.10 shows that the exchangeable and carbonate fractions of Cd, Pb, and As in the Suzhou Creek sediment are more than those in the suspended matter. This fact shows that, in the Suzhou Creek water–sediment system, trace metals bound in the sediments have the possibility and conditions to be transferred to the water again.

(3) **The vertical distribution and pollution of trace metals in the sediment of the Suzhou Creek**

A. The vertical distribution of trace metals in the sediment of the Suzhou Creek

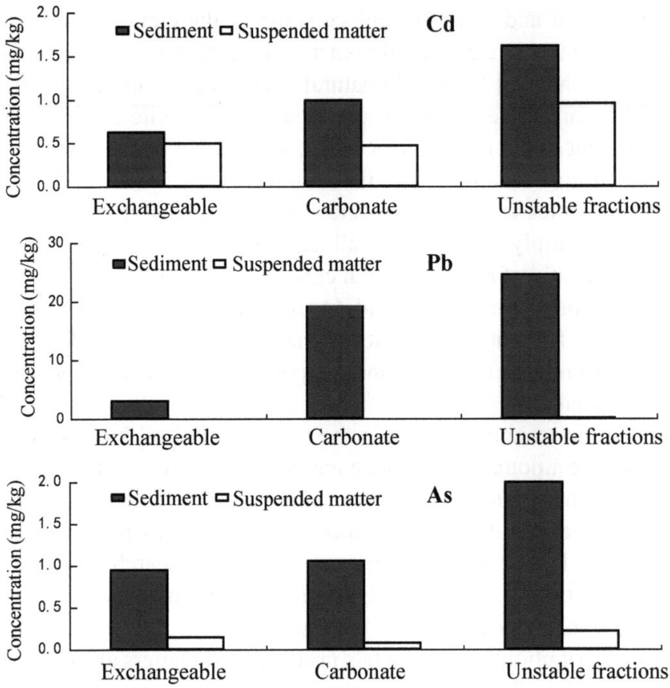

Fig. 3.10 Distribution of unstable fractions of Cd, Pb, and As, respectively, in the suspended matter and the sediment from the Suzhou Creek

The evolution process of water environment can be reflected by the vertical distribution of metal elements in the sediments. In this study, two sampling sites located the main industrial areas, Jiangnan paper plant and Shanghai grease factory, were selected to sample, and the vertical distribution characteristics of five metal elements in the sediments of the Suzhou Creek were systematically researched. The distribution characteristics of Zn, Cu, As, Pb, and Hg in the sediments are shown in Figs. 3.11 and 3.12.

In Fig. 3.11, in the sediments of Jiangnan Paper Mill site, the amount of Zn and As are decreased as depth in the range of 0–25 cm, and Cu, Pb, and Hg increased as depth from the 0–15 cm then decreases as depth in the 15–25 cm with little change. Under 25 cm, the amount of five metals all increase with depth, in which the amount of Zn, Cu, As, and Pb have greatly increased in the range of 50–60 cm, and then reduce rapidly with depth when reached maximum; Hg reaches maximum at the range of 40–50 cm and then reduces rapidly with depth.

Figure 3.12 shows that in the sediments of Shanghai grease plant site of the Suzhou Creek, the concentration of Zn decreases with depth from 0 to 15 cm, increases with depth under 15 cm, and reaches maximum at the range of 50–60 cm. The variation trends for Cu and Zn are generally the same, only from 0 to 30 cm, their concentration remains at the unchanged. The concentration of As increases with depth and reaches

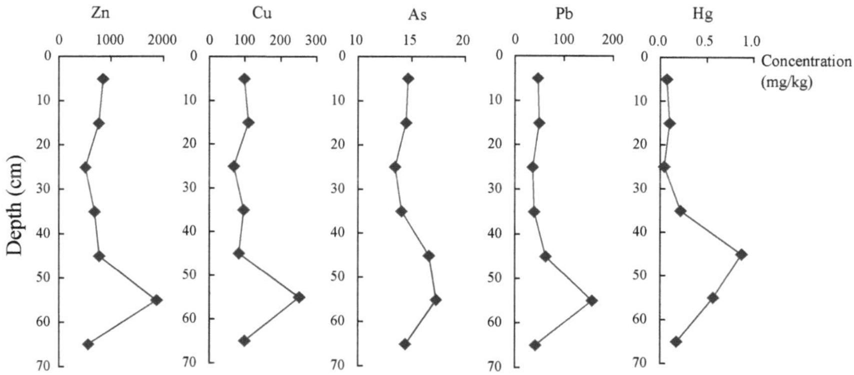

Fig. 3.11 Vertical distribution of trace metals in the sediment from the Jiangnan Paper Mill site

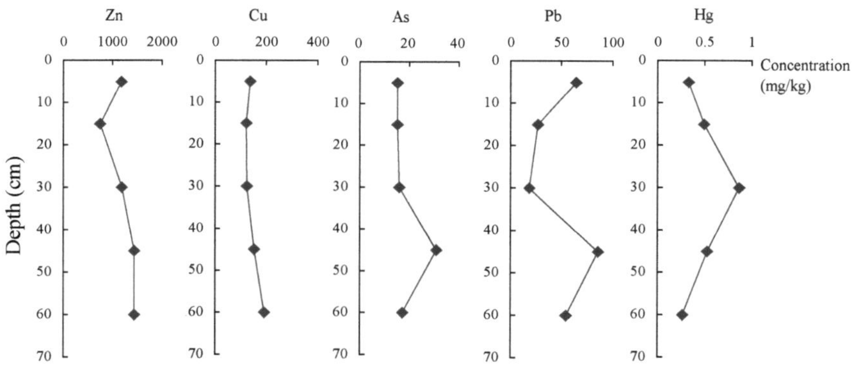

Fig. 3.12 Vertical distribution of trace metals in the sediment from the Shanghai grease factory site

maximum at 40–50 cm depth range, then decreases fast. Pb decreases with depth firstly and then increases to reach the maximum value at 40–50 cm with decreasing rapidly in the deeper sediment below 45 cm. The change trend is same as the Jiangnan Paper Mill's site for Hg, except the highest value appears at the depth about 30 cm of the sediment.

The above situation shows that the concentration of trace metals distribution in the vertical profile of the sediments of the Suzhou Creek is clearly different from the common rivers, which metals increase monotonously from the bottom to the top. There is a slightly lower concentration at the surface and bottom of the sediment, while is higher in the middle section of the sediment column. The accumulation of trace metals in the sediments of the Suzhou Creek is complicated, where the following two factors may be related. On the one hand, the pollution history of the Suzhou Creek, before the first-stage sewage interception project of the Suzhou Creek in 1990s [28, 29], sediments had been the main storeroom of various pollutants in the river, and with the industries and urbanization of high-speed development of

Shanghai, industrial wastewater containing large volume of metals was discharged into the creek so that trace metals had been given priority to with cumulative process, and a characteristic of high concentration of pollution deposition was gradually formed. But since 1993, the sewage interception pipe network of the Suzhou Creek put into operation, the wastewater discharged into the river was aborted largely, then gradually declined in the trace metal concentration in the surface sediments [28, 29]. On the other hand, the oxidation and reduction cycle of Fe and Mn during early diagenesis of the sediments might be related to the features of the metals in the sediment of the Suzhou Creek. Numerous studies have shown that in surface water environment, many trace metals usually associated with Fe–Mn oxides, transported, and transformed along with behavior of Fe–Mn [32]. Fe–Mn oxide is typical sensitive compound for redox conditions. With the change of discharge from production and living along with the Suzhou Creek, a large number of organic matter in the surface sediment of the Suzhou Creek were degraded to cause the water black, hypoxia, and smell, which made the water to be under the reduction condition. Therefore, the part Fe–Mn oxides in the surface sediment occurred the reductive dissolution and released to upside water and deep sediments with pore water [33, 34]. Among them, on the controlling of the matter balance in the pore water, the Fe–Mn materials, with the form of amorphous minerals, diffused into deep sediments and accumulated again in the sediments, while the Fe–Mn released from the sediment could diffuse into overlying water under dissolver form leading to the creek water seasonal pollution of metals from sediment. Under the controlling of this process, various metals combined with Fe–Mn oxides also produced seasonal release and accumulation, which may cause the concentration of trace metals on surface sediments decrease, and accumulation in the middle part of the sediment.

In addition, It is worthy of being noticed that the concentration of Zn in the sediment in Shanghai grease factory site increased gradually over 15 cm, Pb in that increases gradually over 30 cm too, which indicate the pollution in the Suzhou Creek still comes up and the pollution of the metals to remain effect for the creek water quality. From this phenomenon, it would be paid close attention to the trace metal pollution of the Suzhou Creek and the next work to prevent its further deterioration.

B. The characteristics of trace metal pollution in the sediment of the Suzhou Creek

Before 1970, the evaluation of water environment quality was only focused on water phase and biological aspects. With research progress on particulate matter pollution of the river, the importance of particles in water to the water quality assessment is gradually understood. Trace metals which go into water bodies by various channels often quickly are transferred to the particles. Therefore, the extent of trace metal pollution in water body can be more clearly reflected by the content of the trace metals in particulate matters. Comparing to water phase, the indication from sediments to water environment pollution by metals is more stable. Therefore, the method by analysis of the concentration and its change of trace metals in sediment is more accurate and reliable to evaluate the extent of trace metal pollution and ecological crises. This method, sedimental method, could compensate for some shortcomings in the water and bioassessment methods to evaluate water environmental quality,

Table 3.17 I_{geo} and pollution level of the trace metals

I_{geo}	<0	0–1	1–2	2–3	3–4	4–5	>5
Grade	0	1	2	3	4	5	6
Pollution rank	No pollution	No–middle	Middle	Middle–heavy	Heavy	Heavy-extremely	Extremely

for example, the water phase method can usually indicate the transient condition of water quality, and the biological method is generally difficult to reflect the overall situation of the water quality and so on. The degree of water pollution can be more comprehensively reflected by sedimental method, which is of great importance to the potential ecological risk assessment of trace metals in natural water systems.

At present, there are many evaluation methods for trace metals in sediments, each one have specific scope of application and limitation, so the pollution and potential hazards of sediment are difficult to reflect accurately by a single evaluation method. Cumulative index (I-geo) and potential ecological risk evaluation (*RI*) are selected to assess the pollution in this research. These two ways are used to evaluate the pollution of the trace metals in the Suzhou Creek from the metals in the sediments of the creek and then make a preliminary estimation to the pollution level of trace metal in the Suzhou Creek, and according to the results the potential ecological risk of the pollution is estimated.

The geo-accumulation index is a quantitative indicator, which is used for water environment pollution of trace metals in sediment, presented by Müller, used in modern evaluation of trace metal pollution in sediments [35, 36]. The formula is as follows:

$$I_{geo} = \log_2[C_n/(k \times B_n)]$$

In the equation, C_n means the element concentration measured in sediments; B_n means the geochemical background values for the elements in ordinary shale; k is a coefficient which was represented that the lithologic differences from the host rock of the sediment may bring change in the value of the background concentration of the element (it is usually 1.5). The geo-accumulation index is divided into seven levels to represent different pollution levels [37]. Table 3.17 shows the relation between the classification standard for accumulated index of sediments and pollution levels.

The geo-accumulation index and its meaning of pollution level are shown in Table 3.17. The geo-accumulation index at the sampling sites of the Suzhou Creek is as shown in Table 3.18, which is calculated and evaluated according to the formula and grades as Table 3.17. The average geo-accumulation index of the metal elements researched at every sampling site of the Suzhou Creek is, respectively, on behalf of the average geo-accumulation index of the site, which is used to roughly estimate the grade of the trace metal pollution at each sampling site.

Table 3.18 I_{geo} and pollution level of the trace metals in the sediments from the Suzhou Creek

Site	I_{geo}/R^a						
	Pb	Cd	Cu	Hg	As	Zn	AVG I_{geo}^b
Jiangnan Paper Mill	0.67/1	2.57/3	1.45/2	−1.72/0	0.11/1	2.84/3	0.99
Shanghai grease factory	1.11/2	3.51/4	1.94/2	0.53/1	0.40/1	3.32/4	1.80
Zhonghuaxinlu Bridge	1.80/2	1.89/2	2.41/3	3.12 4	1.67/2	3.42/4	2.39
Wuzhenlu Bridge	0.70/1	−0.57/0	2.04/3	−0.42/0	0.52/1	2.74/3	0.84

[a] R means the grade of pollution
[b] AVG I_{geo} means the average index of the six metal elements

As shown in Table 3.18, the grade characteristics of the trace metal pollution in sediments of the Suzhou Creek is indicated generally by geo-accumulation and the pollution grade order for the metal elements were as follows: Zn > Cd > Hg > Cu > Pb > As. Cd and Zn have reached the grade of heavy pollution, Cd is with different pollution levels at different sites, most sites are within medium or heavy pollution grades, while Wuzhenlu Bridge is pollution-free. Zn is within mainly heavy or less heavy pollution grades. Cu, Pb, and As are medium pollution grade, in which Cu and Pb are within medium or less medium pollution grades. As is within medium to light pollution grade, and the pollution grade of Hg is different from different sites. Jiangnan Paper Mill and Wuzhenlu Bridge sites show pollution-free all, Shanghai grease factory shows medium pollution grade, while the Zhonghuaxinlu Bridge site is in heavy pollution grade. Synthesizing from pollution index of the sampling sites, the Zhonghuaxinlu Bridge is in the grade of heavy pollution, Shanghai grease factory is that of medium pollution, Jiangnan Paper Mill and Wuzhenlu Bridge are that of medium pollution.

Potential ecological risk index to quantificationally valuate the pollution is relatively a fast and simple method, suggested by Sweden scholar Hakanson in 1980, which has been used to divide pollution level and potential ecological risk of sediments and contaminants [38]. It can synthetically reflect the ecological environmental impact of trace metals in sediments from analyzing the concentration of contaminants in the sediments. The calculating formula is as follows:

$$C_f^i = C_s^i / C_n^i$$

$$C_d = \sum C_f^i$$

$$E_r^i = T_r^i \cdot C_f^i$$

$$RI = \sum_{i=1}^{n} E_r^i = \sum_{i=1}^{n} T_r^i \cdot C_f^i$$

Table 3.19 Grade of the potential ecological risk, respectively, for single metal element (E_r^i) and multiple metal elements (RI)

Indexes	Risk for ecosystem				
	Light	Middle	Heavy	Very heavy	Extremely
E_r^i	$E_r^i < 40$	$40 \le E_r^i < 80$	$80 \le E_r^i < 160$	$160 \le E_r^i < 320$	$E_r^i \ge 320$
RI	$RI < 150$	$150 \le RI < 300$	$300 \le RI < 600$	$R \ge 600$	

In the formula, C_f^i means the pollution coefficient of the trace metal element, C_s^i means the measured concentration of the trace metal element in sediments, C_n^i means the background value of the trace metal element in the researched region, C_d means the pollution degree of the metal element, T_r^i means the toxicity response coefficient of the trace metal element (Hg = 40, Cd = 30, As = 10, Pb = Cu = 5, Zn = 1) [39], E_r^i means the potential ecological risk coefficient of the trace metal element, RI means the potential ecological risk index of trace metal element in the sediment, and the E_r^i and RI of the metal element can be, respectively, calculated according to the formula. The division standard for potential ecological risk level of trace metal in the sediment of the Suzhou Creek is shows in Table 3.19 [28].

Based on the formula above, potential ecological risk coefficients of trace metals and pollution levels of trace metals are calculated as shown in Table 3.20. It shows that the order of the potential ecological risks of trace metals in the feature of below: Cd > Hg > Cu > As > Pb > Zn. The ecological risks of Cd in Jiangnan Paper Mill site and Shanghai grease factory site are both relatively higher. In the Zhonghuaxinlu Bridge, the ecological risk is extremely higher, in the lower reaches of the Wuzhenlu Bridge there is just slight pollution. The difference in ecological risk of Hg at different sites is obvious, and Jiangnan Paper Mill and Wuzhenlu Bridge sites are in medium pollution degree, the Shanghai grease plant site is in heavy pollution. According the results of calculated by the formula presented above, Cu, Pb, and Zn are characterized by slight risks, As is mainly with a slight risk, only at the Zhonghuaxinlu Bridge it reaches the grade of medium pollution. According to the RI value, trace metal pollution in sediment of the Suzhou Creek is of great ecological risk generally, in the Zhonghuaxinlu Bridge, the pollution of the sediment is most serious, and that of Wuzhenlu Bridge reached the lightest.

In summary, the pollution assessment results of the sediment of the Suzhou Creek in ecological index method are basically the same as that showed in cumulative index, only the pollution levels of trace metals by two evaluations are with slight differences, where for Zn pollution assessment results are different, which may relate to the biological effects of different elements to be concerned more in the ecological index calculating.

Table 3.20 E_r^i, *Cd*, and *RI* of trace metals in the sediment of the Suzhou Creek

Sites	E_r^i						*Cd*	*RI*
	Pb	Cd	Cu	Hg	As	Zn		
Jiangnan Paper Mill	10.5	267.6	24.3	78.8	16.2	10.89	408.29	806.08
Shanghai grease factory	12.6	496.5	30.8	123.6	19.8	15.32	698.62	1384.64
Zhonghuaxinlu Bridge	26.1	166.2	39.8	523.6	47.7	16.03	819.43	1612.76
Wuzhenlu Bridge	12.2	30.3	30.9	44.8	26.3	13.07	157.57	302.94

C. The environmental significance of the distribution of trace metals in the sediments from the Suzhou Creek

In the early twentieth century, the Suzhou Creek was with clear water where fish and shrimp in groups were in the creek. In 1920s, the part reaches of the Suzhou Creek appeared a "black-odor" firstly. According to the data of Shanghai Environment Monitoring Center in 1996, the water quality in the upper reaches of the Suzhou Creek was at the water environmental standard grade V, and in the lower reaches was even worse than that at upper reaches. The main pollution sources of the Suzhou Creek might include following aspects:

a. Industrial wastewater and domestic sewage discharged by coastal enterprises and residential area;
b. Scattered and dumping garbage of loading and unloading along coastal port;
c. The six major tributaries (Pengyuepu Creek, Zhenru Harbor, Mudu Harbor, Xinjing Harbor, Shenji Harbor, and Huacao Harbor) flowed into the Suzhou Creek brought with large quantities of sewage.
d. House refuses, leakage, and emissions of oil from the boats on the creek from about 5000 yachts everyday on the creek, the slurry from workboat emissions;
e. Pollutants released again from the sediment deposited for many years.
f. Pollutants carried by water from upstream, where mainly contain factories and agriculture fields, livestock pollution as well as the sewage and wastewater from the residents living along the creek.
g. Pollutants to the Suzhou Creek carried by urban overland runoff which formed after rain washing roof, air, trees, roads, etc.

The trace metal pollutants in the Suzhou Creek have been caused by long time emissions from the banks of the creek, and they made the sediments of the creek bear a great deal of trace metals outside. From the vertical profile, the sediments of the Suzhou Creek have very clear three-stage sequence structure:

a. Suspended fluid mud layer at the top: This layer is in 20 cm thickness with black flocculation, high moisture. It consists of clayey and fine silt particles, and when the water with slight stir it occurs re-suspension to make the water muddy and black, which have important influence on overlying water quality in the Suzhou Creek.

b. Black middle silty layer: This layer is in 0.5–2.5 m thickness mainly with black clay, silt, and sands. It is in high organic matter, high water with gradually reducing down layer. This layer is loose in the structure and contained large amounts of non-perishable industrial and domestic waste including cinder, gravel, other debris with obviously smell. A lot of garbage buried in the sediments makes the sediment of the Suzhou Creek be dredged harder.
c. Gray sludge on the bottom layer: This layer is grayish-yellow, and mainly with natural muddy sediments of the creek. It consists of muddy materials with lower water, dense texture without peculiar smell.

From the vertical distribution of the trace metals in the sediment of the Suzhou Creek, the trace metal pollution in the Suzhou Creek sediment is mainly arisen in the main stream sides, especially in the convex part of the flow line, the estuaries of tributaries, and the rive bends of abandoned tributaries of the creek. In the middle creek, the sediment is mainly composed of coarse grains and arenopelitic substances with high organic matter concentration and loose structure, which are easy to move with the flow. In the convex part of flow line, contaminated sediment is very thick, while that of the concave is smaller. In the tributaries' estuaries, flow velocity decreases due to the mainstream top support action, leading to sludge accumulation, thus the thickness of contaminated sediment in these parts gets thick. In addition, because of the long-term impaction of the discharge of the building mud, dropped sands in loading and downloading, industrial and domestic waste toppling, etc., from the human activities along the sides of the creek, the distribution of the sludge, sediment, in the Suzhou Creek is distinct local characteristics with the sediment thickness increase at the sewage-draining exit.

The impact of river sediment on water quality is generally from the following actions:

a. The pollutant release by desorption from the particulate adsorbed pollutants during the suspension process of the sediment materials;
b. The pollutant desorption and release from the surficial sediment directly to the overlying water overlying water;
c. The pollutants contained in the interstitial water release and diffusion directly to the overlying water;

As mentioned above, after the effective control for point-source pollution in nearly 20 years, the sediment becomes one of the main sources of water pollution in rivers [40]. During the substance exchange between the sediment and the overlying water, the concentration of trace metals dissolved in water is significantly affected. When sediments are undisturbed, the release of trace metals is spread mainly by pore water and overlying water, and this has little effect on water quality due to the smaller flux. While sediments are disturbed and in re-suspension, the trace metals dissolved in the pore water and adsorbed on suspended particulates will be largely released to overlying water within a short time. And due to the continuous circulation within

Table 3.21 Concentration of PCBs in the surface sediment from the Suzhou Creek

Sampling site	Jiangnan Paper Mill	Shanghai grease factory	Zhonghuaxinlu Bridge	Wuzhenlu Bridge	Average value
PCBs' concentration (μg/kg)	4.4	5.8	14.8	4.9	7.5

overlying water and sediment, therefore, even if all of the external sources has been eliminated, the water quality of the Suzhou Creek would be affected by sediments from the inside source of the pollution in a long time.

Many studies have shown that, when the pollutant concentration in interstitial water of sediments is close to the balance, the action of pollutant mixing with the overlying water is the main effect to water quality, next is static release from sediments at the lower layer, and the release from the suspended particles is relatively small [41]. Statistics showed that the water quality of the Suzhou Creek and concentration of trace metals in sediment was directly proportional [42]. Due to a dynamic balance of absorption and release between the sediment and the water, once the concentrations of water pollutants decreased, the release of pollutants in sediments would increase, and the probability of secondary pollution of water also increased [43]. The pollution in the Suzhou Creek sediment has a significant impact on the environmental quality of overlying water, especially when the creek flow increases, and sediments are disturbed as well as accompanying re-suspension, its effects on water quality will be more prominent. Therefore, the remediation of contaminated sediments in the Suzhou Creek should be an important part of the pollution control for the Suzhou Creek.

3.1.1.4 The Distribution of PCBs in the Sediment of the Suzhou Creek

PCB concentration and variation in the sediment of the Suzhou Creek are shown in Table 3.21. The data show that the content of PCBs in surficial sediments of the Suzhou Creek is in the range of 4.4–14.8 μg/kg, and the max value appears in the tributary, site of the Zhonghuaxinlu Bridge, except this point, the concentration of that in all creek sediments is less than 10 μg/kg with a slight variation.

Because of no national standard for polychlorinated biphenyls in sediments has arisen in China, many scholars at home and abroad had also got different conclusions when PCBs had been included in research. The studies completed earlier by Long et al. suggested that the pollution would reach moderate or severe level if the concentration of PCBs exceeds 50 μg/kg, and the pollution would arise if the total concentration of them were more than 10 μg/kg [44]. In 1995, according to the risk evaluation of pollutants in sediments of the North American coast and estuary, they confirmed the lowest value of the risk assessment ERL (biological toxicity effect under 10%) and mid-value of the risk assessment ERM (biological toxicity effect

Table 3.22 Comparison between different city water systems for PCBs' concentrations in the sediment

States or region	Total PCBs' concentration (μg/kg)	References
Victoria Bay, Hong Kong	6–81	[48]
West Harbor, Xiamen	0.05–7.24	[48]
Liverpool Bay, UK	0.082–38	[49]
Lake Baikal	0.019–0.12	[50]
The Arctic lakes	2.4–39	[51]
Chu Chiang Delta	25.4–485.45	[52]
Second Songhua River	25.4–70.3	[53]
The Yangtze River	3.0–9.5	[54]
The Suzhou Creek (1998)	0.4–3.37	[55]
The Suzhou Creek (2004)	4.4–14.8	This work

under 50%) [44]. The result of this study has been used by US EPA as the national standards of the USA. In this standard, ERL value of total PCBs in sediments is 22.7 μg/kg, ERM value is 180 μg/kg. By comparing different methods of assessment, Macdonald et al. divided the PCBs' toxicity into three bounds, which were the concentration of critical effect (TEC), the concentration of medium effect (MEC), and the concentration of extreme effect (EEC). If the concentration of PCBs is under TEC, the sediment is non-toxic. If the concentration of PCBs is between TEC and MEC, toxic risk would be more than 50%. If the concentration is more than EEC, the sediment is toxic [45]. For sediments in freshwater ecosystems, three threshold values are 35, 340, and 1600 μg/kg. Chinese scholar Zhou et al. proposed that the reference standard value of polychlorinated biphenyls in sediments was 20 μg/kg [46]. Judging from this, the polychlorinated biphenyls in sediment of the Suzhou Creek have not yet reached the level of pollution.

Compared to other urban water worldwide, the concentration of polychlorinated biphenyls in the sediment of the Suzhou Creek is not in heavy. However, compared with historical concentration of PCBs in the sediment of the Suzhou Creek, there is an upward trend, which is an issue of concern of PCBs in the Suzhou Creek, as shown in Table 3.22.

Data in Table 3.21 show that the concentration of PCBs at Zhonghuaxinlu Bridge is significantly higher than the samples from other sites. The sampling site was located in the estuary of Pengyuepu Creek to the Suzhou Creek with more domestic sewage. Because both banks of Pengyuepu Creek are concentrated areas of Shanghai industrial enterprises, this estuary is a serious polluted point of the Suzhou Creek. There is especially with organic pollutants and the COD_{Cr} concentration more than 20 times the class VI grade of China water quality standard, and PCBs are mainly found in organic matter phase of the sediments of the Suzhou Creek [47]. These facts show that the tributary of Pengyuepu Creek is a major contributor to the Suzhou Creek for PCBs.

Suzhou Creek is one of the major shipping channels in Shanghai, fuel spills of shipping vessels, domestic pollution discharge of residents living on board, living activities of residents living on the banks, and scattered cargo in the process of handling and transiting all would affect the water quality of the Suzhou Creek. According to the data of Swedish Environmental Research Institute (2005) survey, the transport ship's fuel and lubricant could release high levels of PCBs, namely 3.2–28 and 9.3–22 μg/kg [56]. In addition to the tributary (the Zhonghuaxinlu Bridge), the polychlorinated biphenyls in the sediments of the Suzhou Creek appear relatively lower with concentration rolling variation, which may be related to the shipping fuel leaked, the volatilization, and deposition of PCBs [57].

To sum up, the concentration of PCBs in surface sediments at urban section of the Suzhou Creek is in the range of 4.4–14.8 μg/kg and has not yet reached the levels of pollution, but there is in the situation with a rising trend.

3.1.1.5 Understanding for the Spatial and Temporal Variation of Trace Metals, PCBs, and Environment Quality of the Suzhou Creek

1. **The distribution and evolution of the trace metals in the sediment–water system of the Suzhou Creek**

Trace metal pollution in the Suzhou Creek based on Zn, Hg, and Cd is more prominent, especially there are higher concentrations in suspended solids and in the surface sediments, which is currently one of the most important problems of trace metal pollution in the Suzhou Creek.

Arsenic, Hg, and Cd pollution in the water phase of the Suzhou Creek is more serious, the water qualities of 100% have met national water quality standard grade III for surface water. Among them, Cd pollution is the most serious. Compared to other cities in China, the trace metal concentration of the water in the Suzhou Creek is higher with concentration of As, Hg, and Cd, and lower concentration of Pb. Seen from longitudinal distribution feature, the elements As, Hg, Cd, and Pb in the total amount of each sampling sites the distribution are not very uniform, from Shanghai grease plant site to Wuzhenlu Bridge site, the trace metals (except Hg) are relatively higher, which may be associated with the banks of a large number of industrial and domestic waste discharging to the water because this section is the concentration area of industrial enterprises and population density.

Among suspended matter samples, except for Pb, other trace metal concentrations are higher, and the older of the concentrations is as the following: Cd 3.19–16.39 μg/kg, As 4.55–6.34 μg/kg, Hg 0.465–1.627 μg/kg, Zn 753.49–2889.34 μg/kg. The distribution of Zn, Cd, and Hg are corresponding to the industrial enterprise distribution along the banks, and among the enterprise sites, at the tributary Pengyuepu Creek on the Suzhou Creek As is contributed greatly.

Trace metals in suspended matter and surface sediment of the Suzhou Creek have shown a good correlation, the total concentration and speciation concentration of trace metals are of obvious relevance, which illustrates that an inner link exists between suspended matter and sediments of the Suzhou Creek for trace metal transportation and transformation under the hydrodynamics and environmental conditions. Apart from the Pb and As, other trace metals in suspended matter are higher than that in sediments, indicating there are important factors of sources of trace metals to affect the water quality evolution of the Suzhou Creek. In these parts of the metals, one part of them with river empties into the Huangpu River would pollute the downstream environment, and the other part under certain conditions would be deposited into the sediment to cause the superposed pollution in the sediment.

2. **The migration and accumulation of As, Hg, Cd, and Pb**

Arsenic, Hg, Cd, and Pb in water mainly consist in soluble form in the Suzhou Creek, Cd might be wide to affect water quality for trace metals into the Huangpu River, which will caused accumulative contribution to the downstream and the China East Sea. As a rough estimate, the contribution of trace metals to the Huangpu River from the Suzhou Creek is as follows: Pb 0.457 t/a, Cd 0.115 t/a, As 1.749 t/a, Hg 0.048 t/a, Zn 24 t/a. Trace metals carried by water in the Suzhou Creek are an important source of trace metals in the Huangpu River, and it would seriously affect the water quality. The highest accumulation in suspended matters and sediments is Hg, followed by As; Cd in minimum degree. Hg accumulated in sediments may be released when the concentration is low or other environmental condition changes. It causes secondary pollution to the water, which would be one of the major risks of the trace metal pollution for the Suzhou Creek.

(3) **The speciation of trace metals in the suspended matters and sediments of the Suzhou Creek**

The Fe–Mn oxides of Pb, Zn, and As are of the main fractions of their speciation in sediments of the Suzhou Creek; it is very easy to release from the sediment to result in secondary pollution of the water when the redox potential decreases or the water is in anoxic. Cd mainly is in the form of exchangeable and carbonate, which indicates that the Cd in the sediment is with a high mobility and biological activity and with a high potential ecological risk.

Various metals of suspensions in the residual fraction are the dominant form in the sediment of the Suzhou Creek. Organic form is a higher level in addition to the residual form, in which Cd and Cu are with the highest concentration in the metals. Lower metal concentration is in carbonate form, Zn is with higher level in the Fe–Ma oxide form, Cd is of the strongest accumulation capacity in the exchangeable and carbonate fractions, with greater mobility and biological effects. Compared with sediments, the exchangeable, carbonates, and Fe–Mn oxide fractions of most metals in suspended matter the proportion have declined, especially that in Fe–Mn oxide decreases very large. The proportion of the residual fraction is significantly relatively higher. This could be the release of trace metal elements to the water during the action of the suspended matter and the water body in the long-term solid–liquid interaction associated with the deposition of more complex geochemical processes (including speciation transformation).

(4) The trace metal pollution in the sediment of the Suzhou Creek

The trace metals in the sediments of the Suzhou Creek show that the metals at surface and the bottom layers are with a slightly lower concentrations and that at the middle layer is relatively higher. It may be shown that there is a low concentration of trace metals in the early in the Suzhou Creek. With the development of the industry and the urbanization on the banks, wastewater contained a large amount of trace metals is discharged into the creek; thus, the accumulation of trace metals is as a dominant process, at last, the metal pollution in the sediment with a high concentration formed increasingly. After the outer source is controlled in the Suzhou Creek, the concentration of trace metals gradually reduces, especially Hg, Cu, and As correspondingly decrease in the sediment. At present, the amount of trace metals at surface of sediment is still in a downward trend, which indicated that outer source pollutions have been effectively controlled. However in the Shanghai grease plant site, the concentrations of Zn and Pb are close to surface's feature in gradually increasing, indicating that contamination factor of Zn and Pb still remains in the Suzhou Creek and had an obvious negative impact on the water quality.

From the preliminary evaluation for the trace metals in sediment of the Suzhou Creek by the cumulative index and potential ecological pollution index, it shows that degrees of the pollution of each metal element are different, the current pollution elements in sediments of the Suzhou Creek mainly are Cd and Hg and this situation for the creek's ecological function should be paid more attention. The section of Zhonghuaxinlu Bridge site of the Suzhou Creek along the urban center is most serious area of the trace metal pollution, the degree of pollution is moderate at the Shanghai grease plant site, and the sites of Jiangnan Paper Factory and Wuzhenlu Bridge are slightly polluted. The pollution distribution feature indicates some corresponding relationship within the industry activities and population sections. Meanwhile in the tributaries of the Suzhou Creek (Pengyuepu Creek), the pollution of trace metals is more serious, which have an important contribution to the Suzhou Creek's pollution.

(5) The distribution and evolution of PCBs in the sediments of the Suzhou Creek

The concentration of PCBs at surface of sediments in the Suzhou Creek ranges from 4.4 to 14.8 μg/kg in the center urban section, which has not yet reached the level of pollution measured by the standard proposed by US EPA, but the PCB level in the sediment increases as time in the Suzhou Creek similar as trace metals. The concentration of PCBs in the sediment of the Suzhou Creek presents the feature of decreasing trend with the distance from the banks.

3.1.2 *The Pollution of Trace Metals and Organic Matter (PAHs) from the Huangpu River, Shanghai*

3.1.2.1 Overview

The Huangpu River, as a major river that flows into the East China Sea in the Taihu Lake basin, is a typical tidal river. The total length is 113 km with 23,800 km^2 basin area. The tributaries, Xietang Creek, the Yuanxie Creek, the Damaogang Creek, etc., meet at the Mishidu Ferry of Songjiang district of Shanghai. The Huangpu River is 28.5 km long from the Mishidu Ferry to the Zhagang Port with roughly flowing direction of east to west, the average depth of the river water is 3.5 m and change within 2–8 m with about 300 m wide. From the Zhagang Port to the Wusongkou Estuary, its length is 54 km with a north–south flowing river within width 77–320 m and depth about 8–17 m. The Huangpu River flows mainly through Minhang, Xuhui, Huangpu, Yangpu, Baoshan districts in Shanghai, respectively. The type of water quality matched the grade III or IV of the China national water quality standard in the main sections [58].

The Huangpu River admits the runoff from the Taihu Lake basin; according to the Mishidu hydrologic station's date from 1954 to 1995, the average runoff was 321 m^3/s, corresponding to amount of water is 101.3×10^8 m^3 per year. Inter-annual variability is large, in highest flow year (1954); the average volume of runoff is 755 m^3/s. In lowest flow year (1979), the average volume of runoff is 153 m^3/s. This suggested that the runoff distribution is not in average every year, and the runoff in the non-flood season was slightly larger than that in flood season [59].

In the early twentieth century, along the Huangpu River there were a number of textile factory, bleaching silk factory, paper mill, metallurgy factory, and others more than 20 factories closely locating on the banks of the river, which discharged industrial wastewater and domestic sewage freely to the water of the river. These industrial enterprises were mainly distributed in the lower reaches of the Huangpu River, so the contaminated water was mainly in the downstream of the Huangpu River. During this period, there was no pollution in the upper stream. At the same time, as affecting by water self-purification function of the river water quality of the river's downstream section also was generally good.

After 1957, with the rapid change of social reforming of China and the development of industry, the region had rapidly increased in the number of industrial enterprises with a fast expansion of scales, the industrial wastewater and domestic sewage increased rapidly in the river. These were large number of untreated wastewater directly dumping into the Huangpu River. Some of the polluted water first poured out into the Yangshupu Port or the Zouma Pond, but the polluted water entered the Huangpu River in the end. Therefore, the Huangpu River became the place of the upstream drainage of sewage and polluted water.

After 1964, the Huangpu River began to appear black and smelly phenomenon called "black smelly," and most tributaries across the downtown including the Suzhou Creek were perennial "black smelly." The pollutants mainly came from factories that were on the banks along the river, and they daily discharged the industrial wastewater about 4.50 × 10^5 t to the river. The second pollutants were sewage that came from 32 sewage outfalls along the Huangpu River, and the wastewater was discharged into the river through drains. Another was the release of the pollutants be dissolved into the river from upstream. During this period, within about 5 km from the wharf of the Qinhuangdao road to the Yangshupu gas plant, water pollution situation was very serious.

In the period of 1980–1990s, the water pollution of the Huangpu River aggravated heavily. Such as in 1978, Shanghai Insulation Material Factory discharged 3000 kg phenolic wastewater into the Huangpu River per day and the phenol amount in this wastewater was as high as 30,000 mg/L, that is equivalent of 90 kg/d phenol into the Huangpu River, and this factory was listed as the one of 33 large wastewater containing phenol sources in Shanghai then. Just in 1982 of the Yangpu district, there were 5.1 × 10^7 t of industrial wastewater without governance and more than 2.90 × 10^7 t of sewage directly discharged into the river. In 1984, electroplating wastewater discharged into the river was 12 × 10^3 t/d, and 70% of this wastewater was from the bicycle factories. It was showed by pollution survey in 1985 that industrial wastewater discharged by 311 enterprises were 1.83 × 10^8, and 1.39 × 10^8 t of wastewater were discharged directly into the Huangpu River or urban sewers, which might account for 76.22% of the total wastewater into the river. In the pollutants into the Huangpu River with the wastewater, 60% of them was in industrial wastewater as organic pollutants, including oily pollutants (1.825 × 10^3 t), chemical oxygen consumption (COD, 3.96 × 10^4 t), biochemical oxygen demand (BOD, 1.09 × 10^4 t), respectively. Furthermore, copper and its inorganic compounds emissions were up for more than 94 t, ammonia nitrogen emissions were 9.72 × 10^2 t. More than 40 large factories (BOD > 5 t/d) discharged out a lots of organic pollutants every day into the Huangpu River. At that time, black smelly days in the Huangpu River were annually more than 110 days and up to 155 days (1985). In the 1990s, that was the most serious period for the Huangpu River water pollution. Such as the 1990 year, 385 enterprises discharged of industrial wastewater 1.39 × 10^8 t/d to the Huangpu River, and 6.3 × 10^7 t of emissions did not fit the national standards of discharging water, account for 45.32% of the total wastewater into the river. It was in a very serious environmental situation for the Huangpu River in this period. After twenty-first century, with a series of environmental protection policy and implements for the Huangpu River, its environmental quality has been becoming more and more obviously improved [60].

3.1.2.2 Samples and Analysis

1. **Sampling and processing**

The sampling in this work was stationed roughly along the Huangpu River trunk stream according to the distance and tried to avoid close to tributaries. Due to busy shipping, many drain outlets, numerous bank residents, and enterprises the point selected for sampling try to avoid outlets of bank enterprise's drainage, etc. Other-

Fig. 3.13 Sampling sites and number on the Huangpu River research

wise, water samples were collected to make as much as possible to reduce the effect of the surface oil contamination of the river and trash instantaneous. The designed sampling depth is 0.5 m below the water surface. In consideration of the Huangpu River flowing turbulence and water being mixed well, sampling point was as far away as possible from the riverbank. Sampling sites were shown in Fig. 3.13.

Influenced by tidal flow backward of the Yangtze Estuary, the Huangpu River is a semidiurnal tidal river with the medium intensity of tides. The Huangpu River Estuary flood tide is two times a day when at high tide period; the river water and sea water always can ascend to the Yangpu Bridge site in the Huangpu River. In order to be representative for the Huangpu River water samples collected, the samples are sampled at low tide within half an hour after the low water level of slack to keep the water representative. The specific sampling time was selected according to the rule of the Huangpu River tidal regular when sampling.

According to the data of marine environmental forecasting on the daily tide at the estuary of the Huangpu River [61, 62], the Huangpu River's high tide time is about 5 h a day with rapid velocity, and ebb tide took about seven hours with slow velocity. The average velocity of tides rose and fell was as the calculation basis to choose the sampling time, and it was about 1.9 m/s. With this velocity, the sampling time at every sampling site was calculated according to the distance from the estuary of the Suzhou Creek, roughly for the creek water to achieve low tide level. The Huangpu River tide-level computation formula was as following:

$$\boldsymbol{T}_2 = \boldsymbol{T}_1 + \boldsymbol{L} \times 1000/3600\ \boldsymbol{V}$$

Here: $\boldsymbol{T}_1$: The Beijing time of the Suzhou Creek Estuary at the lowest tidal level (h); $\boldsymbol{T}_2$: Beijing time of sampling site which was designed of the lowest tide level (h); ***L***: The flow distance from the sampling site to the Suzhou Creek Estuary (km); *V*: The Huangpu River's average flow velocity, 1.9 m/s.

Due to the water of the Huangpu River is pushed-up by the Yangtze River and China East Sea at the estuary of the Suzhou Creek, the ebb and rising tide phenomenon is not obvious here. The sampling time was chosen according to tidal data mentioned above combined with the observed tide then [61]. According to Shanghai Municipal Flood Control and Information Center data (2006) [62], sample sites' tidal level was calculated, respectively, as shown in Tables 3.23, 3.24, 3.25 and 3.26.

Water samples were collected with 10-L Teflon buckets. Before bucket was used to sampling, it was soaked with 10% HNO_3 for over 12 h, washed with deionized water, and dried in nature. Bucket was washed three times with the river water before collecting water sample. Sediment sample was collected by clamshell collector.

After water samples were collected, measuring the pH value and the acidification (with nitric acid to adjust pH < 3 for every sample) were operated before being carried back to the laboratory for preserving under 4 °C in refrigerator for the use next.

Sediment samples were taken out from the sampling clamshell collector, placed into Teflon box to carry back to laboratory with drying naturally in the laboratory and being grinded to $200^{\#}$ with a quartz mortar (Shells, gravel, or branches and other large foreign body were firstly removed before grinding) for use next. The mortar was cleaned twice by distilled water and drying before use.

Every 0.250 g dry sediment sample was put into a 50 mL vitro, then was digested by aqua regia matched up with 37% of hydrochloric acid and 62% of nitric acid and HF, $HClO_4$, and then was disposed constant volume in 25-mL flask for analysis.

2. **Analysis for the relative pollutants**

Analysis items were trace metals (Cu, Pb, Cr, Cd, and Hg), PAHs, and total organic carbon (TOC). Concentrations of trace metal elements were analyzed with ICP-AES (IRIS Advantage 1000, USA), atomic fluorescence analysis (AF610, China), atomic absorption spectrometry (PE5100, USA), solid–liquid mercury measurement instrument (AMA254, China). PAHs was analyzed by GC/MS (Agilent 5973, USA), and TOC analysis used TOC/TN analyzer (Multi N/C3000, Germany). The relevant parameters are shown in Table 3.27.

PAH analysis used the EPA 8270 c standard sample (US EPA 8270c Supelco) to control quality, and the detection limit was less than 2 ng/g with the relative error between 5 and 15%.

3.1.2.3 The Discussion on the Pollution and Evolution Trend of the Trace Metals and PAHs in the Huangpu River

According to Shanghai Environmental Monitoring Center's information from 1991 to 2002, the change trend of wastewater from industry and resident life sewage in

Table 3.23 Tide table at the estuary of the Suzhou Creek (measured at March 2006)

Date	Time	Height (cm)	Time	Height (cm)	Time	Height (cm)	Time	Height (cm)
1	0128	338	0933	5	1345	380	2152	23
2	0214	351	1001	24	1422	383	2241	6
3	0249	352	1039	22	1506	359	2256	26
4	0324	352	1107	50	1535	331	2332	29
5	0405	333	1134	65	1612	294	2339	48
6	0441	308	1158	89	1649	249	2348	71
7	0525	277	1249	111	1740	201		
8	0008	97	0635	245	1443	122	1925	165
9	0126	122	0843	234	1655	103	2220	173
10	0414	118	1021	254	1801	81	2316	207
11	0537	92	1113	281	1847	67	2347	240
12	0635	69	1151	304	1932	59		
13	0013	269	0723	57	1222	321	1955	60
14	0039	291	0756	48	1251	331	2035	54
15	0108	309	0837	42	1318	339	2053	58
16	0131	321	0901	52	1342	340	2133	50
17	0202	330	0936	49	1410	339	2145	53
18	0224	337	0952	60	1431	331	2204	58
19	0250	336	1028	69	1458	317	2231	55
20	0317	334	1045	74	1524	295	2238	63
21	0345	323	1114	89	1558	267	2251	76
22	0422	305	1154	105	1643	232	2316	95
23	0517	279	1308	123	1759	199		
24	0004	120	0702	258	1521	120	2025	190
25	0242	133	0912	271	1701	95	2213	224
26	0449	107	1028	306	1813	71	2307	269
27	0609	78	1121	340	1915	54	2349	310
28	0724	55	1206	364	1959	41		
29	0025	346	0825	41	1234	374	2044	23
30	0109	366	0856	37	1318	381	2136	12
31	0134	376	0941	27	1358	365	2149	30

Table 3.24 Tide table at the Yangpu Bridge site of the Huangpu River (measured at March 2006)

Date	Time of lower tide 1	Time of high tide 1	Time of lower tide 2	Time of high tide 2
1	4.47	12.55	16.75	0.87
2	5.23	13.02	17.37	1.68
3	5.82	13.65	18.10	1.93
4	6.40	14.12	18.58	2.53
5	7.08	14.57	19.20	2.15
6	7.68	14.97	19.82	3.20
7	8.42	15.82	20.67	–
8	3.13	9.58	17.72	–
9	4.43	11.72	19.92	1.33
10	7.23	13.35	21.02	2.27
11	8.62	14.22	21.78	2.78
12	9.58	14.85	22.53	–
13	3.22	10.38	15.37	3.08
14	3.65	10.93	15.85	–
15	4.13	11.62	16.30	–
16	4.52	12.02	16.70	0.55
17	5.03	12.60	17.17	0.75
18	5.40	12.87	17.52	1.07
19	5.83	13.47	17.97	1.52
20	6.28	13.75	18.40	1.63
21	6.75	14.23	18.97	1.85
22	7.37	14.90	19.72	2.27
23	8.28	16.13	20.98	–
24	3.07	10.03	18.35	–
25	5.70	12.20	20.02	1.22
26	7.82	13.47	21.22	2.12
27	9.15	14.35	22.25	2.82
28	10.40	15.10	22.98	–
29	3.42	11.42	15.57	–
30	4.15	11.93	16.30	0.60
31	4.57	12.68	16.97	0.82

Table 3.25 Tide table at the Xupu Bridge site of the Huangpu River (measured at March 2006)

Date	Time of lower tide 1	Time of high tide 1	Time of lower tide 2	Time of high tide 2
1	7.47	15.55	19.75	3.87
2	8.23	16.02	20.37	4.68
3	8.82	16.65	21.10	4.93
4	9.40	17.12	21.58	5.53
5	10.08	17.57	22.20	4.65
6	10.68	17.97	22.82	5.80
7	11.42	18.82	23.67	–
8	6.13	12.58	20.72	1.42
9	7.43	14.72	22.92	4.33
10	10.23	16.35	24.02	5.27
11	11.62	17.22	24.78	5.78
12	12.58	17.85	25.53	–
13	6.22	13.38	18.37	−0.08
14	6.65	13.93	18.85	2.58
15	7.13	14.62	19.30	2.88
16	7.52	15.02	19.70	3.55
17	8.03	15.60	20.17	3.75
18	8.40	15.87	20.52	4.07
19	8.83	16.47	20.97	4.52
20	9.28	16.75	21.40	4.63
21	9.75	17.23	21.97	4.85
22	10.37	17.90	22.72	5.27
23	11.28	19.13	23.98	–
24	6.07	13.03	21.35	2.42
25	8.70	15.20	23.02	4.22
26	10.82	16.47	24.22	5.12
27	12.15	17.35	25.25	5.82
28	13.40	18.10	25.98	–
29	6.42	14.42	18.57	2.73
30	7.15	14.93	19.30	3.60
31	7.57	15.68	19.97	3.82

Table 3.26 Tide table at the Xidu site of the Huangpu River (measured at March 2006)

Date	Time of lower tide 1	Time of high tide 1	Time of lower tide 2	Time of high tide 2
1	10.47	18.55	–	6.87
2	11.23	19.02	–	7.68
3	11.82	19.65	0.10	7.93
4	12.40	20.12	0.58	8.53
5	13.08	20.57	1.20	7.65
6	13.68	20.97	1.82	8.80
7	14.42	21.82	2.67	–
8	9.13	15.58	–	4.42
9	10.43	17.72	1.92	7.33
10	13.23	19.35	3.02	8.27
11	14.62	20.22	3.78	8.78
12	15.58	20.85	4.53	–
13	9.22	16.38	–	2.92
14	9.65	16.93	–	5.58
15	10.13	17.62	–	5.88
16	10.52	18.02	–	6.55
17	11.03	18.60	–	6.75
18	11.40	18.87	–	7.07
19	11.83	19.47	–	7.52
20	12.28	19.75	0.40	7.63
21	12.75	20.23	0.97	7.85
22	13.37	20.90	1.72	8.27
23	14.28	22.13	2.98	–
24	9.07	16.03	0.35	5.42
25	11.70	18.20	2.02	7.22
26	13.82	19.47	3.22	8.12
27	15.15	20.35	4.25	8.82
28	16.40	21.10	4.98	–
29	9.42	17.42	–	5.73
30	10.15	17.93	–	6.60
31	10.57	18.68	–	6.82

Table 3.27 Detecting limit and precision of ICP-AES analysis for metals

	Cd	Hg	Cu	Pb	Cr
Detecting limit (μg/g)	0.0030	0.0002	0.0012	0.0060	0.0210
RSD-8 (%)	0.4781	0.0750	0.0160	0.0369	0.0420

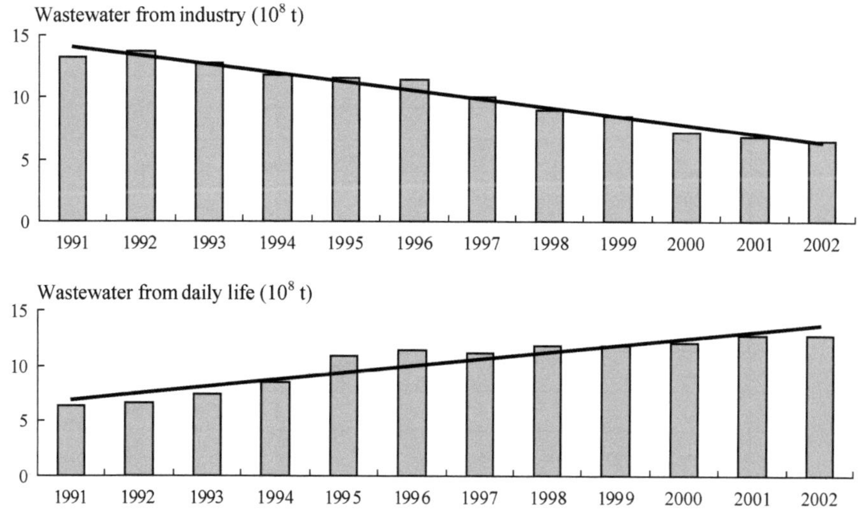

Fig. 3.14 Variation of volumes of the wastewater from Shanghai industry and human daily life

the Huangpu River was with obvious negative correlation, the amount of industrial wastewater was reducing year by year, while the amount of sewage significantly increased, as shown in Fig. 3.14. It could be indicated that the Huangpu River pollution discharge had been significantly changed for the better after decade years of pollution control efforts.

The trace metals and COD monitoring data of wastewater, which is discharged into the river, are shown in Fig. 3.15. The amounts of Cr^{6+} and Cd began to decrease since 1991. In 2002, the amount of Cr^{6+} was equivalent to one seventh of the 1991, and Cd concentration was equivalent to one seventeenth of the 1991. Monitoring data of COD of wastewater was characterized by sharply increased feature. Among them, in 1991–1995, COD was reduced but the amount was significantly increased in 1995–1997. The concentrations and the changes of trace metals in the water and sediment of the Huangpu River were shown in Figs. 3.16 and 3.17.

Figures 3.16 and 3.17 indicate that trace metal concentration in the sediment does not change obviously except at few sample sites (site 5 and site 6) in different sections of the river; however, the PAHs and TOC have a same trend of increase from upstream to downstream, showing the PAHs and TOC concentration increasing gradually with the distance's increase of the river going through the urban center.

The monitoring data for P in the Huangpu River show that the changes of total phosphorus (TP) in the water are with consistent trend with the changes of organic pollutants and residents' living wastewater, as shown in Fig. 3.18. The consistency of changes of the concentration of P and the changes of the Huangpu River's organic pollution indicate that the organic pollution is increasing in the Huangpu River might be mainly related to the living sewage from the residents.

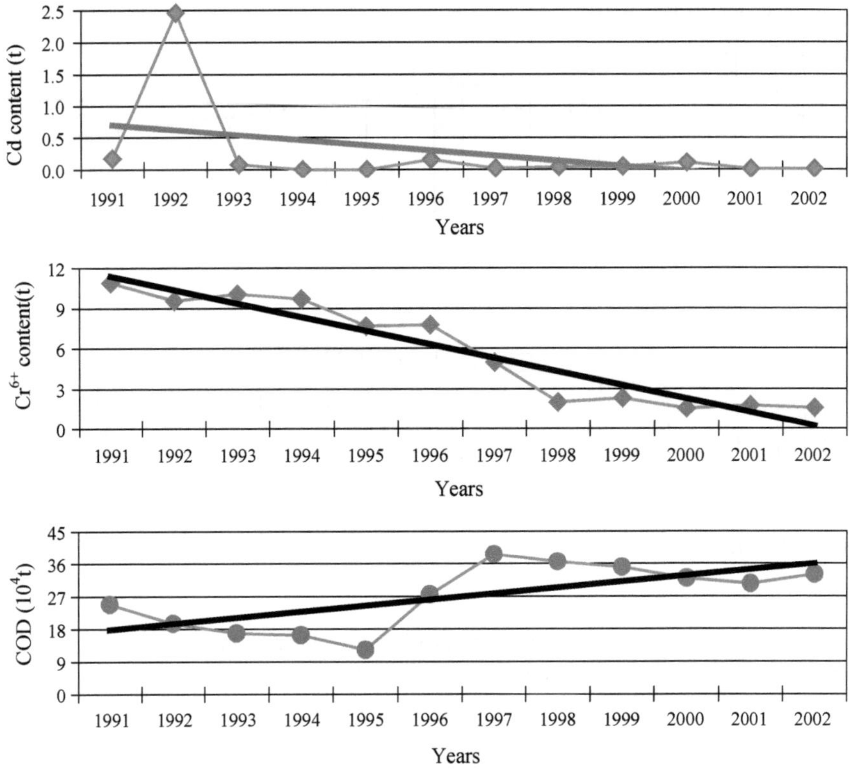

Fig. 3.15 Variation of trace metal and COD concentrations in the wastewater of the Huangpu River, Shanghai

The amounts of pollutants in the river sediment are the results of the pollutant interaction equilibrium between water and sediments [9]. The above change suggests that trace metal concentration in the Huangpu River varies a little with the time. It shows that the trace metal in the Huangpu River (average value of 8 samples) is not a serious problem (Table 3.28), and the water environmental quality is matched the grade III of the water standard specified by China National Environmental Protection Bureau in 2002 [63] and the freshwater standard of US EPA (2006) [64], which adapts to apply to potable, aquaculture, and swimming.

Compared with trace metal concentration in sediments of rivers which flow through the cities and industrial areas in China and other countries in the world (such as Canada and France) and the American sediment environmental standards, the present trace metals in the sediment in the Huangpu River are relatively lower too (Table 3.29).

The data show that the impacts of the water environment protection project for the Huangpu River in Shanghai have been significantly effective except prevention

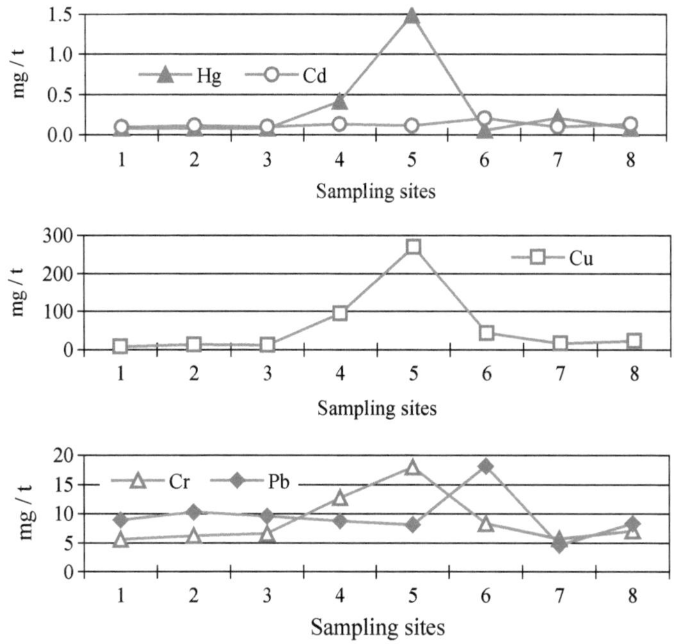

Fig. 3.16 Concentration of metals in water from the Huangpu River

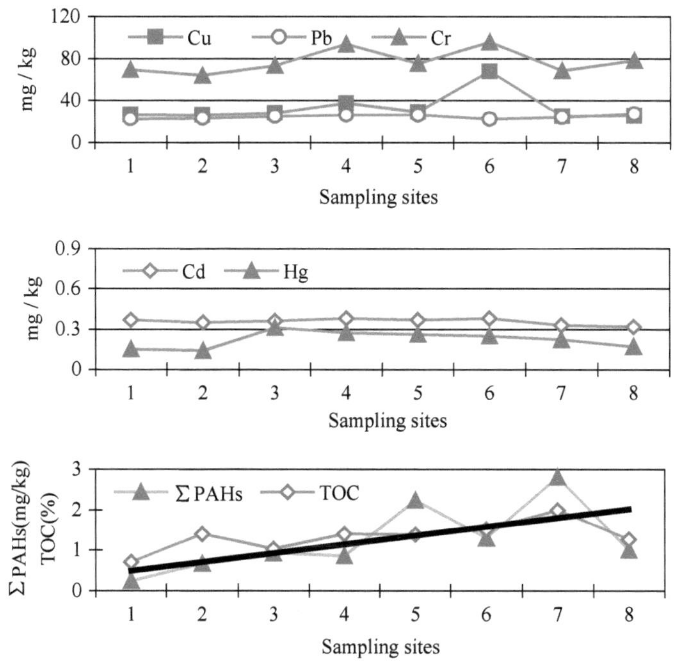

Fig. 3.17 Concentration of metals, TOC, and total PAHs in sediment from the Huangpu River

Table 3.28 Concentrations of metals in the water of Huangpu River (μg/L) and relative criterion

Elements	Criterion [63] [a]	Criterion [64] [b]	Concentration in water of Huangpu River (8 samples)
Cu	1000	13	60.45
Pb	50	65	9.56
Cd	5	2.0	0.12
Cr^{6+}	50	16	8.75 (Total Cr)
Hg	0.1	1.4	0.31
COD_{Cr}	15,000		20,000–30,000

[a]Type III criterion for surface water established at 2002 by China Government
[b]Fresh surface water criterion established by US EPA at 2006

Table 3.29 Metals in the sediment of urban rivers or harbors of countries (μg/g)

	Cu	Pb	Cr	Cd	Hg	References
Canada Sydney Harbor (passing industry region)	19.0–110.0	25.5–408.0	47.0–86.0	0.16–0.94	0.09–0.48	[65]
Seine River (passing Paris)	98.0	107.0	12.0	0.42		[66]
The Sahe river (passing Chendu city, China)	95.71	65.29		0.543	0.283	[67]
The estuary of the Yellow River (passing industry region)	6.7–56.9	13.2–38.1	33.7–88.8			[68]
The Huangpu River (passing Shanghai downtown)	33.43	24.75	77.23	0.36	0.22	This work
Criterion value[a]	390	450	260	5.1	0.41	[69]

[a]Sediment Quality Criterion, US 2006 [64]

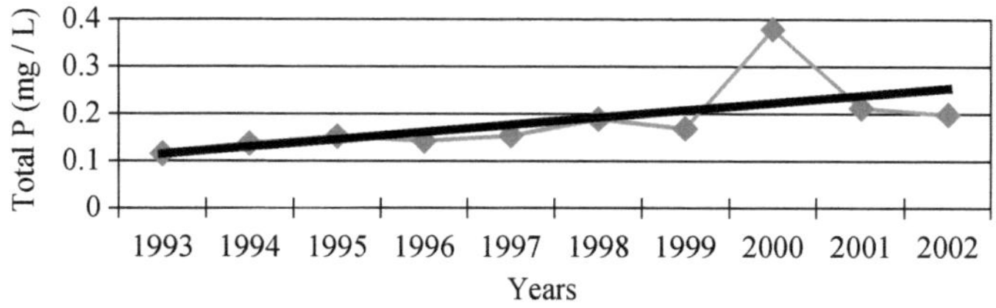

Fig. 3.18 Variations of total P in the water at the sampling site 3 from the Huangpu River

for Hg over the decade. Uncontrollable Hg pollution might be related to the urban atmospheric precipitation and other factors [70, 71].

Between 1999 and 2002, the increase in wastewater from untreated living sewage exceeded that from industry in the Huangpu River (Fig. 3.14), and COD of the wastewater increased with the increase of domestic wastewater (Fig. 3.15). These facts showed that the organic pollutants in the Huangpu River mainly came from living wastewater and organic pollution had the tendency of being serious with time.

The changing trend of PAHs and TOC in the sediment also illustrates the increasing tendency of the organic pollution of the Huangpu River (Fig. 3.17). The average concentration PAHs is 1.26 μg/g, and TOC is 1.34% from 8 sediment samples, respectively. In 2003, the COD of the water from the Huangpu River is changed from 22 mg/L at the upstream to 30 mg/L [72] at the estuary, which is greater than grade III of the national surface water quality standard (15 mg/L). It should be pointed out that the changing trends of these organic pollution parameters imply an gradually arising of the aggravating momentum of organic pollution in the Huangpu River although the PAHs and TOC concentration in the Huangpu River are lower than that of the USA, Canada, and other countries [73, 74], as shown in Table 3.30.

3.1.2.4 The Environmental Quality and Its Evolvement of the Huangpu River

The Huangpu River has gone through a long period of serious pollution and a long governing process. With the continuous strength of governance efforts, the effectiveness has been obvious. But there are also some problems arisen and needed to attach high attention.

Trace metals, PAHs, and related item concentration and their changes in water and sediment of the Huangpu River show that current trace metal pollution has been effectively reduced, but organic pollution has been increased with the increase of domestic wastewater. Comparison with some rivers of China, USA, Canada, and other countries, these pollutants are at a relatively lower concentration, but the fact of the organic pollution increasing shows the potential problems and action direction for the Shanghai's water environment, which is an notable phenomena and direction for the urban river environmental protection worldwide as well as Shanghai [78].

Table 3.30 Concentration of PAHs and TOC in river sediment from different countries

		PAHs (ng/g)	TOC (%)	References
USA	Mill River (passing Connecticut middle region)	590–39,000		[75]
Canada	Sydney Harbor (passing industry region)	4770–246,400	2.91–12.13	[65]
China	The Pearl Diver (passing the industry region of Pearl River Delta)	93.8–4307.0		[76]
	Pearl River (passing industry region)	255.9–16670.3		[77]
	The Huangpu River (passing urban and industry region)	1256.13	1.341	This work

3.1.3 The Trace Metal Pollution in the Nanjing Section of the Yangtze River

3.1.3.1 Samples and Analysis

At the Nanjing section of the Yangtze River, Baguazhou was selected as the research site for sampling, etc. Baguazhou Island is a typical island on the Yangtze River near and downstream Nanjing city, which has formed from long time rushing by the river water. In the research, the sediment of Baguazhou Island was as a record carrier of the pollution of the Yangtze River caused by Nanjing city to understand the pollution history, showed as in Figs. 3.19 and 3.20.

The north beach of the Baguazhou Island was as a research object on behalf of the trace metal pollution features of the Nanjing section of the Yangtze River. The sampling sites were located in the inside of the convex parts of the Yangtze River which is an ideal and the latest sediment deposition of the river. The samples from different depths' sediments as well as the sediment of the beach surface were all yellow-brown sand and silt. The sampling time was at middle March. The section started from the junction of the river and shore to the flood embankment. The sampling line/profile was vertically to the direction of the river flows, showed as in Fig. 3.20.

In this research, the method of combining vertical and horizontal profile sampling for the samples was taken in order to understand the distributed characteristics of trace metals in the sediments and explore the variation of the pollution features and superimposed sedimentary amplitude from the deposition thickness of the sediment

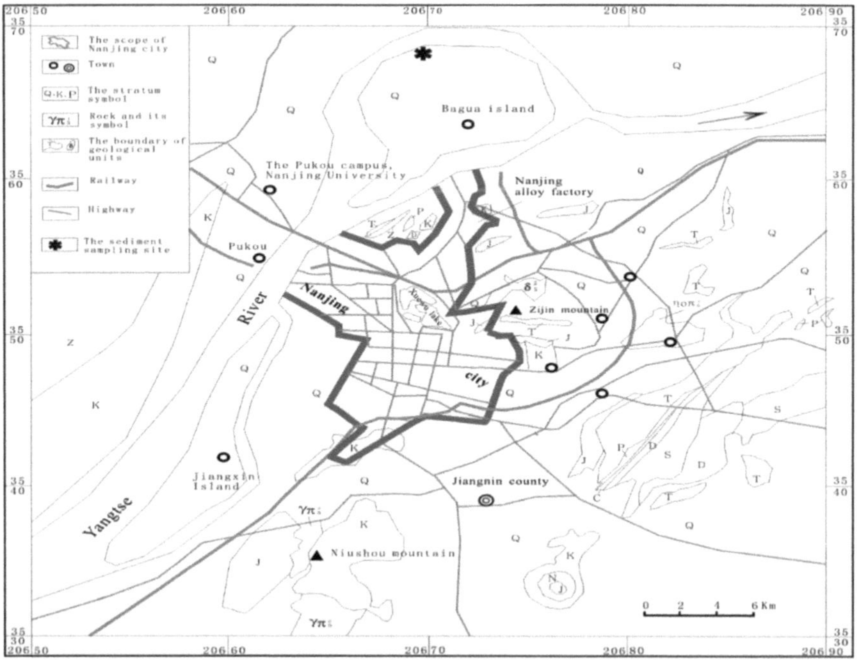

Fig. 3.19 Geochemical background of study area and the sampling sites on Nanjing city research

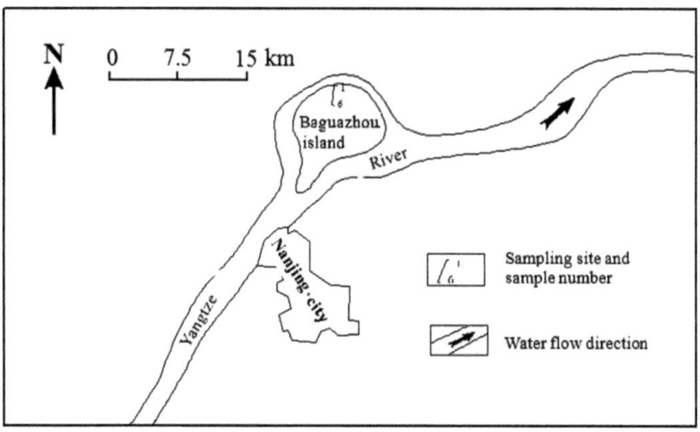

Fig. 3.20 Sampling sites for the sediment from the Baguazhou Island, Yangtze River (Nanjing section)

(i.e., time change). From this work, it could be provided as basis for understanding the concentration variation of the trace metals in the sediment and water of the Yangtze River, which could be used to roughly estimate the pollution development of the trace metals in Nanjing.

The samples were dried naturally and sieved to intercept the part of <50 μm for use in the study. The analysis for As and Cd adopted atomic fluorescence spectrometer (AF-610, Beijing China), and the other elements were analyzed by ICP-AES and ICP-MS (JY38S ICP-AES, France; TJA-1100 ICP-AES, USA; Element II ICP-MS, Germany). All kinds of instruments' testing precision and the analysis quality control indexes showed as in Table 3.31.

3.1.3.2 The Discussion on the Trace Metal Pollution of the Nanjing Section of the Yangtze River

(1) **The trace metal concentration and pollution**

Trace metal concentrations and pollution characteristics in the modern sediment of the Baguazhou Island of the Yangtze River are shown in Table 3.32. The date indicate that Sb, As, Mn, Cd, Pb, Co, Cu, Ni, and V in the surface of sediment had been formed 1–2 grade of Muller pollution grades. Variations of the metal elements studied on the sedimentary plane are shown in Fig. 3.21. The results show that trace metal concentration in the sediments of the Nanjing section of the Yangtze River is slightly higher than its background levels, and the distribution in the same depositing plane is generally uniform and the concentrations' fluctuation was in very small variation range.

The trace metal concentration distributions in the different depths are shown in Fig. 3.22. Figure 3.22 shows that the trace metal concentrations gradually increase with the depth shoaling for the sediment. This implies that the trace metal concentrations in the sediments of the Yangtze River gradually increase with time, but the same sedimentary plane approximately has the same concentrations. According to the change rules of sampling depths and the responded concentrations, the results on the sedimentary rate of the estimation of trace metals in the modern sediment in the Yangtze River can be obtained, shown as Table 3.32. But this estimation is rough one only. The trace metal elements in the sediments except Mn, V, and Hg have the same increase trend in the time range represented by the sampling depth in this research (60 cm vertically). It may be a revelation to the trace metal pollution and evolution trends for the Nanjing city to the Yangtze River. From these facts, the continuous increase of water pollution and ecological risk for the Nanjing city to the Yangtze River is increasing, and the trace metals are accumulating in the sediments with time. The cumulative deposition rate would have a remarkable cumulative effect in future if the current trend continually developed.

Table 3.31 Analytical quality parameters and detect limit of the instruments used in the research

	V	Cr	Mn	Pb	Co	Ni	Cu	Zn	Sb	As	Cd
Element 2 ICP-MS detecting limit (μg/g)									1×10^{-7}		
The rate of data qualification for analysis (%)[a]									100		
JY-38S ICP-AES detecting limit (μg/g)	0.0008	0.00057			0.001	0.001					
The rate of data qualification for analysis (%)	100	100			100	100					
TJA-1100 ICP-AES detecting limit (μg/g)			0.001	0.025			0.002	0.004			
The rate of data qualification for analysis (%)			100	100			100	100			
AF-610 atomic fluorescence spectrometer detecting limit (μg/g)										0.0001	0.0001
The rate of data qualification for analysis (%)										100	100

Note The rate of data qualification for analysis was calculated to be based on the relative deviations within <5% data of every instrument

Table 3.32 Concentration, geo-accumulation index, and added rates of the trace metals in the sediment of the Yangtze River (Nanjing section, concentration unit: mg/kg)

	Cu	Pb	Cd	As	Sb	Hg	V	Co	Ni	Cr	Mn
Background concentration [79, 80]	25.40	22.40	0.23	5.30	4.03	0.06	102.90	9.10	24.70	72.80	426.70
Average concentration in the sediment	59.50	58.70	0.62	20.80	5.70	0.10	193.50	23.80	48.30	98.50	1185.00
Pollution index [35]	0.64	0.80	0.84	1.39	1.88	−3.21	0.33	0.80	0.38	−0.15	0.89
Pollution grade	1	1	1	2	2		1	1	1		1
The rate of adding concentration[a]	0.083	0.067	0.004	0.013	0.0525	0.001	−0.067	0.050	0.067	0.217	−0.667

Note [a]The unit of the rate of adding concentration is mg/kg/cm

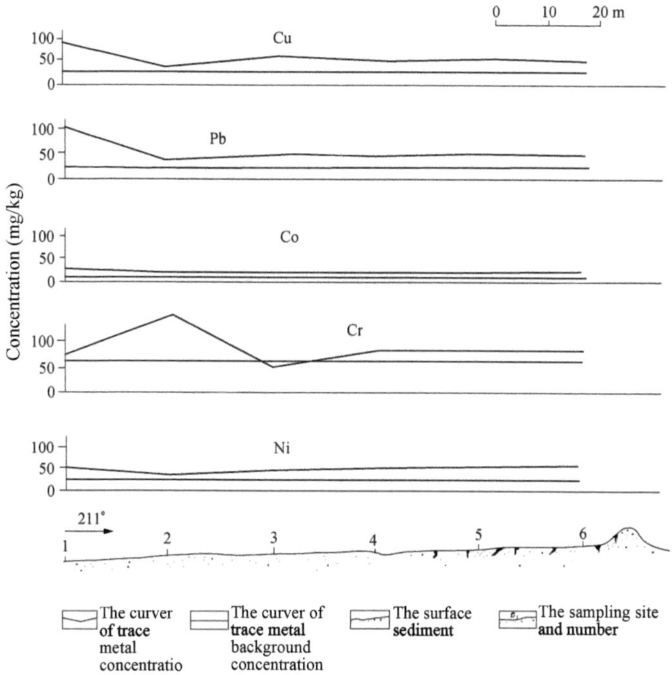

Fig. 3.21 Distribution of the trace metals at same deposition level in the sediment from the Nanjing section of the Yangtze River

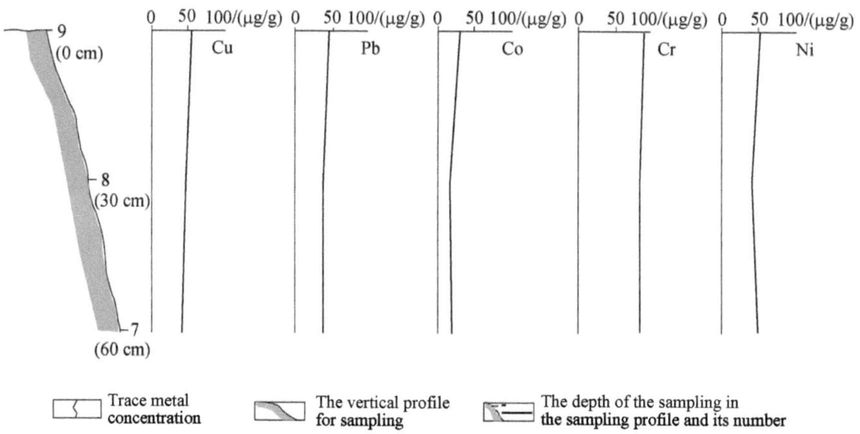

Fig. 3.22 Vertical distribution of the trace metals in the present sediment of the Yangtze River

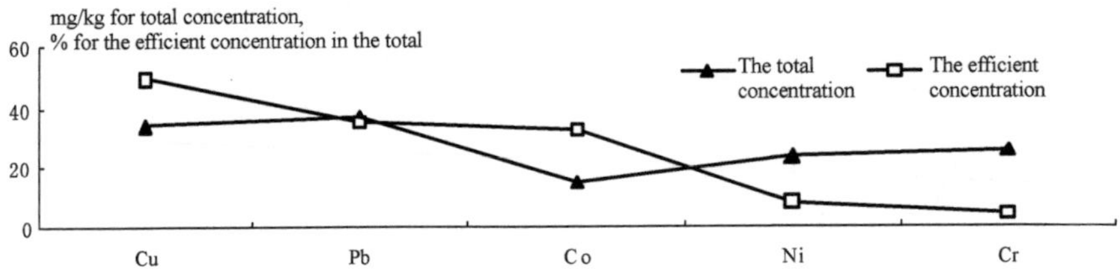

Fig. 3.23 Relationship between the total concentration and the efficient concentration of trace metals in the present sediment of the Yangtze River

2. **The speciation of the trace metals**

Results of chemical species on the trace metals are shown in Table 3.33. The data show that, in effective speciation, V and Cu are mainly in the organic matter and Fe–Mn oxides; Co and Pb are mainly in the exchangeable and Fe–Mn oxidation; Sb and Cd are with small potion; As is mainly in Fe–Mn oxidation, organic and exchangeable; Ni, Cr, and Mn are mainly in Fe–Mn oxides. The order of accumulation of metal elements in its effective speciation is in the following:

$$\text{Exchangeable: Pb} > \text{Co} > \text{Cu} > \text{Mn} > \text{Ni, Cr, and V}$$
$$\text{Carbonate: Cu} > \text{Cr} > \text{Mn} > \text{V} > \text{Co, Ni, and Pb}$$
$$\text{Fe–Mn oxides: Mn} > \text{Co} > \text{Pb} > \text{Cu} > \text{Ni} > \text{V} > \text{Cr}$$
$$\text{Organic: Cu} > \text{Mn} > \text{V} > \text{Cr} > \text{Ni, Co, and Pb}$$
$$\text{Residual: Cr} > \text{V} > \text{Ni} > \text{Co} > \text{Pb} > \text{Cu} > \text{Mn}$$

3. **The cause of formation of the trace metal pollution in the Yangtze River**

In the earlier decades of twenty-first century, the expansion of the production along the Yangtze River had made the sewage to the Yangtze River correspondingly increase by years, which would inevitably lead to the changes of the pollutant concentration in the river water. The fact that the trace metal concentrations gradually increased with the depth from deep to shallow in modern sediments in the Yangtze River is the material record of this change.

While the trace metal concentration of the surface sediments in the beach changed a little with the distance from the river, it suggests that the concentration is restricted by the concentration changes of the river system. The superposition of the trace metals is caused by the metal precipitation carried by the river. And the parts of effective speciation are most likely to move with the water. There is also a positive correlation of the concentrations between the bioavailable fraction of the speciation and superposition amount of the trace metals in the studying area, shown as in Fig. 3.23.

The influence from natural background concentration of the metals by the Yangtze River to this change can be considered in constant. According to this inference roughly, the part of effective speciation of the adding concentration of the trace

Table 3.33 Speciation concentration of the metals in the sediment of the Nanjing section, Yangtze River (mg/kg)

Element	Exchangeable		Carbonate		Fe–Mn oxide		Organic matter		Residue		Total C	Mineral component
	C[a]	CP (%)[b]	C	CP (%)	C	CP (%)	C	CP (%)	C	CP (%)		
Cu	0.68	1.30	0.48	0.90	6.36	12.80	17.50	34.30	25.90	50.70	51.09	Quartz, feldspar
Pb	7.88	15.00	<0.03		10.78	20.50	<0.03		33.90	64.50	52.56	Dolomite (little)
V	0.16	0.10	0.16	0.10	6.80	5.00	3.25	2.40	127.00	92.50	137.37	Calcite (little)
Co	0.64	4.30	<0.03		4.24	28.60	<0.03		9.97	67.10	14.85	Illit (69%)[c]
Cr	<0.05		0.36	0.50	2.86	3.10	0.70	0.80	87.70	95.70	45.74	Chlorite (20%)
Ni	<0.01		<0.01		3.94	8.60	<0.01		41.80	91.40	91.62	Vermiculite (7%)
Mn	1.52	0.20	1.72	0.20	554.90	62.50	32.95	3.70	297.00	33.40	888.09	Kaolinite (4%)

[a]C means concentration
[b]CP (%) means the concentration percent in total (%)
[c]The datum behind mineral name means the relative content of the clay in the sediment

metals in the Yangtze River may be an important contributor. The fact might be the reflection of inherent rule of the concentration changes of metals, and it shows that trace metals increasingly grow in environment with the amount superposition caused by human activities in Nanjing and its upstream.

The differences of the accumulated ability of trace metals in the sediments of the Yangtze River may be more complicated, and the influence factors are also more. The summary of the research on the trace metal speciation is generalized as follows:

(1) **Exchangeable**

According to the experimental procedure, exchangeable trace metals are the concentrations which are soluble in the solution of $MgCl_2$ (pH = 7). Therefore, the dissolved matters in $MgCl_2$ are mainly based on electrostatic interaction. Generally, metals are mainly that combined with the particle surface adsorbing or ion exchange in the environment. According to former study [40, 81] with the facts found by this study, the proportion of changeable trace metals in the river sediments were less than 15% of the total concentrations generally, and most metals among these were less than 5%. Trace metals are relevant with their concentrations in the water and the distribution coefficient between the grain surface and water. The aggregation orders of the metals in the changeable are the results of the interactions of the factors mentioned above. And this order may be the reflection of characteristics of distribution coefficient from different element in the water under conditions of the trace metal concentration of the Yangtze River.

(2) **Carbonate**

In general, the proportion of carbonate metals in the modern sediments of the Yangtze River is less than 1% in the total concentrations, and this may be relevant with less carbonate in the source and carbonate minerals easily to be decomposed in the hot and humid climate conditions in the process of chemical efflorescence.

(3) **Fe–Mn oxides**

Proportions of elements in Fe–Mn oxides are generally larger than 10% in the total concentrations. In addition to combined metals, Fe–Mn oxides have strong adsorption capacity to metal ion (particles) because of its lager surface. Once the condition is fit for precipitation from solution, they will precipitate with carried trace metals. Its concentrations are generally higher with larger proportion of effective speciation. Therefore, this indicator can be as the evaluation standard for evaluating the pollution in river sediments.

(4) **Organic matter**

The difference of the organic proportion of different trace metals is large in the modern sediments of the Yangtze River. The proportion of Cu is up to 34%, with the highest proportion in that of all metals. Proportions of Pb, Co, Ni, and other element are nearly zero. In addition to component difference of anthropogenic sewage, there are other factors also may causing the differences of trace metals in the sediments, such as abundant rainfall in the Yangtze basin, moderate climate, creature booming, causing heavy organic matter in the weathering process flowing into the river, as well as the naturally different combination ability of element with organics.

(5) **Residual**

It was the most common species of trace metals in the sediments of rivers, and it is mainly silicate minerals. As mentioned above, these metals can be as inert elements in the environment, and they are main parts of the natural background. Residual concentration is mainly the embodiment of background factors. For the Yangtze River, this concentration should be relatively stable fraction. All residual element proportions of trace metals are larger than 50% in the total concentrations (Table 3.33) in the modern sediments of the Nanjing section of the Yangtze River. It reveals that the background concentrations of trace metals are major concentrations in the Yangtze River. At present, trace metals have less poison impacts to the creatures in the section. Nevertheless, it should be noted that strongly increasing trend might be a dangerous environmental signal.

3.2 The Pollution of the Trace Metals in Urban Lake, Shanghai

As the case of the trace metal pollution of the urban lake, the Dianshan Lake is selected to reveal the relationship and the mechanism between trace metals and modern metropolis development, Shanghai.

3.2.1 The Distribution, Pollution, Spatial, and Temporal Changes of the Trace Metals of the Dianshan Lake

The Dianshan Lake is only one of protogenesis natural urban lake, and this lake has accompanied almost all development process of Shanghai. The trace metal pollution always has been one of the important environmental issues of the lake.

3.2.1.1 Overview on the Dianshan Lake

Dianshan Lake is the only natural lake in Shanghai, located at the junction of Shanghai, Jiangsu, and Zhejiang provinces, It is located on the east of Taihu Lake and in the western suburb of Shanghai, at 31°04′–31°12′N and 120°54′–121°01′E, with the lake area of 63.8 km^2, about 60 km from the downtown of Shanghai, 18 km from its regulatory agency, Qingpu district, Shanghai. The Dianshan Lake belongs geographically to the Taihu Lake basin, mainly accepting the upstream water from the Taihu Lake basin. Then the water flows into the Yangtze River through the Huangpu

River; finally all flows into the East China Sea. There are 59 rivers around along the lake. The lake belongs to shallow lakes infected by tides, and the water exchange cycle is only about 29 days with the average 2.1 m depth.

The Dianshan Lake is a shallow lake as the Yangtze River moved southward to low-lying alluvial plain, with similar lake-forming causes to the Taihu Lake, the Dongting Lake, and the Poyang Lake in the basin of the Yangtze River.

The range of the Dianshan Lake was originally land. After the formation of the lake, it has experienced repeated changes of the silt shrinkage and expansion by wind waves due to tidal actions, sediment siltation, human reclamation as well as the water and wind wave erosion and unstable lake silt sediment transport from the natural forces.

The Dianshan Lake is one of the important water sources in Shanghai, and the major water sources of it are from the surface runoff discharge of the Taihu Lake, atmospheric precipitation, and the groundwater feeding. The water inlets include Jishui Port, Baishiji Port, Dazhusha Port, and Wangyang Port, from them flowing into about 85% of the total water of the lake. The outlet ports are the Lanlu Port and the Dianpu River. In the tide flood season, the Huangpu River water will flow backward into the lake. In the Dianshan Lake area, the average annual rainfall is about 1000 mm, and the total water storage capacity of the lake is 0.132 billion m^3. The water level is generally stable with the annual average of 2.63 m, which may rise from the middle of June and usually appears the highest level between July and September in the course of a year. According to the data records from the Dianfeng hydrologic station, the water level of the Dianshan Lake in August of 1954 was the highest, 3.7 m. The dry season is from November to March of the following year, with the lowest water level of 1.76 m. The lake flow rate was slow, and water depth was about 1.8–2.5 m. The water level change is relatively small, and the annual change range was less than 0.1 m. There are about 0.2 million acres of farmland irrigated by the lake water. As one of high-quality grain-producing regions in Shanghai, the rice quality of the around lakeside area is usually excellent. According to "The Huangpu River upstream water source protection regulations Shanghai" promulgated in 1985, The Dianshan Lake had been the key water area in water source protection areas of Shanghai. The lands within 5 km from the lake should be designated as water source protection area. Without the approval of the government environmental monitoring department, the production and living of factories and institutions within this area were required to implement the relevant items of preventing water pollution from the "protection regulations." Those who did not meet the requirements of the regulations were closed and stopped for their all activities; some of them might be removed to elsewhere out the lake area. According to the Shanghai municipal people's representatives' inspection in 1987, the water quality of the Dianshan Lake had reached or been close to the situation before 1982, which was in line with national grade II of water quality standard, for drinking in daily life with the background feature of the natural groundwater.

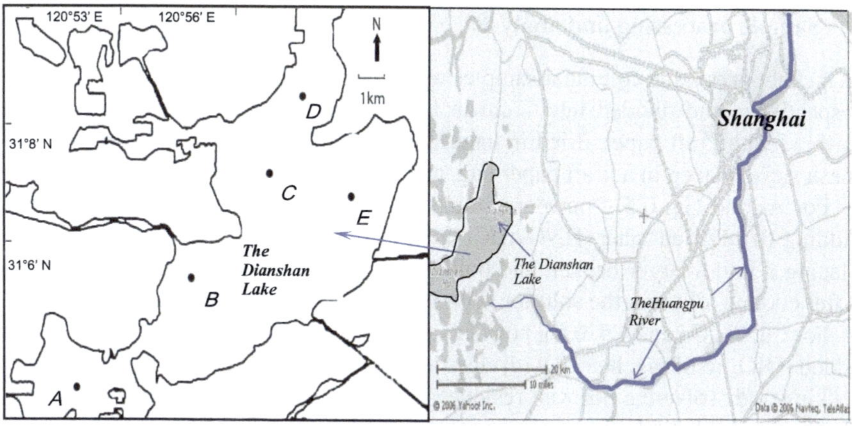

Fig. 3.24 Sampling sites on the Dianshan Lake research

3.2.1.2 Samples and Analysis

In this study, five sampling sites were selected for sampling in the lake area, sites A, B, C, D, and E (Fig. 3.24). The sampling site A was located in the center of a closed lake (Yuandang Lake) in the southwest Dianshan Lake. The sampling site B was located near in the main water inlet (keeping away from the main channel to avoid the possible disturbances). The sampling site C was located in the center of lake, and the sediments were weakly disturbed and relatively stable. The sampling site D was located in Zhaotian area in the north of the Dianshan Lake. The sampling site E was located near the water sports field of the Dianshan Lake, which reflected the frequently influences by human [82].

1. **Samples**

The columnar sediments were collected from five sampling sites in the Dianshan Lake area as mentioned above, using PVC tubes with diameter 8 cm and length 4 m. The PVC tube was directly inserted into the sediment, then taken it out the water and cut off the spare part of PVC tube to take the containing columnar sediment samples PVC tubes back to the laboratory for storage and use next. The length of the columnar samples was 22 cm for No. A, 26 cm for No. B, 19 cm for No. C, and 26 cm for No. D, and 48 cm for No. E. There were differences between colors in the different depth of the samples. For the No. E sample, its color within 0–8 cm was light yellow, dark green within 8–31 cm, and gray-black within 31–48 cm. The other samples were generally gray-black with little differences. At the bottoms of columnar samples, No. A and No. C's were mixed with many shells. From the vertical section, there was gray-black flocculation in the top 1–3 cm of each columnar sample, which had high water content and was mainly clay and fine sandy particles. In the below parts below 3 cm, the colors were gradually light, and the sediments were mainly mud with a little water content and denser texture gradually.

2. **Sample processing and analysis**

In the laboratory, the columnar samples were completely taken out from PVC tubes, respectively, and divided into 1 cm in length fractions with bamboo silver, then covered with kraft paper, dried in natural air, sieved to take the parts under a 200-mesh sieve, placed in a kraft paper bag, and stored under 4 °C for use, respectively.

For As and Hg, 0.5 g powder sample was placed in colorimetric tube (25 ml). Adding 10 ml Aqua water (HNO_3:HCl = 1:3) into the colorimetric tube, capping and placing it into a larger beaker, and boiling for 1.0 h were, respectively, performed. After cooling, diluting the solution with water to 25 ml was for measurement.

For Cu, Pb, Cr, and Cd, 0.5 g powder sample was placed into 50-ml Teflon beaker, added HNO_3 to filling, heated with electrical heating plate to nearly dry (intermittent rocking to dissolve the pot wall residue), added HF to filling, continuously heating to dry, then added HF on the residue to drown the sample and added 2–3 drops of $HClO_4$, heated until white smoke arising, then added 1:10 HCl to the volume of 50 ml for analysis (adding 5 ml concentrated HCl and 45 ml H_2O to fully dissolve the residue in the beaker).

All the reagents were analytical grade, and the water used in the experiments was secondary distilled water prepared by SZ-98 automatic dual water distiller. The polytetrafluoroethylene and glass containers were previously immersed in 10% nitric acid solution for 12 h, and then rinsed with deionized water and dried at low temperature.

(1) **Extraction for the speciation of the trace metals**

A 0.25 g (dry weight) lake sediment sample was used for each step experiment. The four fractions of trace metals were isolated and analyzed by the sequential extracted procedure to assess not only their reactive parts (fractions from the first three steps) but also the experiment analysis quality by the recovery rate between total concentration and the sum of the fractions from the extraction by the formula below:

$$\text{Recovery}\ (\%) = [(\text{Step 1} + \text{Step 2} + \text{Step 3} + \text{Residual})/\text{Pseudo total}] \times 100\%.$$

A. Fraction 1: Exchangeable, water and acid soluble (soluble species, carbonates, and cation-exchange sites)

The sample in tube was extracted with 10 ml of 0.11 mol/L HOAc at 22 ± 5 °C for 16 h with continuous agitation, then centrifuged under 3000 rpm for 20 min. Then the processes were operated as follows: Taking the upper solution into the tube after above process, and holding it at 4 °C for analysis, adding 20 ml water into the residual from above step, shaking for 15 min on the shaker and centrifuging under 3000 rpm for 15 min., decanting the supernatant liquid from centrifuging and discard, taking care not to discard any of the solid residue to be used for next step of extraction.

B. Fraction 2: Reducible (Fe–Mn oxyhydroxide)

The residue from Step 1 was extracted with 10 ml of 0.5 mol/L $NH_2{\cdot}OH{\cdot}HCl$, adjusting pH to 1.5 with HNO_3. At 22 ± 5 °C for 16 h with continuous agitation, then centrifuging at 3000 rpm for 20 min. Then the processes were operated as follows: Taking the upper solution into the tube after above process, and holding it at 4 °C for analysis. Adding 20 ml water into the residual from above step, shaking for 15 min on the shaker and centrifuging under 3000 rpm for 15 min., decanting the supernatant liquid from centrifuging and discard, taking care not to discard any of the solid residue to be used for next step of extraction.

C. Fraction 3: Oxidizable (Organic matter and sulfides)

2.5 ml of 8.8 mol/L H_2O_2 (adjusted to pH 2.0–3.0 with HNO_3) was added to the residue from the Fraction 2 step, reacting for 1 h. Then heating it to 85 ± 2 °C and adding second 2.5 ml 8.8 mol/L H_2O_2 (adjusted to pH 2.0 with HNO_3) to react for 1 h with agitating occasionally to volatilize it to several mls. Adding 12.5 ml/L 1 mol/L NH_3OAc into it to react (adjusted to pH 2.0 with HNO_3) for 16 h at 22 ± 5 °C with continually agitating. Then centrifuging under 3000 rpm for 20 min. Then the processes were operated as follows: Taking the upper solution into the tube after above process, and holding it at 4 °C for analysis. Adding 20 ml water into the residual from above step, shaking for 15 min on the shaker and centrifuging under 3000 rpm for 15 min., decanting the supernatant liquid from centrifuging and discard, taking care not to discard any of the solid residue to be used for next step of extraction.

D. Fraction 4: Residual (this part of metals was mainly in silicates, and the procedure was for most metal digestion except As and Hg, for As and Hg the procedure presented as the addendum below should be handled before the operation).

The residue from the Fraction 3 step was digested with 5 ml concentrated HNO_3 (70%), 2 ml HClO4 (60%), and 8 ml HF (40%) until dry, using the following heating regime: 90 ± 2 °C for 6 h; 120 ± °C for 10 h; 190 ± 2 °C for 6 h. 5 ml of 4 mol/L HCl was added into the remaining material at 70 ± 2 °C for 1 h, then dilute it to 20 ml with 0.3 mol/L HCl and holding it at 4 °C for analysis.

Addendum for the Fraction 4 (Residual) of As and Hg: Adding aqua regia (HNO_3: HCl = 1:3) 10 ml to the residue from Fraction 3 step, coving tubes with covers and heating them for 1 h at under 100 °C (in boiling water). Collecting the solution after treatment above step and holding it at 4 °C for As and Hg analysis. The concentration of As and Hg in Fraction 4 was equal to the total concentration results from these two steps (from Step 4 and addendum step).

(2) **Analysis**

The equipments used in the analysis were sequential injection dual-channel atomic fluorescence spectrometer (AFS-9130, Beijing, China), polarization Zeeman atomic absorption spectrometer (Varian-220z, Varian, USA), high-speed refrigerated centrifuge (Sigma3-18 K, Shanghai, China), thermo-atomic spectrophotometer (Thermo-S4, Germany), electric thermostat oscillation tank (DKZ-2, Shanghai, China), and other processing equipment.

Table 3.34 Working parameters of the atomic fluorescence spectrometer AFS-9130

Element detected	As	Hg
Photomultiplier negative high voltage (V)	270	290
Atomizer height (mm)	8	8
Lamp current (mA)	60	30
Carrier gas flux (mL/min)	400	400
Shielding gas flux (mL/min)	800	800
Time of fetch (s)	7	7
Continual time (s)	1.5	1.5

Table 3.35 Working parameters of the atomic absorption spectrometer

Element detected	Cu	Cr	Cd	Pb
Wavelength (nm)	324.8	357.9	228.8	217.0
Lamp current (mA)	4.0	8.0	4.0	6.0
Ethyne gas flux (mL/min)	1.5	1.5	1.5	1.5
Air flux (L/min)	3.5	3.5	3.5	3.5
Slit (mm)	0.5	0.2	0.5	1.0
Flame height (mm)	13.5	0.0	0.0	0.0

The total concentrations of Hg and As were measured using AFS-9130 sequential injection dual-channel atomic fluorescence spectrometer; the equipment parameters were shown as Table 3.34. The total concentrations of Pb, Cd, Cu, and Cr were measured using Varian-220z polarization Zeeman atomic absorption spectrometer. For the speciation analysis of Pb, Cu, and Cd, the concentrations were measured using thermo-S4 atomic spectrophotometer, and the instrument parameters are shown as Table 3.35. Dilute the samples to 25 times and adding 5% (W/V) thiourea-ascorbic acid as reducing agent before the measurement of As was operated. The other metal elements were directly measured.

Parallel sample determination and recovery experiment of each element were conducted, and the quality parameters for the analysis of each measuring instrument are shown as in Table 3.36. The standard curves of each element, the correlation coefficient, and the fitting equations are shown as Table 3.37.

3.2.1.3 The Distribution and Its Significance of the Trace Metals in the Sediments of the Dianshan Lake

1. **The concentration of the trace metals in the sediments of the Dianshan Lake**

The concentrations of trace metals in the sediments of the Dianshan Lake and the situations of related lakes are shown in Table 3.38. It is revealed that the concentra-

Table 3.36 Working condition and parameters of instruments

Instruments	Elements	Detecting limit (μg/L)	RSD (n = 3, %)	Recovery (%)
AFS-9130 sequential injection dual-channel atomic fluorescence spectrometer	Hg	0.001	4.39	96.95
	As	0.010	2.67	95.47
Varian-220z polarization Zeeman atomic absorption spectrometer	Cu	0.004	3.41	97.36
	Cr	0.010	6.12	103.68
	Cd	0.005	4.20	98.32
	Pb	0.020	7.11	107.90
Thermo-S4 atomic spectrophotometer	Cu	0.003	3.43	101.70
	Cd	0.006	4.22	109.64
	Pb	0.020	5.03	88.06

tions of Cu, Cr, Hg, and As in the sediments of the Dianshan Lake are lower than those in the sediments of Taihu and Dongting lakes in the upper reaches.

Except that the Cu concentration at B sampling site exceeding the soil environmental quality standard grade I [86], the concentrations of other elements and Cu in the other sampling sites are below the soil environmental quality standards. However, the concentrations of Cd and Pb are higher than those in Taihu Lake, Dongting Lake, and Chaohu Lake, as well as higher than the soil environmental quality standards. Moreover, the concentration of Cd is 1.7–4 times of Taihu Lake sediments, and the highest concentration is 14 times as that of the soil environmental quality standard value of China.

From the concentration distribution of each element researched, the element concentrations at C sampling site in the Dianshan Lake are generally lower than those of other sites. This may be related to the fact that the C sampling site is located in the center of the Dianshan Lake and less affected by the pollution source of the coast. The B sampling site is located near a main water inlet and main channel, which is more likely affected by the pollutants and ships of upper reaches. The D sampling site is located near the water amusement park, which is influenced by the artificial disturbance and artificial activity pollution. These factors may cause the high concentration of some metal elements.

2. **The variation of the trace metals with the depth of the sediment**

The Cu concentration in the Dianshan Lake sediments was 16.0–60.7 mg/kg, and the vertical distribution characteristics are shown in Fig. 3.25. The distribution of Cu in the sediments of each sampling site shows a decreasing trend with the increasing depth, and increasing rate of each site was with little change. The highest values from A to E sampling site are 31.83 mg/kg (A, depth 1 cm), 60.731 mg/kg (B, depth 1 cm),

Table 3.37 Parameters of standard curves for trace metal analysis

	Sample number	1	2	3	4	5
Cu	Concentration (10^{-6})	0.0000	0.2000	0.4000	0.8000	1.6000
	Absorbancy	0.0014	0.0248	0.0480	0.0962	0.1892
	Fitted equation and correlation coefficient	$y = 0.1176x + 0.0014$, $R^2 = 1$				
Pb	Concentration (10^{-6})	0.0000	0.2000	0.4000	0.8000	1.6000
	Absorbancy	−0.0038	0.0064	0.0124	0.0267	0.0545
	Fitted equation and correlation coefficient	$y = 0.0357x - 0.0022$, $R^2 = 0.9975$				
Cr	Concentration (10^{-6})	0.0000	0.2000	0.4000	0.8000	1.6000
	Concentration (10^{-6})	−0.0299	0.1577	0.3015	0.5941	1.1096
	Fitted equation and correlation coefficient	$y = 0.7019x + 0.0055$, $R^2 = 0.9966$				
Cd	Concentration (10^{-6})	0.0000	0.0400	0.0800	0.1600	0.3200
	Concentration (10^{-6})	0.0000	0.0144	0.0289	0.0569	0.1075
	Fitted equation and correlation coefficient	$y = 0.3354x + 0.0013$, $R^2 = 0.999$				
As	Concentration (10^{-6})	1.000	2.000	4.000	8.000	10.000
	Concentration (10^{-6})	113.51	242.54	526.66	1152.44	1451.04
	Fitted equation and correlation coefficient	$y = 150.07x - 53.112$, $R^2 = 0.9994$				
Hg	Concentration (10^{-6})	100	200	400	800	1000
	Concentration (10^{-6})	122.01	259.86	538.71	1110.49	1416.92
	Fitted equation and correlation coefficient	$y = 1.435x - 27.88$, $R^2 = 0.9998$				

Table 3.38 Concentrations of trace metals in the sediments of the Dianshan Lake and other lakes in Southeast China

Element	Site A	Site B	Site C	Site D	Site E	Taihu Lake [83]	Dongting Lake [84]	Chaohu Lake [85]	Criterion I of soil environmental quality [86]
Cu (μg/g)	28.69	36.72	17.88	30.77	27.68	41.38	51.30	30.31	35.00
Cd (μg/g)	2.98	3.17	1.34	1.55	2.34	0.79	2.70	0.18	0.20
Cr (μg/g)	37.61	30.26	30.15	38.48	29.66	82.72	87.00	64.66	90.00
Pb (μg/g)	54.69	69.11	44.33	55.97	51.98	41.00	52.80	26.45	35.00
Hg (μg/kg)	53.23	111.69	89.62	115.11	110.41	120.00	220.0	85.47	150.0
As (μg/g)	12.62	9.43	10.71	7.01	7.43	12.43	22.50	4.04	15.00

Note Sites A, B, C, D, and E, respectively, mean the site to be sampled in the Dianshan Lake

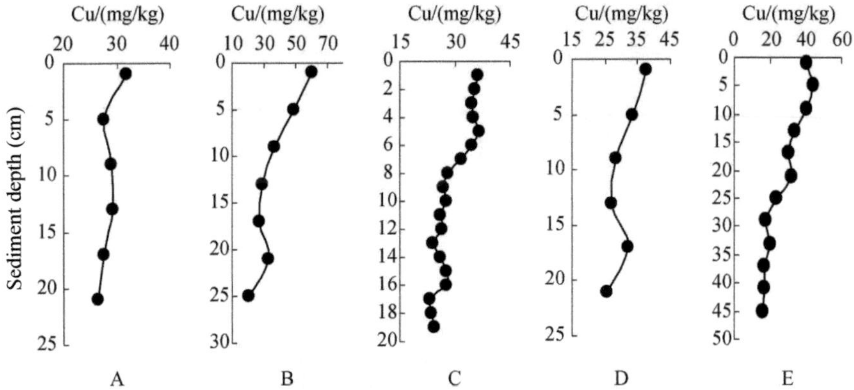

Fig. 3.25 Cu concentration in the core of the sediment from the Dianshan Lake

36.324 mg/kg (C, depth 1 cm), 37.550 mg/kg (D, depth 1 cm), 44.243 mg/kg (E, depth 5 cm), which are all at their surface layers of the sediments. The lowest values are 26.554 mg/kg (A, depth 21 cm), 20.654 mg/kg (B, depth 25 cm), 23.014 mg/kg (C, depth 17 cm), 25.728 mg/kg (D, depth 21 cm), 16.012 mg/kg (E, depth 45 cm), and most of them are at the bottom of sediments. The average concentrations of Cu in each sampling site from large to small in turn are as the order B > D>A > E>C. Except that the Cu concentration in the depth of 0–5 cm at the B sampling site is obviously higher than others in the sediments, there is with a little difference between the concentrations of the different site samples.

The Cd concentration in the Dianshan Lake sediments is 0.12–4.53 mg/kg, and the vertical distribution characteristics are shown in Fig. 3.26. The Cd concentration at B sampling site is relatively stable, and the change is small with different depths. The Cd concentration in C sampling site fluctuates in a little range, but overall it shows a decreasing trend with the increase of the sediment depth. However, the changes of Cd concentration at A, D, and E sampling sites are larger, without obvious regularity. Except D sampling site, the Cd concentrations in the surface (>5 cm) at other sampling sites show increasing trend. The average concentrations of Cd in each sampling site from large to small in turn are in the order B > A>E > D>C.

The concentration of Cr in the Dianshan Lake sediments is 10.47–57.83 mg/kg, and the vertical distributions are showed in Fig. 3.27. The distribution of Cr in the sediments at each sampling site shows an increasing tendency with the sediment depth decreasing. The highest values and occurring places of Cr concentration for the columnar samples, respectively, at sites A, B, C, D, and E are 44.071 mg/kg (9 cm), 40.945 mg/kg (17 cm), 44.831 mg/kg (10 cm), 57.831 mg/kg (5 cm), and 40.119 mg/kg (21 cm). It is showed that all of them were located in the middle and upper sedimentary column. The lowest values and occurring places are 34.200 mg/kg (5 cm), 13.583 mg/kg (25 cm), 25.600 mg/kg (18 cm), 10.474 mg/kg (9 cm), 20.238 mg/kg (45 cm), respectively, which are located in the middle and lower parts of sediments except at site A. The Cr concentration at site A shows an increasing tendency at sections 0–5 and 10–22 cm, and the increasing rates of the

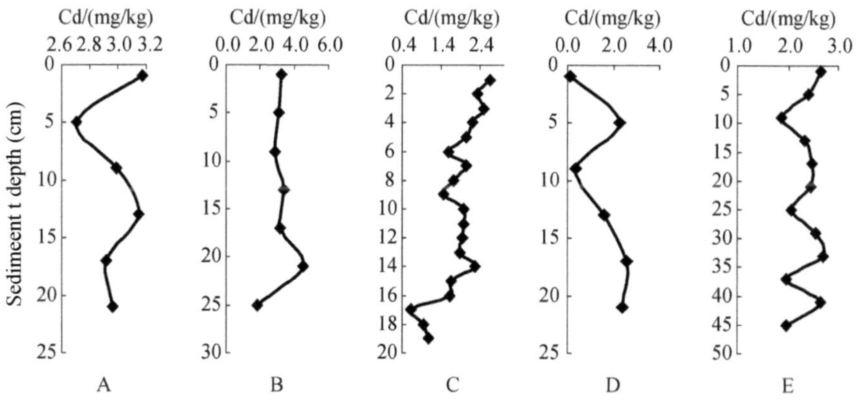

Fig. 3.26 Cd concentration in the core of the sediment from the Dianshan Lake

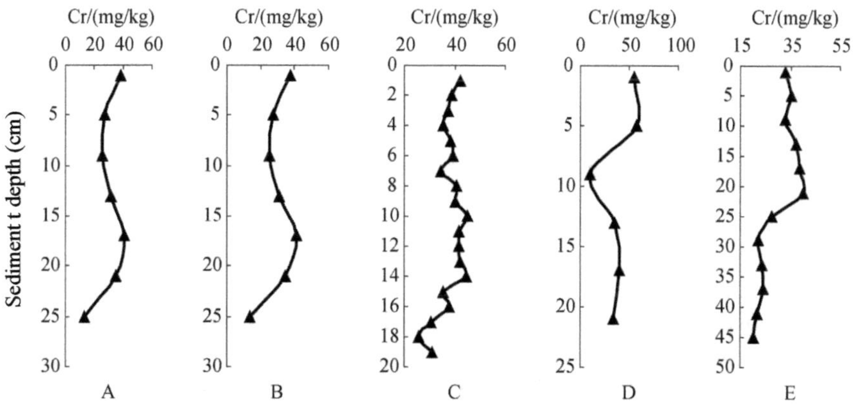

Fig. 3.27 Cr concentration in the core of the sediment from the Dianshan Lake

two sections are also consistent with each other. The Cr concentration at site B also shows an increasing trend at sections 0–10 cm and 17–25 cm, and the increasing rate at 17–25-cm section is relatively faster. There is an increasing tendency at site C from bottom to surface. While Cr concentration at site D shows a decreasing tendency at 10-cm section, and it shows increasing tendency in other depths. The Cr concentration at site E is relatively constant above 20 cm, and it has an increasing tendency below 20 cm. The increasing rate is faster at 20–30-cm section. The order of Cr average concentration of each sampling site from large to small is $D < A < B < C < E$.

The Pb concentration in the Dianshan Lake sediment is 33.97–83.71 mg/kg, and the vertical distribution characteristic is shown in Fig. 3.28. The highest values and occurring places of Pb concentrations in each sampling site from site A to site E orderly are 59.380 mg/kg (17 cm), 83.710 mg/kg (9 cm), 75.635 mg/kg (5 cm), 71.594 mg/kg (1 cm), and 66.526 mg/kg (1 cm), respectively. Most of them are located

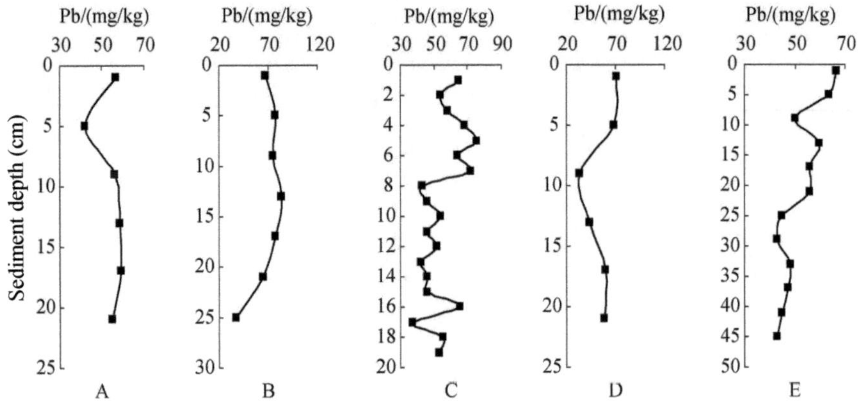

Fig. 3.28 Pb concentration in the core of the sediment from the Dianshan Lake

in the upper part of the sedimentary column. The lowest values and occurring places of Pb concentrations from site A to site E orderly are 41.880 mg/kg (5 cm), 36.909 mg/kg (25 cm), 37.599 mg/kg (17 cm), 33.972 mg/kg (9 cm), and 42.778 mg/kg (45 cm), respectively. Most of them are located in lower part of the sedimentary column. The Pb concentration at site A is relatively constant at section above 10 cm, and decreases firstly and increases later with the decrease of depth at section below 10 cm, and the low value appears at depth of 5 cm. The Pb concentration at site B shows an increasing tendency with the decreasing depth at section below 10 cm and is relatively constant at section above 10 cm. The Pb concentration in the columnar sample at site C fluctuates a little, but had a little increasing tendency with the decrease of depth generally. The average concentration at section below 8 cm (49.252 mg/kg) is smaller than that at section above 8 cm (62.407 mg/kg). The Pb concentration at site D decreases firstly and then increases with the decreasing depth, where the lowest value appears at the point of depth of 9 cm. The Pb concentration at site E shows an increasing tendency with the decreasing depth in general, and the lower values are at depth range of 10 cm and 20 cm. The average concentration of Pb in each sampling site from large to small in turn is B > D>A > E>C.

The Hg concentration in the Dianshan Lake sediment is 43.09–145.66 mg/kg, and the vertical distribution characteristic is shown in Fig. 3.29. The vertical distribution characteristic of Hg is similar with that of Pb. Overall, Hg concentration of each sampling site increases with the decreasing sediment depth. The highest values and occurring places at sampling sites for Hg from site A to site E orderly are 71.06 mg/kg (1 cm), 131.87 mg/kg (1 cm), 140.03 mg/kg (8 cm), 133.07 mg/kg (13 cm), and 145.66 mg/kg (25 cm), respectively. Most of them are located at the top and upper parts of the sedimentary column. The lowest values of that are 43.09 mg/kg (9 cm), 94.46 mg/kg (21 cm), 59.19 mg/kg (19 cm), 91.66 mg/kg (21 cm), and 80.62 mg/kg (33 cm), respectively, basically locating at the bottom and lower parts of the sedimentary column. The Hg concentration at site A is relatively constant at the section

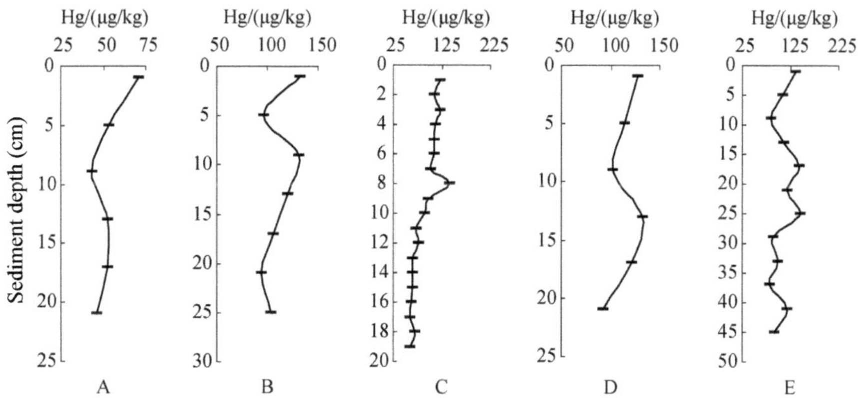

Fig. 3.29 Hg concentration in the core of the sediment from the Dianshan Lake

below 10 cm and increases with the decreasing depth at the section above 10 cm. The Hg concentration at site B increases with the decreasing depth at sections 0–5 cm and below 10 cm; in addition, the increasing rate is faster at the section above 5 cm. The Hg concentration at site C increases at a relatively constant rate with the decreasing depth, and the highest value is at the point of depth 8 cm. The Hg concentration at site D increases as the depth decreases in two sections, respectively (section 0–10 cm and section below 15 cm), and the increasing rates are relatively consistent. The Hg concentration at site E has a great fluctuation, but had an increasing tendency with the decreasing of depth as a whole. The Hg concentration in each sampling site shows the faster increasing tendency at section above 5 cm, overall. The Hg average concentration of each sampling site from large to small in turn is A > C>B > E>D.

The As concentration in the Dianshan Lake sediment is 4.47–14.82 mg/kg, the vertical distribution characteristic is shown as in Fig. 3.30. The As concentration at site A increases firstly and then decreases with the depth decreasing, and the highest value is at the point of depth 12 cm. The As concentration at site B increases with the depth decreasing at sections 0–5 and 20–25 m, respectively, and it is relatively stable at other sections. The As concentration at site C increases with the depth decreasing in general and had an obvious fluctuation with highest values at the point of depth 18, 10, and 4 cm, respectively. The As concentration at site D shows an opposite trend with that of site A, deceasing firstly and then increasing with the depth decreasing. The lowest value appears at the point of depth 13 cm. The As concentration at site E increases at a faster rate below the depth 30 cm, and the highest value appears at the point of depth 30 cm. The concentration generally shows a slightly decreasing trend with the decreasing depth at 10–30-cm section and increases with the depth decreasing in the sediment above the depth 10 cm. The As concentration of each sampling site overall shows an increasing tendency above 5 cm. The average concentrations of As from the sites from large to small in turn are A > C>B > E>D.

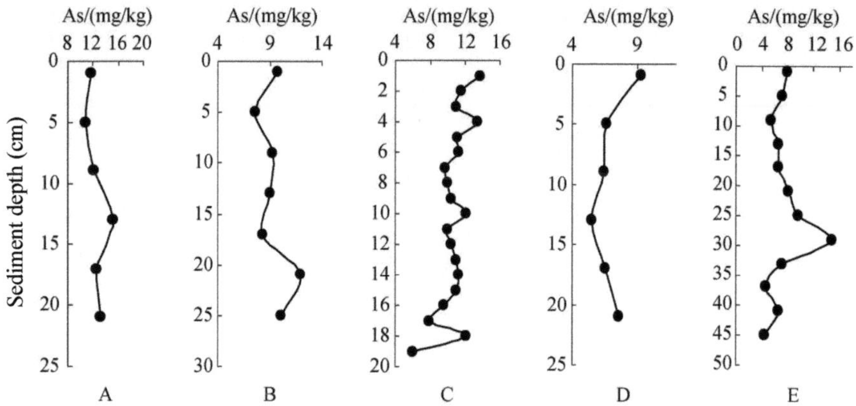

Fig. 3.30 As concentration in the core of the sediment from the Dianshan Lake

From the facts above, the vertical distributions of the metal elements studied in sampling sites suggest following implications for metals in the sediments of the Dianshan Lake:

A. The concentrations of Cu, Cr, Pb, and Hg show the increasing trend with the sediment depth decreasing. The highest values of element concentrations are mostly distributed on the surface and upper part of the sediment, while the lowest are distributed in the bottom and lower parts of the sediment. The vertical distribution of Pb and Hg is similar.
B. The Cd and As concentrations show the great fluctuation with the changes of sediment depth; the changes in sampling sites are not consistent.
C. The concentration of all metals studied in the depth ranging from 0 to 5 cm shows the obvious increasing trend with the sediment depth decreasing, except Cd and Pb at individual sampling site (Cd at site D, Pb at site B), and the increasing rates of most metals as the sediment depth decrease are relatively faster. Thus, it is preliminarily evaluated that the Dianshan Lake is seriously affected by external pollution in recent years.
D. Due to the increasing trend of trace metal concentrations in the surface of the sediments with time, the trace metal pollution in water system of the Dianshan Lake affected by this inertia may continue to increase in the future. Moreover, even if the amount of the pollutants entering the water could be inhibited, the complex exchange between sediments and the overlying water might also cause the trace metals in the sediments to re-enter the water and lead to secondary pollution. The potential impact of the trace metals in the Dianshan Lake is worthy of attention.

Table 3.39 Correlation of metal concentrations in the sediment at site A of the Dianshan Lake

	Cu	Cd	Cr	Pb	Hg	As
Cu	1					
Cd	**0.9534**	1				
Cr	0.0772	0.3744	1			
Pb	**0.7866**	**0.9363**	**0.6763**	1		
Hg	**0.7620**	**0.5312**	−0.5867	0.1997	1	
As	**0.5185**	**0.7523**	**0.8926**	**0.9358**	−0.1585	1

Table 3.40 Correlation of metal concentrations in the sediment at site B of the Dianshan Lake

	Cu	Cd	Cr	Pb	Hg	As
Cu	1					
Cd	**0.9961**	1				
Cr	**0.9305**	**0.8949**	1			
Pb	−0.6996	−0.6344	−0.9127	1		
Hg	0.0043	−0.0831	0.3702	−0.7176	1	
As	0.1827	0.0960	0.5301	−0.8303	**0.9839**	1

Table 3.41 Correlation of metal concentrations in the sediment at site C of the Dianshan Lake

	Cu	Cd	Cr	Pb	Hg	As
Cu	1					
Cd	**0.6501**	1				
Cr	−0.4134	−0.1130	1			
Pb	**0.7173**	0.3799	−0.5194	1		
Hg	0.2028	0.2312	−0.1029	−0.1871	1	
As	0.4708	0.5648	0.2486	0.2853	−0.0332	1

3. **The correlation of the concentration changes of the trace metals in the sediments from the Dianshan Lake**

According to the study of vertical distribution of trace metals in sediments of samples in the Dianshan Lake, it is indicated that the concentration of each metal element in the surface of sediments increased with the depth decreasing. The results of correlating analysis of each element concentration in each sampling site in the depth above 10 cm are shown in Tables 3.39, 3.40, 3.41, 3.42, and 3.43.

The data of Table 3.39 show the correlation better among Cu, Cd, Pb, and As at site A, while the correlation among Cr, Hg and Cu, Cd, Pb, and As is a little less than that. This reveals that there may be partial homology among Cu, Cd, Pb, and As. Table 3.40 shows that the correlation is better among Cu, Cr, and Cd at site B, and the correlation between Hg and As is better too. While there is almost no correlation among the other elements, it suggests that there may be the homology for Cu, Cr, and Cd, as well as between Hg and As. From Table 3.41, it is showed

Table 3.42 Correlation of metal concentrations in the sediment at site D of the Dianshan Lake

	Cu	Cd	Cr	Pb	Hg	As
Cu	1					
Cd	−0.0252	1				
Cr	**0.8816**	0.4495	1			
Pb	**0.9356**	0.3295	**0.9915**	1		
Hg	**0.9961**	−0.1129	**0.8368**	**0.9009**	1	
As	**0.8499**	−0.5483	**0.5006**	**0.6090**	**0.8928**	1

Table 3.43 Correlation of metal concentrations in the sediment at site E of the Dianshan Lake

	Cu	Cd	Cr	Pb	Hg	As
Cu	1					
Cd	0.3761	1				
Cr	**0.9980**	0.3165	1			
Pb	0.4751	**0.9940**	0.4183	1		
Hg	0.1087	**0.9620**	0.0454	**0.9263**	1	
As	0.3841	**0.9999**	0.3247	**0.9949**	**0.9596**	1

that most elements have no correlation at site C, except that individual element had some weak correlations, such as between Cu and Pb, between Cd and Cu. Table 3.42 shows that there is obvious better correlations among Cu, Cr, Pb, Hg, and As at site D, while Cd has no correlation with these elements. This indicates that Cd may have different source with Cu, Cr, Pb, Hg, and As. Table 3.43 shows that there are obvious better correlations among Cd, Pb, Hg, and As at site E, while Cu and Cr have no correlations with other elements. This indicates that Cd, Pb, Hg, and As may have the homology for their cause of formation in the sediment.

According to the correlation analysis of the sediments in individual sampling site, Cd has no correlations with other elements in the sediments for most of sampling sites. This indicates that Cd may have different source with other elements in the sediments. This is also proved by the fact that Cd concentration in sediments of the Dianshan Lake is generally higher (1.34–3.17 mg/kg) than that of others. The correlations among all elements at the site A are better, especially among Cu, Cd, Pb, and As. This indicates that the elements at the site A may have similar source. At the site B, the feature of the distribution of Cu, Cr, Cd, Hg, and As has the homology. At the site D, there are obvious correlations among the concentrations of Cu, Cr, Pb, Hg, and As, which may have similar formation source. At the site E, there are obvious correlations among the concentrations of Cd, Pb, Hg, and As, which may have similar formation source. The correlations of the elements at the site C are generally lower, where the pollution source of these elements may not be consistent. And the concentrations of the metal elements at the site C are mostly lower than that at other sampling sites. It is showed by the comprehensive analysis that trace metals in the sediments show slight pollution at the site C, and their sources or processes may be not the same. This may be related to the fact that the site C is in the center of the Dianshan Lake and is slightly polluted by channels and peripheral point sources. Therefore, compared to the sites A, B, D, and E, the trace metal concentrations in the sediments at the site C are more representative of the features of trace metal pollution in Yangtze River Delta region.

4. **The features of the speciation of the trace metals in the sediments of the Dianshan Lake**

After trace metals to enter the sediments, they exist in different compound species with different energy states. Their migration abilities in sediments are different, and the migration abilities determine the biological effectiveness of trace metals and the harmful degree to the ecological environment. Therefore, in order to study the environmental effects of trace metals in water system, it is necessary to study the speciation of trace metals in sediments and the change or transformation dynamics of the speciation. At present, researches on the speciation distribution of trace metals in sediments are mainly concentrated on the extraction methods and the proportions of trace metals in different speciation, and there are few researches on the speciation transformation of metal elements. The speciation transformation direction of trace metals in sediments and the changes with time could provide the information about trends of speciation transformation of trace metals in sediments and the impacts on the biological effectiveness, which is of great significance for understanding the regulations of migration and transformation as well as the biological effectiveness of trace metals in sediments.

According to the analysis results of total amount for trace metals in different sampling sites, the site C, which is the most typical in all sampling sites of the Dianshan Lake in this study (the most stable sediments and the minimum interference), is selected for analyzing the speciation from the sample core column with intervals of 4 cm. The weak acid-extractable, the reducible, oxidizable, and residual As, Pb, Cd, and Cu are extracted using modified BCR continuous extraction method. The concentrations are determined, and the percentage of each species in total amount of trace metals is calculated. Most of the speciation concentration of Cd is below the detection limit of the instrument, which would not be discussed here.

The extraction procedure operated for 1 g dry sediment sample is as follows:

A. Weak acid-extractable specie: Add 40 ml 0.11 M HOAc into the sample, oscillate overnight at 20 °C, centrifuge at 1500 rpm for 20 min, and then get the supernatant to be analyzed.
B. Reducible specie: Add 40 ml 0.1 M $NH_2OH{\cdot}HCl$ in the extracted residue in the previous step (adjust pH to 2 by HNO_3), oscillate overnight at 20 °C, centrifuge at 1500 rpm for 20 min, and then get the supernatant to be analyzed.
C. Oxidized specie: Add 10 ml 8.8 M H_2O_2 in the extracted sediment of the above step (adjust pH to 2 with HNO_3), oscillate 1 h at room temperature, add 10 ml 8.8 mol/L H_2O_2 again (pH = 2), oscillate 1 h at 85 °C, evaporate the solution to several milliliters, finally add 50 ml 1 M NH_3OAc (using HNO_3 to adjust pH to

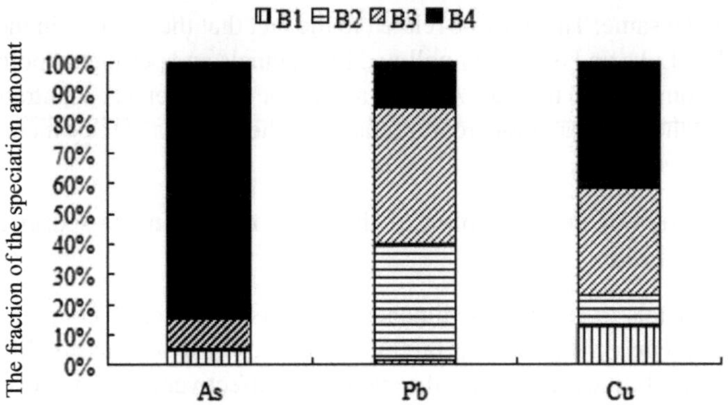

Fig. 3.31 Distribution of the speciation of trace metals in the sediment from the Dianshan Lake. B1—acid soluble, B2—reducible, B3—oxidizable, B4—residual

2), oscillate overnight at 20 °C, centrifuge at 1500 rpm for 20 min, and then get the supernatant to be analyzed.

D. Residual specie: Dissolve the extracted residue from the above step and get the digestion solution to be analyzed.

(1) **The speciation distribution of the trace metals**

The distribution of speciation for As, Pb, and Cu in the sediments of the Dianshan Lake is shown in Fig. 3.31. It is showed that As in the sediments of the Dianshan Lake was mainly combined in the residue species (85.46%), followed by reducible species (Fe–Mn oxide), weak acid-extractable species (water-soluble, exchangeable carbonate species), and oxidized specie (organic and sulfide species). The element As has a strong migration capability and mainly combines with organic matter in sediments and soil. In acidic conditions, As is easy to combine with Fe and Al to form compounds, and combine with Ca to form compounds in the soil of being rich in calcium carbonate [87]. Arsenic in the sediments of the Dianshan Lake is mainly combined in residue specie, which is consistent with the study on the speciation and distribution of As in the sediments of Miyun Reservoir by Liu et al. [88].

Pb is mainly in oxidized specie (44.76%), followed by reducible specie (38.51%), residue specie (14.92%), and weak acid-extractable specie. Pb in the sediments is mainly combined with organic complexes and sulfides, while Pb concentrations are low in forms of exchangeable and carbonate species. This indicates that Pb concentration in dissolved species is lower, which is consistent with the results in the relevant foreign literatures [89, 90]. The concentration of Pb in the Fe–Mn oxide specie is also higher, only second to the oxidized specie. It has been reported that the Pb concentration in the Fe–Mn oxide is generally lower [91], but recent studies show that the Pb concentration in the Fe–Mn oxide specie can be relatively higher, which might be due to the high concentration of Fe–Mn oxide with the strong adsorption capability of Pb [92, 93].

Cu is mainly combined in residue species (41.65%), followed by oxidized specie (35.26%), and then weak acid-extractable specie (12.72%) and reducible specie. Cu is mainly combined with sulfide in nature, and always exists in the montmorillonite and clay minerals under reducible conditions. In this study, Cu mainly exists in the form of organic and sulfide species in addition to residual species, and the high stability of organic copper compound contributes to forming the stable complex between Cu and organic matter [94]. Researches by Weisz et al. also found that the solubility of Cu increased under oxidative conditions [87], which is the same as the conclusion from Tokalioglu et al. on Cu, generally in organic species in sediments [95].

(2) **The vertical variation of the speciation of As, Pb, and Cu in the sediment of the Dianshan Lake**

The differences of geochemical behaviors of trace metals and changes of sedimentary environment can lead to the differences of the migration and transformation of trace metals in sediments. The vertical variation characteristics of the speciation of As, Pb, and Cu in the sediments of the Dianshan Lake are shown as in Fig. 3.32.

Trace metals in the exchangeable and carbonate species in sediments can easily migrate and transform, which are susceptible to external conditions such as salinity, temperature, and pH of the system. Changes of ionic composition could cause the changes of exchangeable species, and pH changes could cause changes of carbonate species [96]. The concentrations of As, Pb, and Cu with the form of weak acid-extractable species in sediments of the Dianshan Lake have increasing trends with the decreasing sediment depth, as shown in Fig. 3.32.

In the sediment of the Dianshan Lake, it is shown by Fig. 3.42 that the proportion of weak acid-extractable element is relatively small, and the proportion of residual species was relatively stable, so the total amount in Fe–Mn oxides, organic, and sulfide species is with small changes. As an example for Pb, the concentration in Fe–Mn oxides decreases with the increasing of the sediment depth, while the concentrations in organic matter and sulfide increase with the increasing of the sediment depth. This might be due to the dissolution of the hydroxide under reducing conditions, and ions absorbed and coprecipitated on these substances would enter pore water of sediments, leading to transform to other forms (organic and sulfide species) or directly enter to the water body as free ions. In anoxic conditions, sulfide anions in sediments can react with these ions released from the compound to form the corresponding metal sulfides. Conversely, if oxidation reaction occurred in sediments, the sulfide was oxidized to sulfate, and the original ions combined with sulfide would be released too [97]. The facts in Fig. 3.32 show that there may exist transformation relations among Fe–Mn oxides, organic and sulfide specie in sediments.

The comparison of elements in the form of Fe–Mn oxides, organic and sulfide species is shown in Fig. 3.33. The total amount of Pb in the Fe–Mn oxides and the organic and sulfide species has the increasing trend below 14 cm of the sediment depth, which may be caused by the increase of the total Pb due to the external pollution. The total amount of Cu in the Fe–Mn oxides, organic and sulfide species has a decreasing trend in the section above 14 cm of the sediment depth. Due to the high stability of the combination between copper ions and organic compounds,

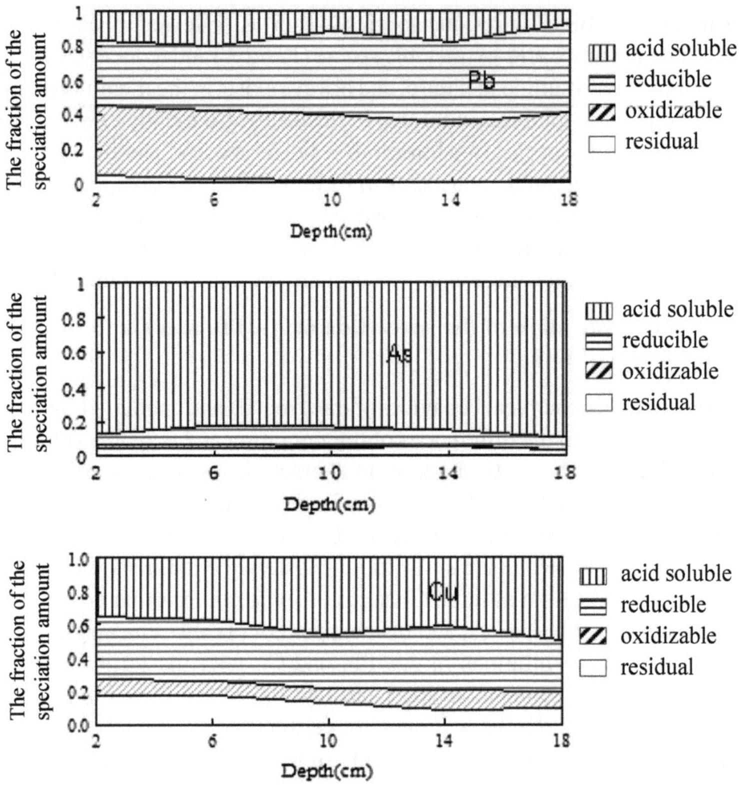

Fig. 3.32 Variation of the speciation amount of the trace metals versus the depth of the sediment from the Dianshan Lake

there is the weak transformation between organic copper and other forms of copper. The proportion of weak acid-extractable Cu is relatively large (12.72%), and this might be due to both Fe–Mn oxide species and organic and sulfide species of copper all might transform to more unstable acid-soluble copper. The ratio between weak acid–extractable Cu and reducible Cu is shown in Fig. 3.34. When the concentration of weak acid-extractable Cu decreases, the concentration of reducible Cu increases and vice versa. There is a clear decreasing–increasing relationship between the two species, where there may exist mutual transformation.

The residual species of metals is present in mineral crystal, which is relatively stable and is not easily released into system as free ions under natural conditions. The concentrations of residual As and Pb have the increasing trend with the decreasing sediment depth, while residual Cu has decreasing trend. This indicates that, if the exogenous factors are same for the effects, the residual species may be transformed to unstable species in the sediments. This transformation would be different for different elements, which is quite complex and may be related to the geochemical nature of the elements.

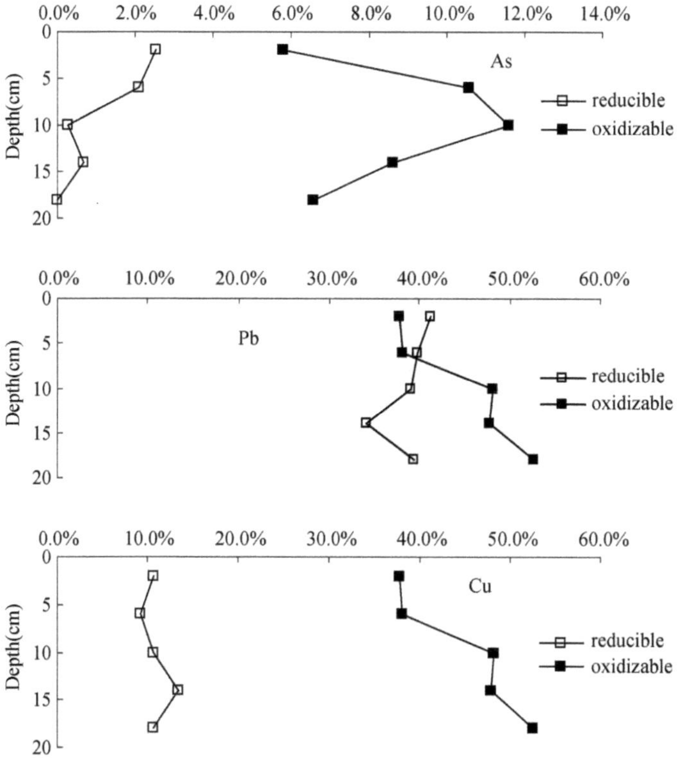

Fig. 3.33 Distribution of the fraction of reducible and oxidizable concentration in the total concentration of the metals versus the sediment depth of the Dianshan Lake

Fig. 3.34 Distribution of the ratio of the acid-soluble and the reducible fractions in the total concentration of Cu in sediment of the Dianshan Lake

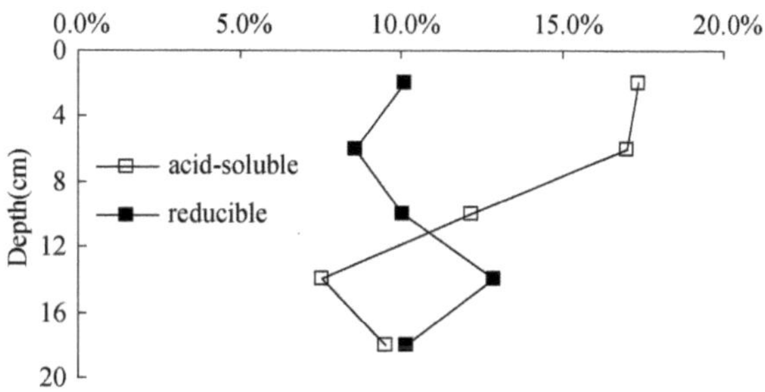

Changes of oxidation and reduction conditions in the sediments may cause the dissolution or precipitation of Fe–Mn oxides, and metal elements combined with Fe–Mn oxides may be subsequently released or adsorbed. The oxidation–reduction potential of the sediments may decrease rapidly below the depth of about 20 cm, while Fe and Mn are rich in the sediments at the depth of about 10–15 cm [87]. With the development of the sedimentation, surface oxides are buried under the oxidation-reduction interface and became unstable. If hydrated Fe–Mn oxides are reduced, the trace metals precipitated or adsorbed on these components may be released. The released trace metals will firstly enter the pore water of sediments. When there are tiny amounts of sulfur ions or organic matters in the pore water, these metal ions can combine with them. As long as the concentration of sulfur ions can reach 10^{-9} M, Pb, Cu, and other elements can be fixed on the reductive sediment particles in the pores [97]. The proportion of Pb in Fe–Mn oxides decreases along the depth increase, possibly due to the reason that the reduction of the redox potential causes the reduction of part of Fe–Mn oxides, and Pb ions combined with Fe–Mn oxides are released, while the increasing proportion of organic and sulfide Pb indicates that some released Pb ions may be combined with organic matter or sulfur ions. Therefore, the reduction of the redox potential leads to the speciation conversion of Pb from Fe–Mn oxides to the organic and sulfide species. In the sediment of the Dianshan Lake, there is a low concentration of Pb in the Fe–Mn oxide specie below depth of 14 cm, which may be located near the redox boundary. The proportion of Cu in the Fe–Mn oxides decreases below 14 cm, while that in organic and sulfide species and carbonate specie increases. When the Fe–Mn oxides are reduced, the Cu ions combined with them are released, and the various matrixes in the sediments would compete for the surface binding sites of Cu ions. When copper ions are combined with carbonates, the concentration of carbonate Cu increases, and the concentration of oxidizable Cu increases when Cu ions are combined with organic matter and sulfides. Under the oxidation conditions of the shallow sediments, the CuS in the sediments is easily oxidized by Fe (III), and the released Cu ions are redistributed to other species. The concentration of As with Fe–Mn oxide species decreases with the increasing sediment depth, while those with the organic and sulfide species increase accordingly. This result is in accordance with the transformation regulation between reducible and oxidizable species. However, under the sediment depth of 14 cm, the total concentration of the two previous species, reducible and oxidizable, of the metals relatively decreases, and the concentration of the organic species decreases obviously. It may be related to the decreasing of the total trace metal concentrations, but it cannot exclude the possibility of transition to other more stable species.

5. **The assessment for the potential ecological risks of the trace metals in the Dianshan Lake**

(1) **The evaluation of geological accumulation index**

Geological accumulation index is a parameter to determine the degree of pollution with the total concentration and the background value of trace metals, which can intuitively give the pollution level of trace metals. The geological accumulation

Table 3.44 Average index of I_{geo} of the metals in sediments from the Dianshan Lake

Elements	Cu	Cd	Cr	Pb	Hg	As
I_{geo}	−0.300	3.177	−2.007	0.885	−1.310	−0.574
Grade	I	V	I	II	I	I

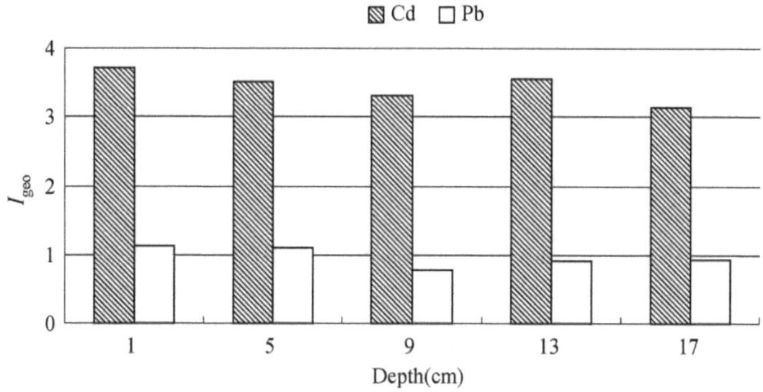

Fig. 3.35 Variation of the geo-accumulation index (I_{geo}) of Cd and Pb in the sediment versus the sediment depth of the Dianshan Lake

index analysis not only considers the effect of background value caused by the natural geologic processes (e.g., diagenesis), but also pays more attention to the influence of anthropogenic activities on the trace metal pollution. Therefore, the index can reflect the natural variation properties of trace metal distribution and could identify the influence of human activities to the environment too, which is an important parameter to distinguish the effects of human activities. Results of the trace metal geological accumulation indexes in the Dianshan Lake sediments are shown as in Table 3.44. Among the elements in this study, the pollution indexes of Cu, Cr, Hg, and As are smaller than zero, indicating that these elements are still at the pollution-free level. Pb and Cd show different degrees of pollution. The pollution degree of each element from heavy to light is generally in the order below: Cd > Pb > Cu > As > Hg > Cr.

The vertical distribution of pollution indexes of Cd and Pb is shown as in Fig. 3.35, and two element pollution indexes have increasing trends with the decreasing of the sediment depth. It is indicated that the pollution degree of Cd and Pb gradually increases in recent decades, and the increasing rate is more obvious with the sediment depth above 9 cm.

(2) **Evaluation with the concentration ratio of the metals, respectively, in the secondary phase and primary phase (RSP)**

The water-soluble and exchangeable trace metals can be released under neutral conditions of the system, which are most likely to cause an effect in the environment. Trace metals with carbonate species are sensitive to pH value in the environment and easily released under acidic conditions. Trace metals in Fe–Mn oxides in natural water are present on the particles in the forms of Fe–Mn nodules or condensates, and some oxides in the gel-film form are covered on the particles, and they are excellent adsorbents for trace metal ions. The reducible metal (iron–manganese oxide species) is the encapsulated part of the metal associated with the Fe–Mn oxides, or the part of the hydroxide precipitation, so it is not easy to be released. When the redox potential of the water decreased, that is when the water was deficient in oxygen, the trace metals with this kind of binding species might be reduced. The oxidizable species (organic and sulfide species) is the binding species with trace metal ion as center ion and with active groups of organic matter as the ligand, or the sulfide-binding species when sulfur ions and trace metals produce insoluble substances, and the substances with oxidizable species could be decomposed only in strong oxidizing conditions [98].

According to the classification proposed by Mao [99], the bioavailability classified for all forms of trace metal elements, the water-soluble, the ion-exchange, and the carbonate-binding species are classified as the effective speciation, the Fe–Mn oxide, organic (humic acid-binding species and organic species) and sulfide-binding species are classified as the potentially effective species, and the residual species is classified as unavailable species. As in sediments of the Dianshan Lake, the metals studied are mainly in form of unavailable species (85.46%), the proportions of effective species and potentially effective species are all relatively lower. Pb existed in the form of potentially effective species (83.28%); the proportions in the form of unavailable species and effective species are lower. The proportion of Pb with effective species shows an increasing trend with the decreasing sediment depth, which has potential hazard to the environment. Cu is mainly in the form of potential effective species (45.63%) and had obvious increasing trend above 10 cm of the sediment depth. The proportion of Cu with effective species is 12.72% and also increased with the decrease of the sediment depth. Therefore, Cu has greater potential hazard on the environment, as shown in Fig. 3.36.

Under uncontaminated conditions, most of the trace metals are distributed in the mineral lattice or present in the Fe–Mn oxide as the coating film of particulates. However, under polluted condition, trace metals from anthropogenic sources are mainly present on the surface of particulates in the adsorbed form, or combined with organic matter, and exist in various weak binding phases such as carbonate and organic species [43]. According to the origin of geochemical facies and the sources of trace metals, Hakanson et al. divided the sediments into primary and secondary phases [38], and proposed the method of distribution ratio in secondary phase and primary phase to estimate the pollution degree. That is, the ratio of the total mass fraction of trace metals in secondary phase to the mass fraction of trace metals in the primary phase was used to reflect and evaluate the sources and pollution levels of trace metals in the sediments. The method is used to evaluate As, Pb, and Cu pollution in the Dianshan Lake sediments, and the calculation formula shows as follows:

$$\mathrm{RSP} = M_{\mathrm{sec}}/M_{\mathrm{prim}}$$

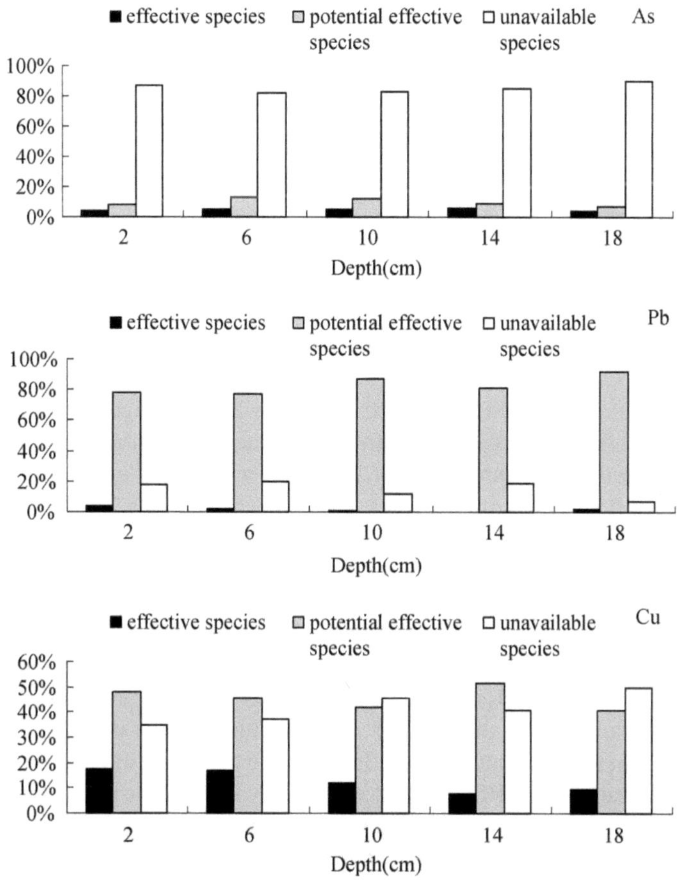

Fig. 3.36 Ratio of the effective speciation concentration in the total concentration of the metals versus depth of the sediment of the Dianshan Lake

where RSP represents the grade of pollution, M_{sec} represents the trace metal concentration in the secondary phase, and M_{prim} represents the trace metal concentration in the primary phase. RSP < 1 means at no pollution grade, 1 < RSP < 2 means at light pollution grade, 2 < RSP < 3 means at moderate pollution grade, and RSP > 3 means at severe pollution grade. The secondary phase is calculated by the sum of concentration of weak acid-extractable species, reducible species, and oxidation species. The concentration in the primary phase is calculated according to the concentration of the residual metal. The distribution proportions of As, Pb, and Cu in the secondary phase and primary phase are shown as in Fig. 3.37.

The ratios of the secondary phase to primary phase for As at different depths all are smaller than 0.3. From the vertical distribution of the RSP value, it is larger in the surface sediments than that of bottom sediments, indicating that although As concentration had not yet reached the pollution level, the pollution trend gradually

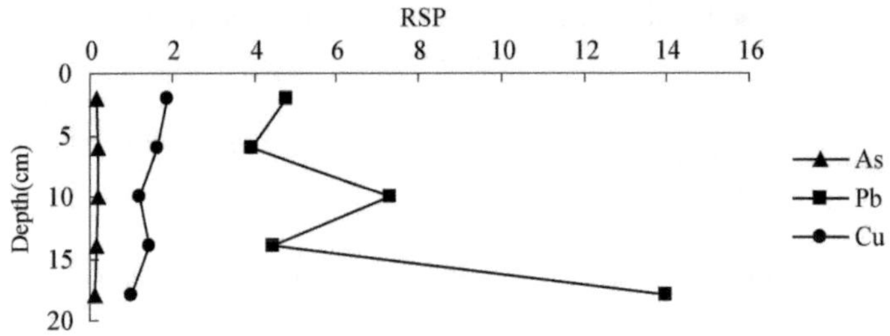

Fig. 3.37 Ratio of the secondary phase to the primary phase (RSP) of the metals in the sediment of the Dianshan Lake

increased. The average RSP value of Cu is 1–2, which belonged to light pollution. There is an obvious increasing trend for RSP value from bottom to surface for Cu, indicating that the pollution degree of it showed a gradually aggravating tendency with time. All RSP values of Pb are larger than 3, which belonged to severe pollution grade, and there is an obvious decreasing trend of RSP value from bottom to surface of the sediment, indicating that the trend of Pb pollution is weakened with time. Nevertheless, due to the more serious pollution of Pb, it will still be in the pollution level for a long time, and there will be with a largely potential environmental risk.

Comparing the result of ecological risk assessment based on the species (RSP) to that based on the total amount (I_{geo}), the two methods are consistent for the assessment results of As and Pb, and the evaluation results of Cu are slightly in difference from the different method. Based on the definition of the pollution grade, the threshold value of the RSP method is relatively lower. This is due to the fact that the latter distinguished the natural and anthropogenic sources of trace metals from the human activity and the geological diagenesis, and took into the chemical activity and bioavailability of the trace metals. Therefore, the ecological risk assessment based on the total concentration can more intuitively evaluate the pollution degree of the trace metals. In contrast, the ecological risk assessment based on the species can more deeply evaluate the migration characteristics and ecological hazards of the trace metals.

In fact, urban lakes are reservoirs of water pollutant, and in the sediments of the lakes it is preserved for the pollution records of the lakes in different historical periods. The trace metals in the sediments may also be re-released and cause the "secondary pollution" as the change of environmental conditions. According to the research on the environmental quality evolution of trace metals based on lake sedimentary records and the understandings of the evolution trend of trace metal pollution in the Dianshan Lake with its surrounding environments, the following points can be drawn:

A. The concentrations of the trace metals in the sediments of the Dianshan Lake are 16.012–60.731 mg/kg for Cu, 0.119–4.532 mg/kg for Cd, 10.474–57.831 mg/kg for Cr, 33.972–83.710 mg/kg for Pb, 43.088–145.658 μg kg^{-1} for Hg, 4.473–15.281 mg/kg for As. Cd and Pb concentrations are higher than those of the surrounding lakes, exceeding the relevant environmental quality standards too.

B. The trend of vertical distribution of the trace metals shows that the concentrations of Cu, Cr, Pb, and Hg increase with the decreasing of the sediment depth. The maximum values of the elements concentration mostly appear on the surface or the upper part of the sediment, and the minimum values are generally distributed at the bottom or lower part of the sediment. The concentrations of Cd and As vary greatly along the sediment depth, and the variation regulations at different sampling sites are different. However, at the sediment depth ranged from 0 to 5 cm, almost all of the metals show the increasing trend with the decreasing of the sediment depth. Thus, there can be a preliminary judgment that the Dianshan Lake has been affected by external pollutants in recent years, and the trace metal pollution is with the aggravated trend as the time then. According to the correlation of the trace metals in the sediments of each sampling site, Cd in the sediments at most of sampling sites has no correlations with other elements, indicating that Cd may have different source with other elements in the Dianshan Lake.

C. Based on the potential ecological risk assessment by the geological accumulation index in the Dianshan Lake, the results show that Cu, Cr, Hg, and As are still at pollution-free grade, while Pb and Cd appear on different degrees of pollution. The pollution indexes of Cd and Pb show the increasing trend with the decreasing of the sediment depth, indicating that the pollution degrees of Cd and Pb in the sediments increase gradually in recent years, especially in the surface sediments above 9 cm.

D. Results of the RSP analysis show that the As concentration has not yet reach the pollution grade, but the pollution trend is increasing with time. The Cu pollution is slight but the degree of pollution is obviously increasing with time. The Pb pollution is severe, but the trend of its pollution is in the weakened feature gradually with time. Due to the heavy pollution of Pb, it will have largely potential harmful impact on the environment for a long time.

E. The speciation proportions of the trace metal elements distribute in the sediments of the Dianshan Lake in order as: Residual > reducible > acid extractable > oxidizable for As, oxidizable > reducible > residual > acid extractable for Pb, residual > oxidizable > acid extractable > reducible for Cu. The concentrations of acid-extractable As, Pb, and Cu in the sediments of the Dianshan Lake decrease with the decreasing of the sediment depth, and the total concentration of the reducible and oxidizable As, Pb, and Cu is in more relatively stable range. It is indicated that there may be transformation reaction between these two species. The concentrations of residual As and Pb show the increasing trends with the decrease of the sediment depth, while that of the residual Cu shows a decreasing tendency with the decrease of the sediment depth. It can be concluded that the change of redox potential in the sediment may cause the transformation of trace metal elements among the Fe–Mn oxide, organic species, and the carbonate species.

F. The arsenic in the sediments of the Dianshan Lake is mainly in the form of unavailable species (85.46%), and the proportions of effective species and potentially effective species are relatively lower, which is less harmful to the environment. The Pb mainly exists in the form of potentially effective species (83.28%), and the proportions of the effective and biounavailable species are relatively smaller. The proportion of effective Pb increases with the decreasing of the sediment depth, which has a greater potential hazard to the environment. Cu exists mainly in the form of potentially effective species (45.63%). For the section above 10 cm of the sediment, the Cu concentration shows obvious increase with the decrease of the sediment depth. Relating to the proportion of effective Cu was 12.72%, all of these situations indicate that Cu also will have potential hazards to the environment of the Dianshan Lake.

3.2.2 The Coupling Relationship of Levels of Trace Metals, N, P, C, and Pesticides in the Dianshan Lake

3.2.2.1 The Features and the Evolution of the N in the Dingshan Lake

With the rapid economic development for decades, an increasing amount of agricultural and industrial wastewater has been produced and discharged into the Yangtze River, which accelerates the accumulation of contaminants in freshwater lakes in the lower reaches, causing the aggravation of eutrophication. An example was the blue-green algae bloom in the Taihu Lake, China in 2006. The Dianshan Lake, downstream of the Taihu Lake, is a key water source of Shanghai. Its water quality is associated with the safety of drinking water for a population of 24 million as well as the city's economy and industries. As a result, it is both important and urgent to study the water quality trends and its influencing factors in the Dianshan Lake.

Water quality greatly depends on the levels and behaviors of nitrogen in aquatic systems, which is a major contributor to eutrophication. Variation of nitrogen in water can be deduced by investigating the levels of total nitrogen (TN), ammonium nitrogen, and nitrate nitrogen in the sediment and their transformation to one another, which is of great significance in studying the sources and behaviors nitrogen in the Dianshan Lake. Based on the studies of concentration and species of the nitrogen, associated with the issues of phosphorus, organic carbon, trace metals, and pesticides in the Dianshan Lake, it is performed to discuss the status of water quality and pollution evolution trend of the Dianshan Lake and provide scientific basis for the environmental protection of the lake.

1. **Sampling and analysis**

(1) **Sampling**

The water exchange period of the Dianshan Lake is 29 days. With a large area but only one main entrance, water quality varies across the lake. The lake is therefore divided

Fig. 3.38 Sampling sites for N research in the Dianshan Lake

into several characteristic sections for sampling. The Wusong Estuary downstream is over 100 km away from the Dianshan Lake, so far that the effects of the East China Sea tides can be ignored here. Samples were taken from the sampling sites 4 and 5 in this work, as shown in Fig. 3.38 [100].

The site 4 is at the center of the lake, with wide open spaces void of fisheries and shipping, so the samples taken from here could well indicate the normal water quality of the lake. The site 5 is located in the north of the lake, which was in Jiangsu province. The semi-enclosed bay features fish farming using seine nets, and the water exchange is rather slow. So this sampling site is representative of the northern part of the lake.

The sediments were collected using a 4-cm-long PVC pipe with a diameter of 8 cm. It was uniformly punctured on the side to allow for drainage. Compass intersection method was adopted to determine the sampling locations. The PVC pipe was inserted into the sediment and retrieved after reaching the hard bottom. The segment which contained the sediment was cut off, sealed with kraft paper tape, and brought back to the laboratory for next use.

The segment of the pipe was further cut into 1-cm-thick slices. The samples, first naturally dried, were ground and sieved through a 200-mesh sieve. Then they were stored in the refrigerator at the temperature of 4 °C until use. Substances such as shells, stones, and branches were removed from the samples before grinding. The mortar was cleaned with double-distilled water before use.

(2) **Analysis**

Given that lake sediments share similar properties with soils, measurement methods for nitrogen in soil were applied to the sediment samples in this study.

A. Total nitrogen (TN)

Sample extraction techniques mainly include the Walkley–Black method, sulfuric–perchloric acid digestion, and the Kjeldahl method. Studies have shown that quite a few oxidants, such as potassium permanganate ($KMnO_4$), perchloric acid ($HClO_4$), hydrogen peroxide (H_2O_2), potassium dichromate ($K_2Cr_2O_7$), all perform well in digestion. Sulfuric–perchloric acid digestion was employed here. It is a lot less time-consuming and more cost-effective in comparison with the widely used Kjeldahl method while achieving satisfying results.

Upon completing the digestion of the sediment, the quantity of total nitrogen present in the digest was determined using the alkaline potassium persulfate method. Sodium hydroxide solution (NaOH) was added to the digest to avoid the interference of metal ions released from the sediment, such as ferric ions (Fe^{3+}). The pH was adjusted to 4 to ensure the complete precipitation of Fe^{3+} while retaining all the ammonium sulfate ($(NH_4)_2SO_4$).

In the oxidation of the sample by perchloric acid ($HClO_4$), organic nitrogen was decomposed into ammonia, which reacted with concentrated sulfuric acid and produced ammonium sulfate ($(NH_4)_2SO_4$), as the following chemical equations represents.

$$NH_3 - CH_2COOH + 6\,HClO_4 \rightarrow NH_3 + 2CO_2 + Cl_2 + 9O_2 + 4H_2O$$
$$2NH_3 + H_2SO_4 \rightarrow (NH_4)_2SO_4$$

At 60 °C, potassium persulfate in the aqueous solution can be decomposed into potassium bisulfate and atomic oxygen. A certain amount of hydrogen cations is released from potassium bisulfate dissolving, so an alkaline NaOH environment can help promote the decomposition to the fullest. At the temperatures of 120–124 °C, atomic oxygen will oxidize nitrogen compounds to nitrates; meanwhile, organic compounds will also break down. The absorbances of the solution at the wavelengths of 220 nm and 275 nm, denoted by A_{220} and A_{275}, can be measured using a UV spectrometer. The corrected absorbance can be determined as such.

$$A = A_{220} - 2A_{275}$$

The experiment proceeded as follows:

a. 0.5 g sample was weighed and put into a 150-ml Erlenmeyer flask.
b. 5–10 drops of distilled water was added into the flask to dampen the sample. 1–2 min later, 5–10 ml concentrated sulfuric acid was added and mixed by gently shaking the flask, which was then left standing for 30 min.
c. A drop of 70% perchloric acid was added and a funnel was placed above the flask. The flask was gently shaken to allow for sufficient reaction.
d. The flask was heated using an electric furnace for digestion for 5–8 min. For this process, the temperature should not be too high and excess smoke caused by heating sulfuric acid should be avoided. Then the flask was taken off from the furnace to cool down. A drop of perchloric acid was added into the flask, which was heated again until the sample turned gray, when the digestion completed. If the sample were still blackish or brownish, the flask could be temporarily removed from heat for adding another drop of perchloric acid. But perchloric acid should be used in moderation to prevent nitrogen loss.
e. After cooling down, the digest was transferred to a 50-ml volumetric flask to make up a solution.
f. After suspended solids finished settling, 1–2 ml of the supernatant was pipetted into a 50-ml color comparison tube. Ammonia-free water was added to the tube until the level reached about 10 ml. An additional 5 ml of alkaline potassium persulfate solution was put into the cube as well. The pH was adjusted to 4 using NaOH solution to avoid the interference of Fe^{3+}. Then the tube was sealed with the stopper, secured by gauze and elastic bands to prevent it from popping out. The cube was heated by a high-pressure steam sterilizer at 121 °C, which was about 0.11 MPa, for 40 min straight.
g. The temperature was lowered, and the valve was opened for deflation. After the tube cooled to room temperature, 1 ml 10% hydrochloric acid was added to the solution, which was subsequently diluted to a volume of 25 ml with ammonia-free water.
h. The solution was transferred into a dry centrifuge tube. It was spun at 3500 r/min for 10 min, and the precipitates (ferric hydroxide) deposited were removed.
i. The supernatant was taken and placed in a 10-mm quartz cuvette, afterward whose absorbances were measured at 220 and 275 nm, respectively, with a UV spectrometer. Ammonia-free water was used as the reference solution.

The total nitrogen concentration in the sediment sample can be determined with the following equation.

$$C = \left(m / V_{\text{sample}}\right) \times \left(V_{\text{diluted}} / M\right)$$

where C is the concentration of total nitrogen (TN) in the surface sediment (mg/g), m is the nitrogen content read from the standard curve, computed using $m = [A_{220}-2A_{275}]/10.9970$ (mg), V_{sample} is the volume of the supernatant taken from the diluted digest and put into the color comparison tube (mL), V_{diluted} is the total volume of the diluted sample digested in the volumetric flask (mL), and M is the mass of the sediment sample weighed (g).

B. Ammonium nitrogen (NH_4^+–N)

The ammonium nitrogen in sediment is in the form of the exchangeable ammonium ion, and the NaCl or KCl solution is always used to extract the ammonium ions from the sediment. Under the condition of pH > 7, ammonium ion was combined with Nessler's reagent to become yellow component, and it could be measured with colorimetric. The exchangeable and combining chemical equation as follows [92]:

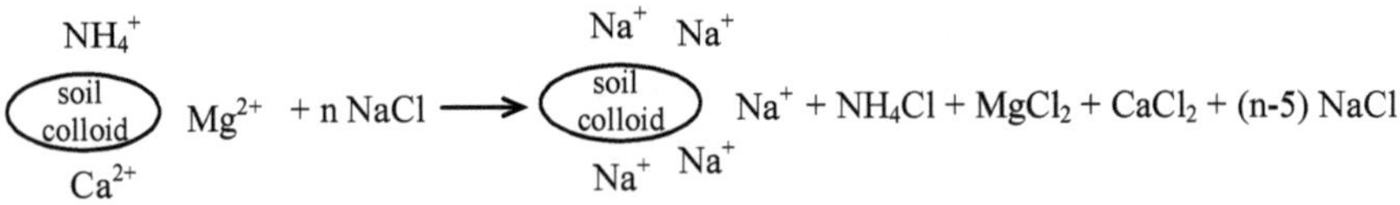

$$NH_4Cl + NaOH \rightarrow NaCl + NH_4OH$$
$$NH_4OH + 2K_2HgI_4 + 3KOH \rightarrow HgOHgNH_2I + 7KI + 3H_2O$$

The 1–2 drops of sodium potassium tartrate were added into the solution for avoiding the interference from the ion Ca and Mg in the solution; chemical equation as follows [101]:

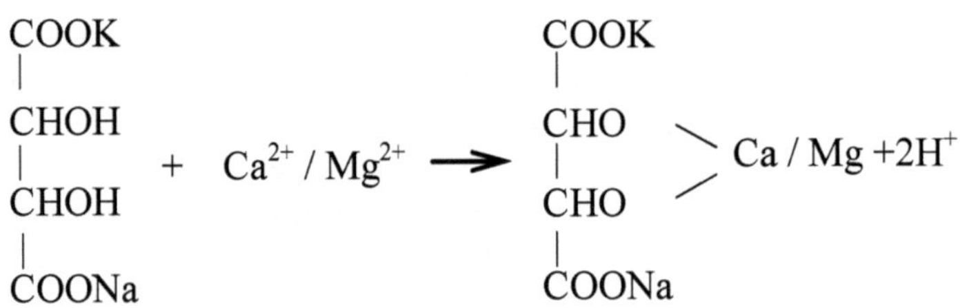

The operating steps for the measurement were as follows:

a. 1–2 g sample in the conical flask with 20% NaCl solution was shocked for 30 min and was filtered with qualitative filter paper.
b. The 5 ml filtrate was taken into the 50-ml colorimetric tube and diluted into 20 ml with water, then added 1 ml 50% potassium tartrate solution stewing for 5 min after shaking up.
c. 1.5 ml $HgCl_2$–KI–KOH agent was added and metered volume to 50 ml for stewing for 10 min after shaking up to be colorimetric, and the colorimetric operation was operated under 420 nm with reference water in 10-mm glass cuvette.

C. Nitrate nitrogen (NO_3–N)

The concentration of nitrate nitrogen can be obtained from measuring the absorbance value of the sample at the points of 220 nm wavelength of the UV light because of the absorbing taking place at the point of 220 nm for the anion of nitric acid. The absorbance value of the sample at the points of 275 nm wavelength of the UV light should be measured for rectifying the effects from organics, and result from rectifying was as the real absorbance value of the nitrate nitrogen in the sample.

Due to the interference substances in the sediment, the key point for measuring the nitrate nitrogen in the sediment by ultraviolet spectroscopy was to choose the extracting agent and restrain the interferences from organics and nitrite nitrogen. The extraction agent should extract out the nitrate nitrogen quantificationally from the sample and let the extraction object being clear colorless with no any interferences for the analysis.

In the measurement for nitrate nitrogen in soil, it had been generally acknowledged that the species of NO_3^-, O_2^- were the main forms of the nitrate nitrogen and nitrite nitrogen in the systems, and they were absorbed little by soil colloids so as to they could be extracted in a short period of time by water from vibration in a short period of time (5–10 min). To avoid the extraction object muddy and colored in the above extraction procedure, the water used in the extraction should contain the flocculant for flocculating the soil colloids and precipitating the organisms in the soil.

In this study, the samples were lake sediment, and the methods of nitrate nitrogen measurement of soil were referenced. The saturated solution of calcium sulfate was used as the extraction solution, and the sulfamate solution was added into the extraction solution to restrain the interferences from nitrite nitrogen [102].

The schedule of the operating for nitrate nitrogen measurement was as follows:

a. The 15 ml saturated solution of calcium sulfate was added into the 1.5 g sample taking in a 150-ml conical flask and filtered by qualitative filter paper after vibrating for 15 min.
b. 5 ml filtered solution was added into a 50-ml colorimetric tube, and added 15 ml non-ammonia water, 1 ml HCl (10%), 0.1 ml sulfamate solution (0.8%), then metered volume to 25 ml with shaking up and stewing for 5 min for measurement.
c. Using the quartz cuvette with 10 mm optical path length, the absorbance value was measured, respectively, at 220 nm and 275 nm points by ultraviolet spectrophotometer, referenced by ammonia-free water in the measurement [103].

D. The parameters and analysis quality control

The instrument used in the analysis was ultraviolet and visible spectrophotometer (UV-2102, UNICO, Shanghai), and the standard curve method was used in the measurement. The main technical parameters of the instrument are as shown in Table 3.45.

a. Total nitrogen analysis

The 0.0, 0.1, 0.3, 0.5, 0.7, 1.0, 3.0, 5.0, 7.0, and 10.0 ml $C_N = 10$ mg/L potassium nitrate standard solutions were added into the 50-ml colorimetric tubes, and metered volumes to 10 ml with ammonia-free water respectively. Then they were analyzed as the total nitrogen measurement schedule presented above.

The two parallel operations were carried out, and the average result from them was used to drawn the standard curve for use in calculate, shown as Table 3.46, Fig. 3.39. From the analysis data of the samples, the working curve of revised absorbance value from fitting ($A_{220}-2 \times A_{275}$) versus the concentration of the potassium nitrate standard solution is:

Table 3.45 Parameters of UV-2102 ultraviolet–visible spectrophotometer

Items	Parameters
Optical system	Single beam, 1200 strip/ mm diffraction grating
Wavelength	200–1000 nm
Wavelength accuracy	±0.5 nm
Wavelength repeatability	0.3 nm
Scope range	0–125% (luminousness), −0.097–2.5 (photometric parameter), 0–1999 (reading concentration), 0–1999 (slope)
Luminosity accuracy	±0.5% (luminousness)
Luminosity repeatability	±0.5% (luminousness)
Stray light	≤0.2% (luminousness) at 220 nm and 340 nm point
Bandwidth	0.5, 1, 2, 4 nm
Stability	±0.002 A/h at 500 nm point
Baseline flatness	±0.005A
Data output	Serial port

Table 3.46 Data of standard curve for total nitrogen measurement (potassium peroxodisulfate method)

Sample number	1	2	3	4	5	6	7	8	9	10
Solution N (mL)	0.0	0.1	0.3	0.5	0.7	1.0	3.0	5.0	7.0	10.0
N concentration (mg)	0.000	0.001	0.003	0.005	0.007	0.010	0.030	0.050	0.070	0.100
Adjust $A_{220\text{-}2} \times A_{275}$	0.000	0.016	0.052	0.072	0.101	0.109	0.384	0.546	0.786	1.071

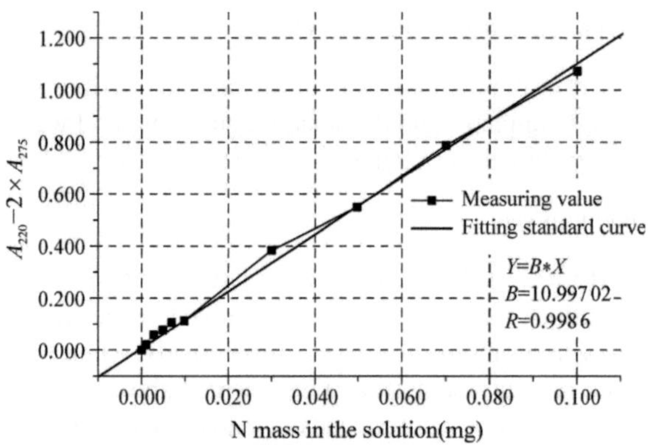

Fig. 3.39 Standard curve for the total nitrogen analysis

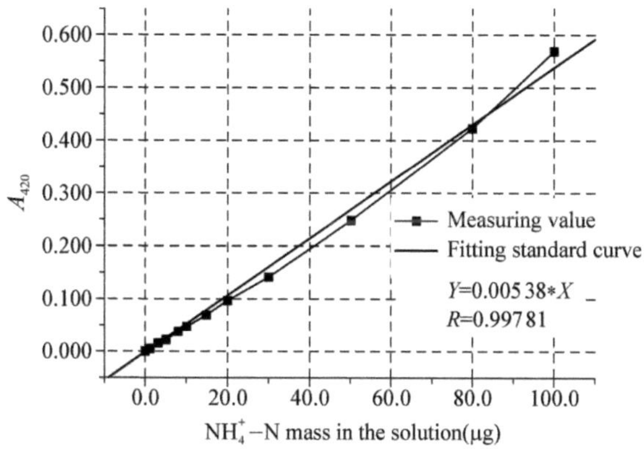

Fig. 3.40 Standard curve for NH_4^+–N analysis

$$Y = 10.9970 \times X\left(R^2 = 0.9972\right)$$

In the equation, Y means the revised absorbance value (A_{220}–2 × A_{275}) and X means the concentration of nitrogen (mg).

b. Ammonium nitrogen analysis

The 0.0, 0.1, 0.3, 0.5, 0.8, 1.0, 1.5, 2.0, 3.0, 5.0, 8.0, and 10.0 ml $C_N = 10$ μg/L NH_4Cl standard solutions were added into the 50-ml colorimetric tubes, and metered volume to 20 ml with ammonia-free water respectively. Then they were analyzed as the ammonium nitrogen measurement schedule presented above.

From the analysis data of the samples, the working curve of revised absorbance value from fitting (A_{220}–2 × A_{275}) versus the concentration of the NH_4Cl standard solution is:

$$Y = 0.0054 \times X\left(R^2 = 0.9956\right)$$

In the equation, Y means the revised absorbance value (A_{420}) and X means the concentration of ammonium nitrogen (μg). The measurement results are as shown in Table 3.47, and the working curve used in calculating is as shown in Fig. 3.40. The formula used in the calculating for the ammonium nitrogen is:

$$C = m \div V_{\text{sample}} \times V_{\text{total}} \div 10M$$

In the equation, C means the concentration of nitrate nitrogen in the sediment (10 μg/g), m means the nitrogen concentration showed by the standard curve ($m = A_{420}/0.0054$) (μg), V_{sample} means the volume of the extraction agent in measurement

Table 3.47 Data of standard curve for NH_4^+–N measurement

Sample number	1	2	3	4	5	6	7	8	9	10	11	12
Solution N (mL)	0.0	0.1	0.3	0.5	0.8	1.0	1.5	2.0	3.0	5.0	8.0	10.0
NH_4^+–N content (μg)	0.0	1.0	3.0	5.0	8.0	10.0	15.0	20.0	30.0	50.0	80.0	100.0
A_{420}	0.014	0.019	0.028	0.035	0.051	0.060	0.082	0.108	0.154	0.262	0.436	0.581

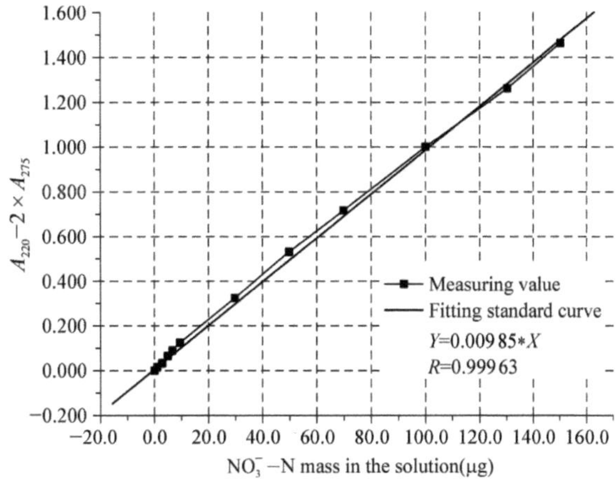

Fig. 3.41 Standard curve for NO_3^-–N analysis

(mL), V_{total} means the extraction agent volume added to the sample in the measurement (mL), M means the mass of the sample, and 10 is as a conversion coefficient.

c. Nitrate nitrogen analysis

The 0.0, 0.1, 0.3, 0.5, 0.7, 1.0, 3.0, 5.0, 7.0, 10.0, 13.0, and 15.0 ml $C_N = 10$ mg/L KNO_3 standard solutions were added into the 50-ml colorimetric tubes, and metered volume to 15 ml with ammonia-free water respectively. Then they were analyzed as the nitrate nitrogen measurement schedule presented above. The measurement results are as shown in Table 3.48, and the working curve used in calculating is as shown in Fig. 3.41.

From the analysis data of the samples, the working curve of revised absorbance value from fitting (A_{220}–2 × A_{275}) versus the concentration of the potassium nitrate standard solution is:

$$Y = 0.0099 \times X\left(R^2 = 0.9993\right)$$

In the equation, Y means the revised absorbance value (A_{220}–2 × A_{275}) and X means the concentration of nitrate nitrogen (μg). The formula used in the calculating for the ammonium nitrogen is:

$$C = m \div V_{\text{sample}} \times V_{\text{total}} \div 10M$$

In the equation, C means the concentration of nitrate nitrogen in the sediment (10 μg/g), m means the nitrogen concentration showed by the standard curve ($m =$ [A_{220}—2 A_{275}]/0.0099) (μg), V_{sample} means the volume of the extraction agent in measurement (mL), V_{total} means the extraction agent volume added to the sample in the measurement (mL), M means the mass of the sample, and 10 is as a conversion coefficient.

Table 3.48 Data of standard curve for NO_3^-–N measurement

Sample number	1	2	3	4	5	6	7	8	9	10	11	12
Solution N (mL)	0.0	0.1	0.3	0.5	0.7	1.0	3.0	5.0	7.0	10.0	13.0	15.0
NO_3^-–N content (μg)	0.0	1.0	3.0	5.0	7.0	10.0	30.0	50.0	70.0	100.0	130.0	150.0
Adjust A_{220}–$2A_{275}$	0.000	0.013	0.031	0.063	0.085	0.119	0.320	0.528	0.711	0.991	1.260	1.460

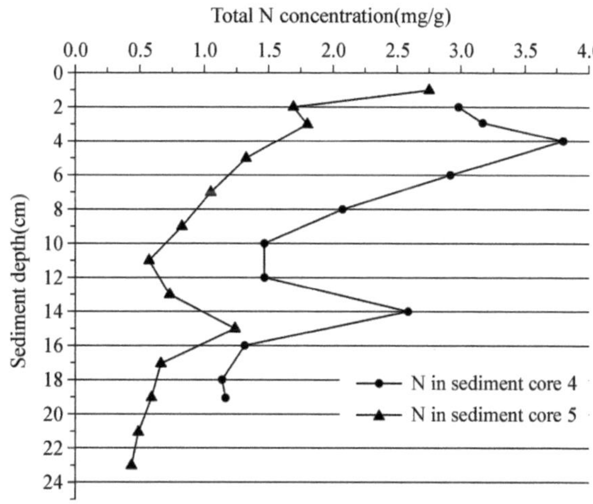

Fig. 3.42 Distribution of TN in the sediment at sites 4 and 5

2. **Results and discussion**

(1) **The distribution of the total nitrogen in the sediment from the Dianshan Lake**

The concentration of the nitrogen in the sediment was obtained from the measurements for 2–3 parallel samples. The distribution of the total nitrogen (TN) in the sediment of the Dianshan Lake is shown as Table 3.49 and Fig. 3.42.

It is shown by Fig. 3.42 that at the 19–10-cm section of the Sample No.4, the concentration of the TN increases slowly between 1.0 and 1.5 mg/g, except at the point of 14 cm; between 10- and 4-cm section, the concentration of TN increases fast, and to the top concentration (3.804 mg/g) with strait increasing trend to imply that nitrogen importing to the Dianshan Lake is increasing continually and fast in the period; at the 4–2-cm section TN concentration begins to decrease. Corresponding to the TN concentration top value point of Sample No.4 at 14 cm, the TN concentration top value of Sample No.5 is at the point of 15 cm, and is with rising at the 11–3-cm section then decreasing after 3 cm depth. The concentration of TN in the core of Sample No.5 is to 2.75 mg/g, highest at top layer sediment of this core sample. The concentration of TN in the two sample sites, No.4 and No.5, is corresponded well with same variation trend to imply that the nitrogen content in the Dianshan Lake is increasing with time, in other words the nitrogen present to the Dianshan Lake is in increasing. Moreover, the TN concentration in Sample No.5 is obviously less than that in Sample No.4, this caused maybe from the differences of nitrogen inputting differences at different sites, and the sampling site No.4 is near at the main channel which is with more surface runoff relatively. It can be seen that, except the impacts to the TN in asynchronism in different deposit depths or time at different deposit places, the external input is in same trend generally, and this may be the basic cause to form the TN content status in generally same variation in the Dianshan Lake.

Table 3.49 Distribution of total nitrogen (TN) in the sediment of the Dianshan Lake

Sample number	Site	Depth (cm)	Sample amount (g)	TN (mg/g)	Sample number	Site	Depth (cm)	Sample amount (g)	TN (mg/g)
4-2	4	2	0.507	2.979	5-2	5	2	0.606	1.694
4-3	4	3	0.509	3.168	5-3	5	3	0.498	1.799
4-4	4	4	0.499	3.804	5-5	5	5	0.527	1.323
4-6	4	6	0.501	2.924	5-7	5	7	0.516	1.048
4-8	4	8	0.502	2.078	5-9	5	9	0.516	0.823
4-10	4	10	0.497	1.469	5-11	5	11	0.489	0.573
4-12	4	12	0.507	1.463	5-13	5	13	0.508	0.725
4-14	4	14	0.493	2.585	5-15	5	15	0.609	1.238
4-16	4	16	0.51	1.316	5-17	5	17	0.598	0.657
4-18	4	18	0.513	1.141	5-19	5	19	0.631	0.585
4-19	4	19	0.506	1.165	5-21	5	21	0.606	0.487
5-1	5	1	0.519	2.753	5-23	5	23	0.701	0.437

The fact that the TN in the sediment of the Dianshan Lake is increasing as the depth decrease of the sediment demonstrates that the nutrient substance into the lake is in growing as the development of the urbanization process as well as the corresponding growth of the production and life development as time in recent decades.

It is shown that the rate of the deposit is about at 4 cm/a with a little increment in recent years [104]. The samples in this work were taken at 2006, and according to TN variation as the deposit depth shown as Fig. 3.42 and the deposit rate the TN change as time in the core of the sediment is in obvious three trend stages. In 1960–1980 (18–10-cm section), the TN in the sediment of the Dianshan Lake was relatively slowly increased with time, corresponded that the well-balanced agriculture at the starting stage (1949) after just now establish the new state with no fast development in production all over the country; and in near 1970s, the TN of the lake occurred increase in speediness implied that it might be from some heavy effects continued in short period to the lake water so as to affect little to the trend of the TN of the lake; to the economic reformation of the state from 1980 (corresponding the section of the sediment core 10 cm above, after 1980) the production and economic development in fast speed in China and Shanghai, the TN recorded in the sediment of the lake shows a straight climb, demonstrates that the TN to the Dianshan Lake is in increase sharply, caused by industry and agriculture development around the lake in fast speed to form the TN into the lake shooting up. In the end of 1990s and after (about above the core 5-cm section) the TN in the lake turned to decrease trend, it might be caused by the impact of environment protection actions along with the economic development inspired by the consciousness of importance for environment quality to the society sustainable development. Generally, the pollutant discharge has been decreasing as the environmental protection policies to be complete oneself.

(2) **The distribution of the nitrogen (ammonia nitrogen, nitrate nitrogen, and organonitrogen) in the sediment of the Dianshan Lake**

The nitrogen in the sediment is most in the form of organonitrogen, and mineral nitrogen is mainly ammonia nitrogen and nitrate nitrogen. The nitrite nitrogen is little in sediment and is often to be lost in sight. So, the organonitrogen concentration in the sediment could be obtained from the total nitrogen concentration to be deducted the mineral nitrogen.

The concentration of the every chemical form of nitrogen in the sediment of the Dianshan Lake is shown as in Table 3.50. The variation of the chemical forms of nitrogen in the sediment of the Dianshan Lake is as shown in Fig. 3.43.

These results show that the concentrations of organonitrogen and mineral nitrogen in the No.4 sample core are decreasing as the depth increases, generally corresponded with the trend of TN concentration change in the Sample No.4. The concentration of nitrate nitrogen is in certain range with no obvious variation except in relatively higher concentration near the interface of water and sediment. These suggest that the TN in the Dianshan Lake is affected mainly by the levels of ammonia nitrogen and organonitrogen, and little affected by that of nitrate nitrogen. In the Sample No.4, the concentration of the nitrogen is mainly in the form of organonitrogen, 1.086–3.745 mg/g, and accounting for 95.2–98.4% of TN. The concentration of

Table 3.50 Distribution of N species from the sediment of the Dianshan Lake

Sample number	Site	Sampling depth (cm)	NH_4^+–N concentration (mg/100 g)	NO_3^-–N concentration (mg/100 g)	Organic N (mg/g)
4-2	4	2	6.742	0.791	2.904
4-3	4	3	4.944		
4-4	4	4	5.309	0.589	3.745
4-6	4	6	5.253	0.590	2.866
4-8	4	8	4.129	0.559	2.031
4-10	4	10	4.916	0.535	1.414
4-12	4	12	5.309	0.530	1.405
4-14	4	14	4.129	0.636	2.537
4-16	4	16	4.719	0.627	1.263
4-18	4	18	4.804	0.674	1.086
4-19	4	19	3.961	0.672	1.119
5-1	5	1	14.158	0.928	2.602
5-3	5	3	13.681	0.688	1.655
5-5	5	5	14.102	0.676	1.175
5-7	5	7	14.664	0.697	0.894
5-9	5	9	13.653	0.554	0.681
5-11	5	11	13.709	0.527	0.431
5-13	5	13	7.500	0.573	0.644
5-15	5	15	7.051	0.619	1.161
5-17	5	17	6.405	0.688	0.586
5-19	5	19	9.017	0.580	0.489
5-21	5	21	8.708	0.675	0.393
5-23	5	23	10.225	0.593	0.329

ammonia nitrogen is in the range of 3.961–6.742 mg/g, accounting for 1.40–4.21% of TN. The nitrate nitrogen is few in the sediment with the stable change trend in no obvious variation, average accounting for 0.3% of TN only. In the Sample No.5, the concentration of organonitrogen is in the range of 0.329–2.602 mg/g, accounting for 75.22–94.52% of TN. The concentration of ammonia nitrogen is in the range of 0.064–0.147 mg/g, accounting for 5.14–23.40% of TN. Being similar to the case of Sample No.4, the nitrate nitrogen in the Sample No.5 is in lower concentration level and in no obvious variation as the deposit depth change, and with average accounting for 0.63% of TN. These facts show that there are differences between two samples for the concentration of nitrogen chemical forms, shown as in Fig. 3.44.

In the most cases, nitrogen gathered in living body, and this may be the cause of formation of the nitrogen status in the Dianshan Lake. For the mineral nitrogen, the NH_4 + can be adsorbed by colloids in the system to be with relatively higher

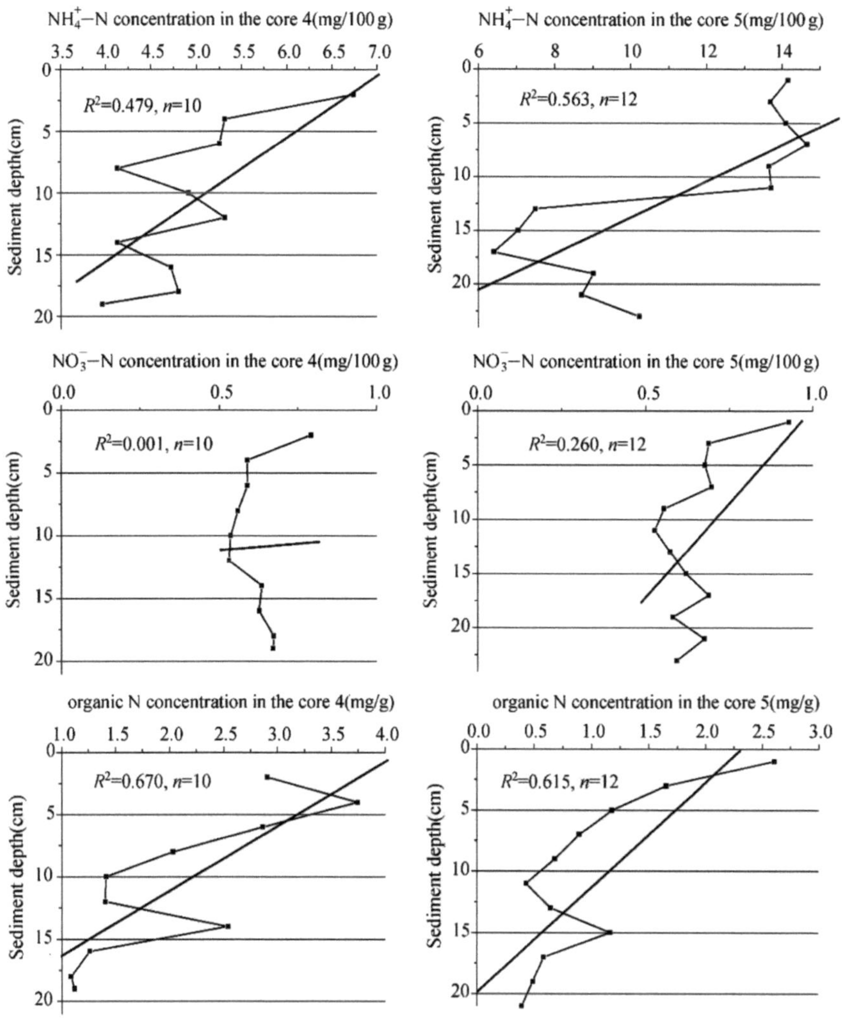

Fig. 3.43 Nitrogen species change in the sediment from the Dianshan Lake

concentration, and for the nitrate nitrogen its NO_3^- is with relatively little possibility to be absorbed by the colloids in the system so as to be lost and hard to deposit in the sediment. These may be the reason that nitrate nitrogen is in relatively lower level and has no obvious gathering as the sediment depth variation. The correlation between total nitrogen and nitrogen forms is shown as in Fig. 3.45.

As presented above, the total nitrogen in the sediment of the Dianshan Lake is affected mainly by ammonia nitrogen and mineral nitrogen, and little by nitrate nitrogen. Because it is limited for nitrate nitrogen to be absorbed by the colloids in the sediment system, it is showed by this work that the nitrate nitrogen is with no obvious

Fig. 3.44 Ratio of nitrogen species in the concentration of TN from the sediment of the Dianshan Lake

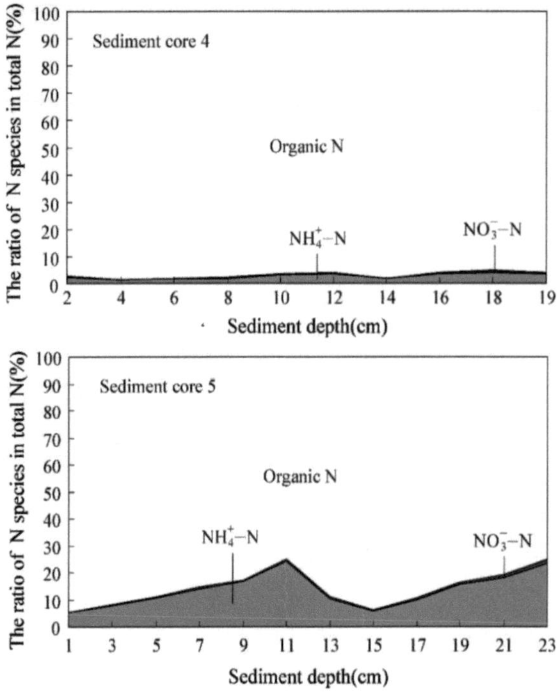

variation in different depth of the sediment of the Dianshan Lake. The correlation index (R^2) between total nitrogen, ammonia nitrogen, and organonitrogen within Sample No.4 and No.5 all are more than 0.99, implying that it is mainly caused for nitrogen gathering in the Dianshan Lake by organonitrogen to be assemble in the water in the process of the lake evolution and human activity around the lake.

(3) The relationships between the nitrogen distribution, the human activities, and the water eutrophication of the Dianshan Lake

In the period of the Sixth Five-Year Plan of China, The East China Normal University carried out a research titled "Study on the Capacity and Plan for the Water Environment of the Dianshan Lake" according to the state key project, The Comprehensive Plan of Prophylaxis and Treatment, cooperated with relative departments to investigate all-sided for the hydrology, water quality, and aquatic organism of the lake, and it concluded that the Dianshan Lake had been polluted with middle to high level of the eutrophication which was in development of deteriorating [105].

A lot of work has suggested that the components of the sediment of a lake usually reflect the environmental status of the lake and the water bodies around it [106–113]. The pollutants in the sediment of a lake are the savings for the particles taking with the pollutants to precipitate or to be absorbed by the sediments. Therefore, the geochemical change of the components of the lake sediments is with significant value for describing the eutrophication status in its history evolution of the lake ecosystem.

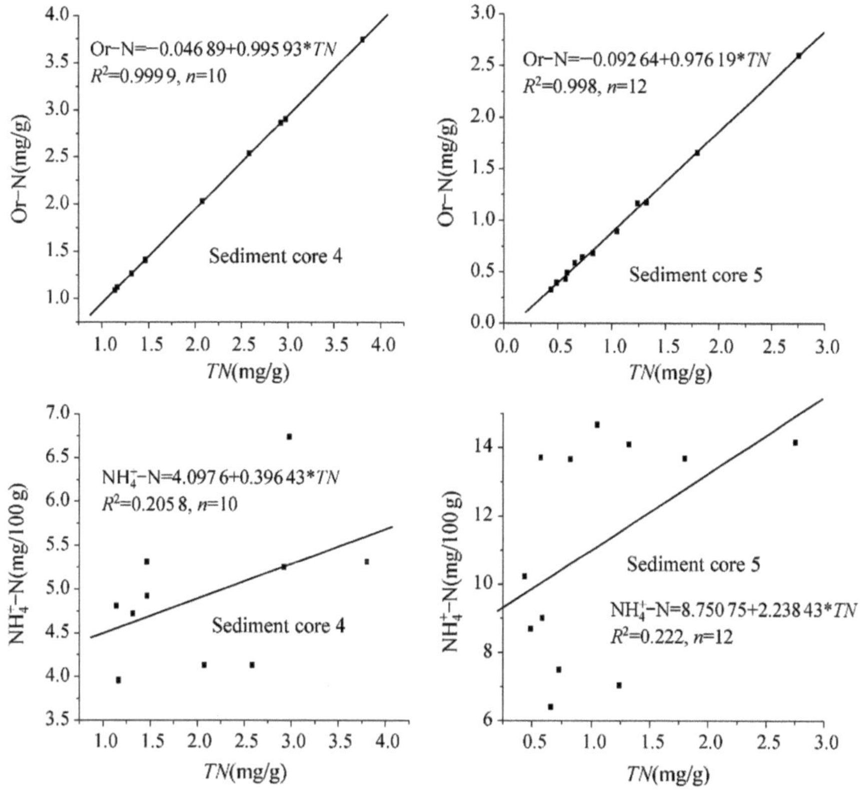

Fig. 3.45 Correlation between TN and its species in the sediment of the Dianshan Lake

Based on the discussion for the nitrogen history status associated with relative data, the following items can be in summary from the research on the nitrogen in the sediment of the Dianshan Lake:

A. The variation trend of nitrogen from the sediment samples of the Dianshan Lake, No.4 and No.5 is in generally same, and the concentration range of nitrogen, respectively, is 1.141–3.804 mg/g for Sample No.4, 0.437–2.753 mg/g for Sample No.5. The top concentration of Sample No.5 is at the surficial layer, 2.75 mg/g.
B. The concentration of nitrogen at sampling site No.5 is generally lower than that at the sampling site No.4, and the trend of the nitrogen variation in the sediment is similar, implying that the site of No.4 is near the main channel and the effects from the external source import such as nitrogen brought by runoff heavier.
C. The fact that nitrogen in the sediment of the Dianshan Lake is in decrease as the deposit depth increase demonstrated that the pollutants and nutrient substances are increasing in the recent decades as the fast urbanization and development of agriculture and industry around the lake. According to the rate of depositing of

the sediment at the Dianshan Lake, 4 cm/a, in the period of 1960–1980 nitrogen concentration in the lake varied a little with slower increase status responded the slow economic development in the area around the lake. After 1980 (the economic reformation of China beginning), the nitrogen into the lake was sharply increasing as the urban economic, industry, and agriculture values. Among this period, in the end of 1990s the TN in the lake was to be in decreasing, implying that the impacts from environmental protection actions and policy as well as the environmental protection awareness of people to reduce the direct discharges in the lake area were reflecting.

D. Comparison for the concentration variation trend of the speciation of nitrogen to that of the total nitrogen in the Dianshan Lake, the ammonia nitrogen and organonitrogen are decreasing as the deposit depth increase corresponding to the variation trend of TN in the sediment; and the concentration of nitrate nitrogen is in lower with no obvious change except in relative higher at the interface of the water and sediment, implying that the total nitrogen in the Dianshan Lake is mainly affected by the concentration of ammonia nitrogen and organonitrogen and affected weaker by that of nitrate nitrogen.

E. The fact that 95% of the nitrogen in the total nitrogen is organonitrogen in the sediment of the Dianshan Lake and in mineral nitrogen NH_4^+–N is account for most portion in the total nitrogen all implied that it is associated with ions to be absorbed easy by the colloids in the sediment system. From the relationship between the total nitrogen, ammonia nitrogen, organonitrogen, and nitrate nitrogen, the nitrate nitrogen concentration is in lower with no obvious change as the deposit depth variation in the sediment may be caused by the anion characteristics of NH_4^+–N difficult to be absorbed easy by the colloids in the sediment system and easy taking away by water so as to be deposited few in the sediment.

F. The results on the variation of the nitrogen in the sediment of the Dianshan Lake above show that the trend of nitrogen change corresponds to the economic variation trend of Shanghai, especially corresponds to the agriculture development and chemical fertilizer use increasing. It is shown that the process of the eutrophication is speeding with the change of the Dianshan Lake natural evolution affected by human activities. In recent years, the increased trend is somewhat diminished as the environmental protection police and actions carrying out with the progress of science and technology of the field. In the whole, the potential crises of nitrogen increase and eutrophication of the Dianshan Lake caused mainly by the human activities around the lake, and it is the significant event in the sustainable development of Shanghai to enhance the protection for the Dianshan Lake from diminishing and forbidding discharge the pollutants to the lake by means of developing science and technology in the production links, as well as governing the pollution formed in history.

3.2.2.2 The Concentration and Evolutionary Trend of Phosphorus and Total Organic Carbon (TOC) in the Dianshan Lake

Lake eutrophication and blue-green algae blooms are one of the major environmental problems, which all lakes are facing in China. Among the major factors for the lake eutrophication, phosphorus is the most common restrictive factor that causes the blue-green algae blooms. Due to the relationship among the contaminated factors such as metal pollutants, nitrogen, phosphorus, and others, it is necessary to research the bio-geochemical behavior and the evolution trend of phosphorus in the lake. These researches are important to reveal the mechanism of lake eutrophication and pollution prevention [114].

In the research of bio-geochemical behavior of phosphorus, the research on the sediment has gradually becomes one of the hot spots in China and abroad. Sediment is the assemble place of the nutrients in river basin, and it has played an important role in the research on the ecological evolution process of lakes and phosphorus geochemical cycles. Accompanying with the migration, chemical species transformation, and the mineralization of phosphorus organic matter in sediment with the process of sedimentary burial and diagnosis, the profile of phosphorus in the sediments may change in its distribution. Different species and concentration of phosphorus in the sediment are different in effects on exchange capacity of phosphorus in the interface between sediment and water, on the concentration of phosphorus in the overlying water, and on the process of lake eutrophication [115]. So, the research of phosphorus species in the sediments can offers an insight into the source, geochemical cycle, and buried features of phosphorus to reveal the mechanism of eutrophication and its factors. Sediment is a more stable recording medium for the space–time distribution of phosphorus in the lake than water. Meanwhile, in order to understand and predict the evolution trend of phosphorus well, it is necessary to research the accumulation of phosphorus in different depth of sediments.

Eutrophication is usually caused by multiple factors. A comprehensive study shows that phosphorus is one of the control factors that caused the lake eutrophication, and the geochemical cycle of phosphorus became the key point for understanding the behavior and impacts of phosphorus in lakes [116, 117]. According to the researchers' hypothesis, the critical ratio of carbon, nitrogen, and phosphorus is 106:16:11 and phosphorus may become to limit the growth of algae when its value is lower than this critical ratio. Someone thought that the lake eutrophication is limited by phosphorus account for 80% in all eutrophication lakes worldwide [117]. The concentration of phosphorus is usually considered as an indicator of lake eutrophication. It is found by the biological analysis that in the Great Lake, USA, the claw crescent algae is sensitive to phosphorus, but insensitive to nitrogen [118]. In the research conducted in northwestern Ontario, it is found that when it is without nitrogen, the addition of phosphorus can lead to cyanobacteria outbreak, which improves that it may significantly enhance the primary productivity of ecosystem [119]. Therefore, it is generally believed that phosphate is the limiting factors for plants to grow in the freshwater system of earth's surface.

The process of change of water quality in a lake usually needs thousands of years as the lake get nutrition from natural systems under the influence of natural factors. However, this process can be accelerated by human activity. Ostorm et al. (1998) found that because of complicated function of natural and artificial factors, the effect of eutrophication in the Lake Erie in North America in 25 years was equivalent to the effect of that from natural factors in 15,000 years [120].

Most of the lakes in middle and lower reaches of Yangtze River are shallow lakes that always may theoretically not be oligotrophic. Before the influence of human activity, these nutritious lakes did not be in eutrophication mainly from the flourishing growth of a large number of wetland and aquatic vegetation. The existence of aquatic plants not only reduces exogenous nutrient in the lake effectively, but also contains the nutrient salt stored and released in and from the sediments. The lake eutrophication has an obvious relation with human activity. On the substances, lake eutrophication is mainly marked with the concentration of TN, TP in lake to be exceed its self-purification capacity. In order to promote economic development, on the one hand lakes are polluted by human discharge the pollutants into the lakes, on the other hand a lot of aquatic plants are destroyed from the pollution from the discharges and various agricultural activities performed in the coastal belt, thus it will be significantly reducing the self-purification capacity of lake again. All the above-mentioned activities accelerate the process of lake eutrophication. The sewage from production, living, and the agricultural drain all contains large amounts of nitrogen, phosphorus, and other inorganic salts. Some sewages that are dumped into lakes lead to an increase of nutrients. These nutrients promote the large number of algae growth rapidly and prevent other algae growth. A large number of blue-green algae are a sign of eutrophication algae reproduce rapidly, and its growth cycle is very short. When they die, the aerobic microorganisms degrade them as the consumption for the dissolved oxygen, or by the anaerobic microbes with the production of hydrogen sulfide gas. Due to these results, many aquatic organisms would die. In the process of algae and other plankton residue decomposition, many nutrients deposited in these organism bodies would be released into the water for the use of a new generation of algae. So the lake eutrophication is a chain process, and it is very hard to clean thoroughly.

The indexes for water eutrophication evaluation are currently including in the items such as the concentration of the nitrogen in the water (more than 0.2–0.3 mg/L), biochemical oxygen demand (COD) (more than 10 mg/L), the concentration of phosphorus in the water (more than 0.01–0.02 mg/L), the total bacteria count in freshwater (more than 100,000/mL when pH is 7–9), and the concentration of chlorophyll-a of algae (more than 10 μg/L). Before 1980s, the water quality of shallow lakes in the middle and lower reaches of the Yangtze River was good, except for some urban and suburban lakes. Since the late of 1980s, most of the lakes had been in the status of middle or middle more eutrophication [121, 122], even had been in that of super-eutrophication for some lakes, such as Chaohu Lake in Anhui province and East Lake in Wuhan province. [123, 124]. In addition, some lakes in mid-nutrition had reached mid-eutrophication indicated by the monitoring at 2000, such as Gucheng Lake [125].

With the fast increase of economy in China, shallow lakes in the middle and lower reaches of the Yangtze River have eutrophicated. With the increasing of pollutants, water quality in the Dianshan Lake gradually becomes poor. The Dianshan Lake is one of the main water sources of Shanghai; the water quality is directly linking to the existence and development of Shanghai. Therefore, the study on the water quality evolution and factors of the Dianshan Lake is an important environmental issue. There are many researches on the concentration of total phosphorus in sediments, such as the relationship between phosphorus concentration in sediments and the soluble phosphate in overlying water, and the effects of total organic carbon, total iron, pH on the concentration of total phosphorus. Because of the transformation and migration phosphorus in the sediment system as presented above, the distribution profile of phosphorus in the sediments would be in variation. The migration and transformation of phosphorus in sediment and water is an important factor for the trophic level of a lake, and the behavior of phosphorus depends on its chemical species. Therefore, the speciation study of phosphorus is significant for understanding the geochemical behavior of phosphorus in sediments and water.

The studies of having been carried out on the evolution of the phosphorus in the sediment of the Dianshan Lake is in lack, and the trace metals are often as an emphasized issue. This study mainly focuses on the change of species and total concentration of the phosphorus in the sediment of the lake to understand the evolution behavior and mechanism of this eutrophication factor in the water source of Shanghai. Moreover, from this work the migration and transform of phosphorus in the Dianshan Lake will be explored to reveal the effects of human activities on the phosphorus cycle and the relationship between phosphorus and lake eutrophication in the process of urban development. From this work, the important consults can be effectually provided for the prediction for the change trend of the water quality of the Dianshan Lake and water environmental protection in Shanghai.

1. **The fractions of phosphorus speciation and extraction from the sediments**

The extraction for the speciation fractions of the P has been used in soil research since many years ago, and the researchers put forward the initial classification scheme according to the different dissolving capacity of the different chemical reagent to the P in minerals. In 1957, the phosphorus speciation fractions were classified into labile or loose-bound P, occluded P, and organic P (called C-J method P). Weng extracted phosphorus from the sediments of rivers and lakes of Washington using the fixed C-J method modified and suggested that the different P combined in different species were mainly inorganic phosphorus (Fe–P, Ca–P, Al–P), solid-state Fe–phosphorus, and Ai–phosphorus (Fe–P, Al–P) [126]. Other researchers improved this method and divided the speciation of phosphorus into apatite phosphorus (AP), no-apatite phosphorus (NAP), and organic phosphorus. In order to overcome re-adsorption use NaOH for the organic phosphorus extraction, some researchers suggested use NH_4Cl as extracting agent. This method can remove the carbonate when extracting the instable phosphorus and it is important for the P extraction from the alkalis sediments. In 1985, Psenner et al. put forward another extraction method for phosphorus in sediments, which was used to analyze the phosphorus in the sediments from Piburger Sea in Austria [127]. This method divided the phosphorus speciation into water-soluble phosphorus (WSP), reducible water-soluble phosphorus (RSP), combined iron and aluminum state of phosphorus, Ca–P, and inert phosphorus. In 1988, Peterson et al. had used this method to extract the phosphorus in sediments in Balaton Lake in Hungary [117], and found that the extracting agents were interfered by the lake's high basicity and the volume of $CaCO_3$. In the later researches, the method which was put forward by Ruttenberg was most effective in the methods suggested. Ruttenberg found that washing the samples with $MgCl_2$ and H_2O, respectively, between the steps of the extractions could minimize re-adsorbing, and based on the procedure the SEDEX method (named R method) was developed [128]. The main steps of this method were $MgCl_2$ extracting exchangeable phosphorus, citrate-sodium bicarbonate-sodium dithionite (CBD) extracting Fe–P, sodium acetate extract carbonate fluorapatite phosphorus, HCl extracting phosphorite phosphorus, and the podzolic residue to be extracted by HCl to estimate the concentration of organic phosphorus. Baldwin then modified this method for enhancement of the extraction efficiency with repeating each step extraction until the P concentration reaching the pointed value. The method suggested first for the separation of the speciation of original detrital phosphorus and the calcium phosphorus-combined, but this method did not apply to the separation of other forms of phosphorus satisfactorily [129]. Li et al. used this method to classify of aluminum phosphorus, iron phosphorus, and occluded phosphate of soil system, and the results obtained were with more clearly geochemical meaning [130]. Golterman proposed another method for P speciation extraction that using nitrilotriacetic to extract Fe–P and Ca–P did not destroy the clay crystal combined phosphorus or organic phosphorus [131, 132]. This method extracts more Fe–P and organic phosphorus [133, 134], and it might be that the hydrolysis from some organic phosphorus makes the values of Ca–P higher when the NaOH/HCl extracted Fe–P/Ca–P, because of the existence of OH^-/H^+. In 1996, Golterman proposed using Ca–EDTA (pH 9.0) instead of Fe–P to extract Fe–P, using Na–EDTA (pH 9.0) to extract Ca–P [131]. This method reduced the times of extraction cycles effectively. The main extraction methods for phosphorus in sediments are shown in Table 3.51.

This work goal is to shed light on the vertical distribution characteristics of the total phosphorus and the different fractions of phosphorus in the sediment of the Dianshan Lake, based on field investigation and combining with regional economic development and population growth, to discuss the history and evolution trends of phosphorus in the Dianshan Lake.

The items of the work have been carried out are as following:

a. Study on the concentration, source, and evolution trends of total phosphorus in sediments at different depths;
b. Study on the distribution characteristics of the speciation fraction of phosphorus in sediments at different depths; discussion on the relationship within various speciations of phosphorus with the coupling relationship and the regulation of

Table 3.51 Procedures of sequential abstraction for P in sediment

Methods	Extractant	The components extracted
C-J method (1957)	a. NH_4Cl 1.0 mol/L	Labile—P
	b. NH_4F 0.5 mol/L	Al–P
	c. NaOH 0.1 mol/L	Fe–P
	d. HCl 0.5 mol/L	Ca–P
	e. CBD	RSP
	f. NaOH	Refractory P
W method (1976)	CBD 0.22/1.0/1.0 mol/L	NAP
	NaOH 0.1 mol/L	AP
	HCl 0.5 mol/L	
H-L method (1980)	a. NH_4Cl 1.0 mol/L	Labile P
	b. NaOH 0.1 mol/L	Fe, Al–P
	c. HCl 0.5 mol/L	Ca–P
P method (1985)	H_2O	WSP
	BD 0.11 mol/L, 40 °C	RSP
	NaOH 1.0 mol/L	Fe, Al–P
	HCl 0.5 mol/L	Ca–P
	NaOH 1.0 mol/L, 80 °C	Refractory P
G method (1990)	a. Ca–NTA 0.02 mol/L + $Na_2S_2O_4$ 0.045 mol/L (Tris buffer, pH 8.0)	Fe–P
	b. Na_2EDTA 0.05 mol/L, pH 8.0	Ca–P
R method (1992)	$MgCl_2$ 1.0 mol/L pH 8.0	Exchangeable P
	CBD 0.3/1.0/0.144 mol/L	CFAP
	NaAc–Na_2CO_3 1.0 mol/L, pH 4.0	FAP, Ca–P
	HCl 1.0 mol/L	FAP
	550 °C ashing, HCl1.0 mol/L	OP
G method (1996)	a. Ca–EDTA 0.05 mol/L + 1% $Na_2S_2O_4$ 0.045 mol/L (Tris buffer, pH 7.0–8.0)	Fe–P
	b. Na_2EDTA 0.1 mol/L, pH 4.5	Ca–P
	c. H_2SO_4 0.5 mol/L	ASOP
	d. NaOH 2.0 mol/L, 90 °C	ROP
Sequential extracting method (1998)	$MgCl_2$ 1.0 mol/L pH 8.0	Adsorption and soluble P
	NH_4F 0.5 mol/L, pH 8.2	Al–P
	NaOH–$NaCO_3$ 0.1/0.05 mol/L	Fe–P
	CBD 0.3/1.0/1.125 g, pH 7.6, stirring for 15 min	O–P
	NaOH 0.5 mol/L, 8 h	

(continued)

Table 3.51 (continued)

Methods	Extractant	The components extracted
	NaAc–Na_2CO_3 1.0 mol/L, pH 4.0, vibration for 6 h	Authigene and apatite P
	HCl 1.0 mol/L, vibration for 16 h	Inorganic and chipping P
	550 °C ashing, HCl1.0 mol/L, vibration for 16 h	OP

speciation fraction transformation between organic matter and phosphorus in the sediments;

c. Analysis and discussion on the history of phosphorus pollution and eutrophication process in the Dianshan Lake associated with the economic development of Shanghai.

(1) **Samples**

The Dianshan Lake belongs to the Taihu drainage basin which is a depression area forming by river alluvial and ancient lagoon deposits. The formation of the Dianshan Lake experienced a shrink then expanding by water and wave scouring. The formation mechanism of the Dianshan Lake is same with lakes of Taihu plain as the characteristics and evolution of lakebed terrain of the lakes of the Taihu plain. The main formation mechanism is strata subsidence to take the lake basin, and then associated with the sediment filling. The Dianshan Lake is a soft water and eutrophic lake, and phosphorus is the main factor for the eutrophication. The other main contaminants are COD, trace metals, petroleum, etc. According to the "Rule of Shanghai Huangpu River Upstream Water Protection" by Shanghai Government, the Dianshan Lake as well the 5 km areas around the lake is an important water-protected area.

The samples of this study were taken at the center of the Yuandang Lake, the southwest branch lake of the Dianshan Lake, shown as in Fig. 3.46.

The sediment samples were sampled with a PVC tube sampler (length 4 cm, diameter 8 cm), then the tube section loaded with sample was cut off and taken to laboratory immediately after sealing by adhesive tape with kraft paper lining. Then the samples were taken from the tubes and sliced into 1-cm sections respectively by bamboo and dried by air, smashed, sieved (mesh 200), blended before use in the next research schedules.

(2) **Analysis**

Determination of moisture concentration is the basis of the analysis and calculation. The sample was placed in an oven to dry to constant weight, all the moisture was dried up but ensure organic matter were not decomposed with the temperature remained below 110 °C, to prevent lose the crystal water (e.g., $CaSO_4 \cdot 2H_2O$), gases adsorbed by particles, organic matter decomposition products, and so on. In the experiment, the instruments and equipment used were 101-A digital display electric heating constant temperature oven, drier with $CaCl_2$, JA1003 N electronic balance, etc. Experimental steps were as follows:

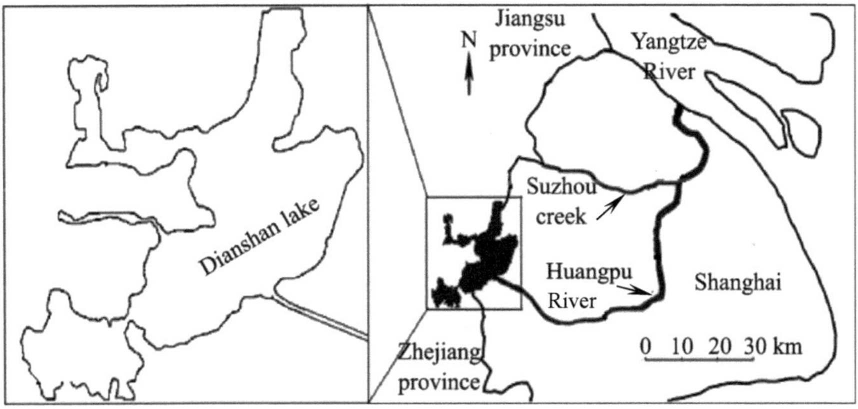

Fig. 3.46 Sampling site on P research in the Dianshan Lake

a. Placed the covered glass plate in the oven to dry (105–110 °C), then it was cooled and weighed (W_1).
b. Took 2.0 g sample into the plate (W_2).
c. Placed the plate with sample into the oven, dried for 6–hours (105–110 °C), and then it was cooled in the drier than weighing. The plate with sample was dried again for 2 h then cooled again and weighted. Repeat this cycle until the weight of the plate with sample remained same (W_3), and ensure that the difference from two weights was less than 0.02 g.

Water concentration (W) was calculated by following formula:

$$W\% = \frac{W_2 - W_3}{W_2 - W_1} \times 100\%$$

The pH value as an important parameter to affect phosphorus release directly, it will affect phosphorus rapidly release from the sediments easily at lower than that at higher. In the study, the pH value was determined by glass electrode and the detailed process was as follows: Put 2 g of dry sample in a beaker then added distilled water 15 ml and stirring for 1 min, then let the sediment to deposit at the beaker bottom for 30 min. Inserted the glass electrode in the turbid liquid and stirred the water. Get the pH value repeatedly until it was at a stable value.

(3) **The extracted concentration of total phosphorus and checking**

The sediment sample was digested by the $HClO_4$–H_2SO_4. This method is easy to operate and very efficient (97–98% of total P can be extracted out). $HClO_4$ is a strong oxidant that can form complexes with trivalent iron when the organic matter and ore materials are decomposed and colloid SiO_2 is dehydrated. It can suppress the interferences from iron and silicon in the colorimetric test. H_2SO_4 can increase the temperature of digestive solution and prevent solution to dry out. After digestion, all kinds of phosphorus were converted to soluble orthophosphate.

Molybdenum-antimony anti-spectrophotometric was used to measure the concentration of phosphorus. The orthophosphate reacts with ammonium molybdate and antimony potassium tartrate under acidic conditions to produce molybdophosphate heteropoly acid that can be reduced by ascorbic acid. The solution was determined by spectrophotometric ($\lambda = 700$ nm), dipotassium phosphate as the standard liquid in the measurement (the limit of determination was 0.01–0.6 mg/L).

In above measurement, the 100-ml volumetric flask, 50-ml Erlenmeyer flask, funnel, induction cooker, 50-ml colorimetric tube, phosphate-free filters (MN619G), UV–visible spectrophotometer (UV-2450), etc., were used, respectively.

Reagent preparation was as follows:

H_2SO_4–Mo–Sb–Vc: After adding 13 g ($(NH_4)_6Mo_7O_{24}\cdot 4H_2O$) into 100 mL solution, adding 0.35 g ($K(SbO)C_4H_4O_6\cdot 1/2H_2O$) into 100 ml solution, adding the ammonium molybdate solution into 300 ml$(1 + 1)H_2SO_4$, and adding the antimony potassium tartrate solution with stirring, the mixed solution was stored in the brown bottle for use.

10% ascorbic acid: After adding 10 g ascorbic acid into 100 mL solution, it was stored in the brown bottle under 4 °C for use, and 4 mol/L NaOH, H_2SO_4 (2 mol/L), 50–60% $HClO_4$, phenolphthalein indicator (adding 1 g phenolphthalein into 100 mL alcohol), phosphorus standard solution (phosphoric acid stock solution into 250-ml volumetric flask to ensure concentration of phosphorus is 2 mg/L) were, respectively, prepared for use in the measurement.

The schedule of measurement for the total phosphorus in the sediment was as follows:

a. Making of the phosphorus standard curve: After adding phosphate standard solution (2 mg/L) into colorimetric tubes, 0, 0.50, 1.00, 1.50, 2.00, 2.50, 3.00, 3.50, 4.00, 5.00, 7.50, 10.0, 12.5, 15.0 ml, respectively, first, then adding deionized water to the constant volume to 50 ml for every tube after adding the coloring agent, respectively, for each solution, then phosphate solutions with concentrations (0, 0.02, 0.04, 0.06, 0.08, 0.10, 0.12, 0.14, 0.16, 0.20, 0.30, 0.40, 0.50, 0.60 mg/L) were achieved. Next, adding 1 ml 10% ascorbic acid solution into the solutions, respectively, for reaction 30 s, then adding 2 ml molybdate solution into every solution, the absorbance was determined under 700 nm frequency band after coloration for 15 min. The work completed as above procedure, the phosphorus standard curve were obtained as the concentrations and responded absorbances measured from above solutions.
b. After adding 0.1–0.3 g sample in 50-ml Erlenmeyer flask, adding 3 ml concentrated H_2SO_4 after the sample was wetted by a little water dropwise, then 20 drops of $HClO_4$ were added and proceeded by the digestion on the electric stove. It was to continue digestion for 20 min until the color of the solution changes to white. The control experiment was proceed same time with no adding sample. The digested solution was put into 100-ml volumetric flask after it was cool

Table 3.52 Data of phosphate standard curve in analysis

Concentration of phosphate (mg/L)	Absorbancy (A)[a]	Concentration of phosphate (mg/L)	Absorbancy (A)[a]
0	0	0.14	0.069
0.02	0.01	0.16	0.081
0.04	0.019	0.2	0.101
0.06	0.031	0.3	0.153
0.08	0.042	0.4	0.203
0.1	0.051	0.5	0.253
0.12	0.06	0.6	0.307

[a]A value was the value after deducting the blank value

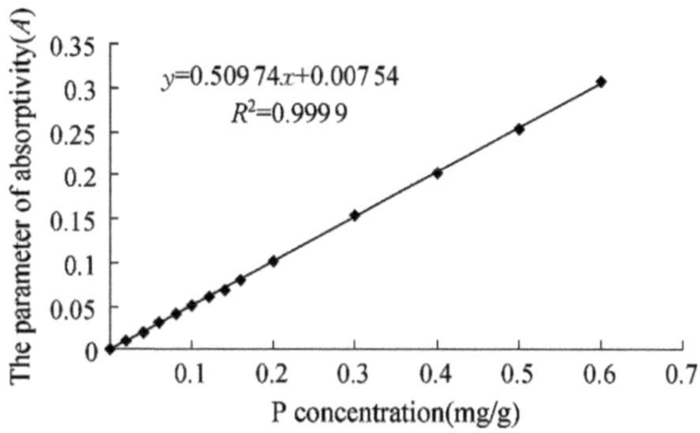

Fig. 3.47 Standard curve for phosphate analysis

down, and then water was added to be constant volume for the solution before filtrating treatment by phosphate-free filters.

c. 10 ml filtrated solution was sucked to 50 ml volumetric and adding 1 drop phenolphthalein indicator to the flask together with 20 ml water. Then the NaOH (4 mol/L) was added dropwise into the solution until its color changes to red, then dropwise adding H_2SO_4 (2 mol/L) until the red color fades away.
d. Mixing the ascorbic acid solution and molybdate solution, then water was added to the constant volume and mix it. The absorbance under 700 nm was measured after every solution sample coloration for 15 min.

The experiment quality control is shown as Table 3.52 and Fig. 3.47. The measurement accuracy is shown as in Table 3.53.

(4) **The extraction and measurement for the speciation of phosphorus**

In this study, phosphorus in sediments is classified into inorganic and organic according to the classification method suggested by Li [130]. Inorganic phosphorus includes

Table 3.53 Parameters of the analytical accuracy for total phosphorus from the sediment of the Dianshan Lake

	The depth of sampling (cm)				
Sample	6	11	15	20	25
Repeating measuring 1	0.3392	0.3274	0.2987	0.2926	0.2732
Repeating measuring 2	0.3443	0.3213	0.3069	0.2981	0.2783
Average value	0.3418	0.3244	0.3028	0.2954	0.2758
Relative mean variation	0.0075	0.0094	0.0135	0.0093	0.0092
Standard deviation (s)	0.0036	0.0043	0.0058	0.0039	0.0036
Variable coefficient (CV)	0.0106	0.0133	0.0191	0.0132	0.0131

diffluent phosphorus, Fe–P phosphorus, Al–P phosphorus, and Ca–bound phosphorus. In the experiments of this work, the speciation of the P was extracted orderly for diffluent P, Fe–Al bound P, occluded P, Ca–bound original P, Ca–bound detrital P, and organic P, modified from references of Li and others work.

Using Mg_2Cl to extract diffluent P in the sample was firstly processed, and then using NH_4F to extract the P was processed when F^- and Al^{3+} complexing under pH 8.2 so that can separate Fe–P and Al–P. The third extraction step was using NaOH due to the phosphate radicals in the Fe–P were released with the hydrolysis reaction between NaOH and Fe–P. Then the occluded phosphate (Oc–P) was extracted by sodium citrate and sodium hydrosulfite. This part of P was iron phosphate and aluminum phosphate encapsulated by Fe_2O_3 glue film. The Oc–P cannot be released under normal reducing condition. Iron oxide is reduced into ferrous oxide under the strong reducing action of sodium thiosulfate, and then it is complexed by sodium citrate to let the encapsulation of ferrous oxide stripped constantly. In the extraction solution, Fe_2O_3 and Fe–P could dissolve in the extracting solution, but Al–P could not. Therefore, to extract with NaOH sequentially the Al–P, remainder Fe–P was released all, and then all the Oc–P would be extracted. Ca–bound P, or original Ca–P was extracted by NaAc–HAc (pH 4) solution. The chippings Ca–bound P were extracted by HCl. For organic P, HCl was used for extraction after the sample to be ash for 5 h under 550 °C.

It was found that the extraction effects of Al–P were not clear when NH_4F was used. However, it was not the point in this study, so NaOH was use to extract the Al–P and Fe–P together in the experiment.

The PVC centrifuge tube, beaker, 50-ml cuvette, 15-ml porcelain crucible, induction cooker, Constant Temperature Vibrator (JA1003 N), centrifugal machine (Anke TDL-40B), ultraviolet spectrometry photometer (UV-2450), muffle were used in the experiment. The reagents were used as follows: 1 mol/L $MgCl_2$, 0.1 mol/L NaOH + 0.05 mol/LNa_2CO_3, Na_3Cit (0.3 mol/L) + $NaHCO_3$ (1 mol/L) + $Na_2S_2O_4$ mix extract solution (pH = 7.6), 4 mol/L NaOH, 2 mol/L NaOH, 30% H_2O_2, 1 mol/L HAc–NaAc buffer solution (pH = 4), 1 mol/L HCl. All the reagents used in the experiment were analytically pure grade quality. The speciation of the phosphorus extraction steps is shown as Fig. 3.48.

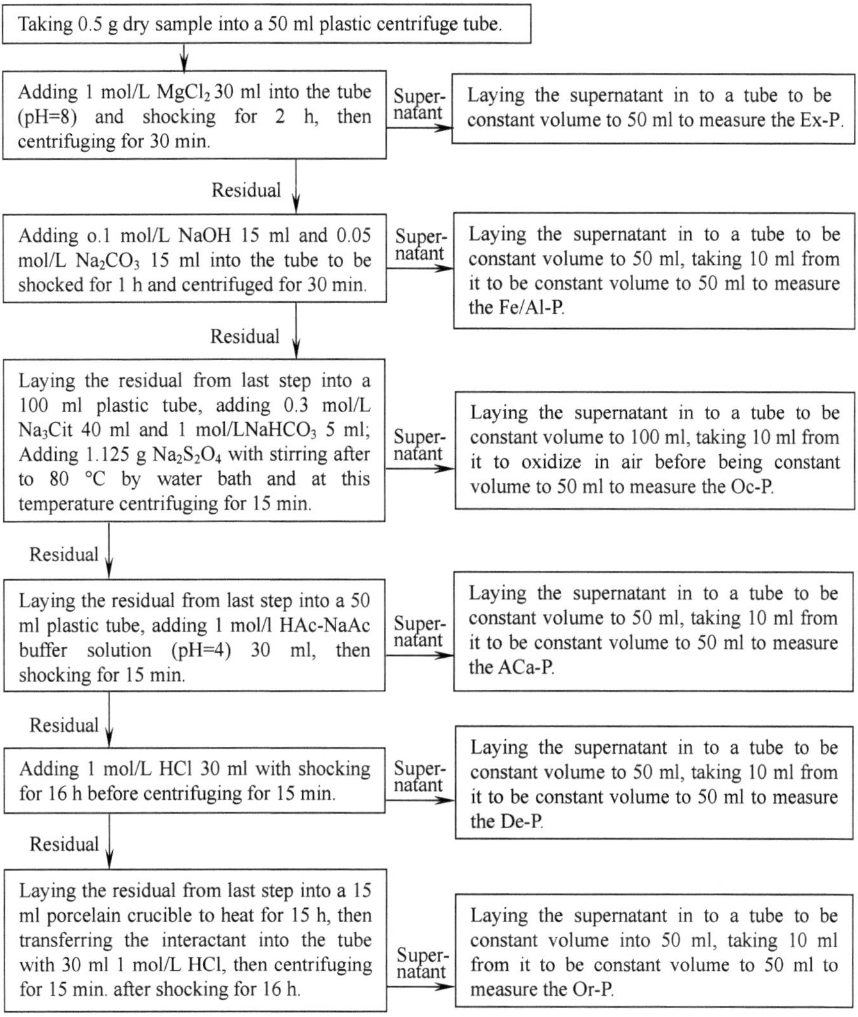

Fig. 3.48 Schedule of sequential extraction for the speciation of phosphorus from the sediment of the Dianshan Lake under 25 °C with continual shocking

Due to acids and alkalis were used in the extraction process, all the supernate after centrifuge in the experiment requires to regulate pH to neutral or weak acid. The acidity regulating method as follows:

Regulating the solution for acidic extraction: After orderly adding 1 drop of phenolphthalein indicator, dropwise adding NaOH (4 mol/L) into the solution until its color changed to red, adding H_2SO_4 (2 mol/L) dropwise until the red color to be faded with shaking up, then 1 drop of phenolphthalein indicator was added into the solution.

Table 3.54 Parameters of the analysis accuracy for phosphorus species from the sediment of the Dianshan Lake (unit for Ex–P is μg/g, and others all are mg/g)

Sample	Ex–P	Fe/Al–P	Oc–P	ACa–P	De–P	Or–P
Repeating measuring 1	0.7296	0.0564	0.1084	0.0170	0.0456	0.0630
Repeating measuring 2	0.6747	0.0541	0.1087	0.0175	0.0531	0.0660
Repeating measuring 3	0.8612		0.1273			
Average value	0.7552	0.0552	0.1148	0.0172	0.0493	0.0645
Relative mean variation	0.0936	0.0207	0.07259	0.0142	0.0762	0.0236
Standard deviation (s)	0.0958	0.0016	0.01083	0.0004	0.0053	0.0022
Variable coefficient (CV)	0.1269	0.02932	0.09431	0.02014	0.1078	0.0333

Regulating the solution for alkaline extraction: After orderly adding 1 drop phenolphthalein indicator letting the solution change to red, dropwise adding H_2SO_4 (2 mol/L) until the red color to be faded, then 1 drop of H_2SO_4 (2 mol/L) was added into the solution with shaking up.

As the supernate produced in above step, sodium thiosulfate was contained in the solution and it was in strong reducibility. The Mo–Sb–Vc color agent was disturbed under the pH condition so that the solution was red and not blue in the colorimetric process above, the oxidized operation had to be processed as follows: 10 ml solution was added into 150-ml Erlenmeyer flask with heating, then 20 ml H_2O_2 (30%) was added into the flask. After the oxidation was completed (no bubbles were in the solution), the solution was distilled to nearly dry. Adding 10 ml NaOH (2 mol/L) and boiling for 5 min in the water bath, then it was moved into 50-ml volumetric flask and regulate acidity as mentioned above.

The solution after regulating acidity was used to measure the concentration of soluble orthophosphate in sediment. The accuracy of the measurement is shown as Table 3.54.

2. **The distribution of the phosphorus in the Dianshan Lake sediment**

(1) **The total phosphorus distribution**

The distribution characteristics of phosphorus in the sediments of the Dianshan Lake are shown in Figs. 3.49 and 3.50. The results indicate that the total phosphorus concentration in the sediments of the Dianshan Lake gradually decreases with depth growth and the phosphorus concentrations are very close to the eutrophication levels of lake.

It has been shown that the average deposition rate of the Taihu basin reached 0.41 cm/a calculated by ^{210}Pb by CIC model [104]. As one of lakes of the Taihu basin, this sedimentary rate can be used to roughly estimate the sedimentary rate for the Dianshan Lake. Calculated from this deposition rate, the sediment of 25 cm in the sediment of the Dianshan Lake can roughly represented the deposition conditions of the Dianshan Lake in the recent 60 years. The phosphorus concentration in sediment cores from 0 to 25 cm of the Dianshan Lake ranges from 0.2776 to 0.5197 mg/g with

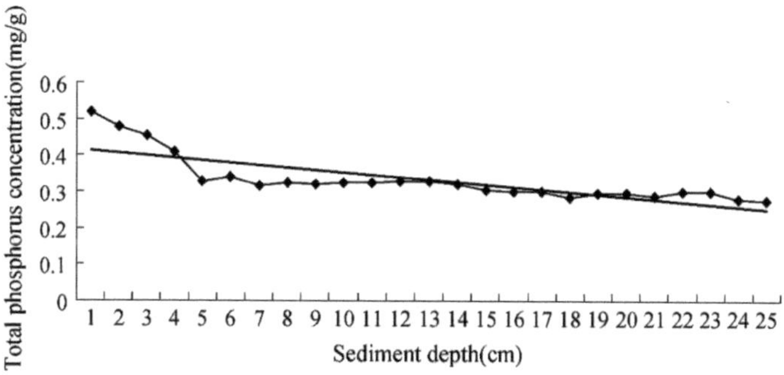

Fig. 3.49 Distribution of phosphorus in the sediment of the Dianshan Lake

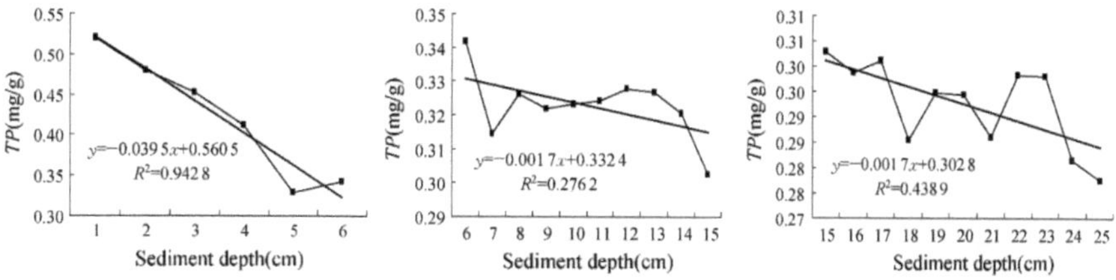

Fig. 3.50 Distribution of the total phosphorus (TP) from the sediment of the Dianshan Lake

Table 3.55 Features of total phosphorus concentration from the lakes, China

Lakes	Total P in sediment (mg/g)	Degree of eutrophication
Moshui Lake, Wuhan	4.505	Abnormal eutrophication
Liuhua Lake, Guangzhou	1.793	Heavy eutrophication
Xuanwu Lake, Nanjing	2.161	Heavy eutrophication
Dian Lake, Yunnan province	1.716	Heavy eutrophication
Hongfeng Lake, Guizhou province	1.569	Eutrophication
Chaohu Lake, Anhui province	0.332	Eutrophication
Erhai Lake, Yunnan province	1.093	Heavy eutrophication
Taihu Lake, Jiangshu province	0.580	Eutrophication
Dianshan Lake, Shanghai	0.3335	Eutrophication

average value 0.3335 mg/g. The result of comparison to the phosphorus concentration of the Dianshan Lake with the concentrations of some lakes of China is shown as in Table 3.55.

The above data suggest that the phosphorus concentration in the sediment of the Dianshan Lake has a downward trend with the depositing depth growth. The concentration of P in the depth from 0 to 7 cm shows a sharp adding trend, in the depth from 7 to 25 cm shows a slower decline trend with generally stable concentration between the 7 and 14 cm depth (Fig. 3.50). It may be deduced that the phosphorus in the Dianshan Lake has experienced a slow and fluctuant increase to a gradually stable

Table 3.56 Feature of the phosphorus distribution in the sediment of the Dianshan Lake

Depth (cm	Ex–P (mg/g)	Fe/Al–P(mg/g)	Oc–P (mg/g)	ACa–P (mg/g)	De–P (mg/g)	Or–P (mg/g)	TP (mg/g)
1	0.003229	0.06716	0.1274	0.04407	0.1817	0.09070	0.5146
5	0.003028	0.04940	0.1000	0.02590	0.06468	0.08559	0.3286
9	0.002936	0.05048	0.0973	0.02579	0.05700	0.08249	0.3159
13	0.002641	0.05048	0.1117	0.02187	0.06042	0.08105	0.3281
17	0.002338	0.05523	0.1148	0.01721	0.04558	0.08038	0.3155
21	0.002041	0.05310	0.1197	0.01530	0.04628	0.07718	0.3136
25	0.000755	0.04344	0.1177	0.01631	0.04637	0.06452	0.2891
Average	0.002424	0.05276	0.1126	0.02378	0.07172	0.08027	0.3436

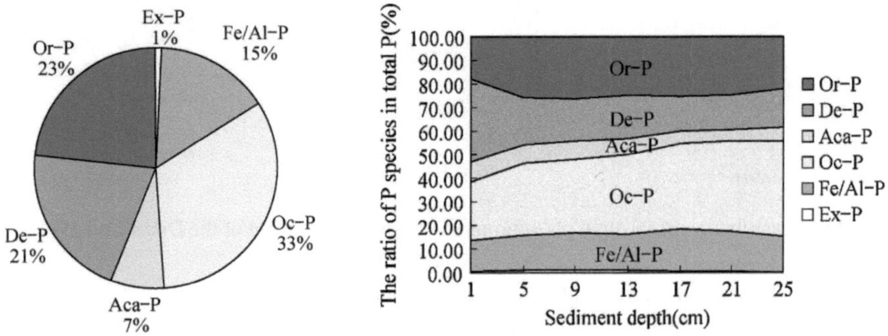

Fig. 3.51 Ratio of species of the phosphorus in the total phosphorus from the sediment of the Dianshan Lake

concentration and finally a sharp rise in recent years. Contrast to the total phosphorus concentration in the sediment of some drawing attention lakes in China, the water quality of the Dianshan Lake is in gradually reaching the eutrophication level and the concentration of P is increasing rapidly. If appropriate measures to prevent the increase of total phosphorus are not to be taken, in the next few decades or even in few recent years, the Dianshan Lake would be subjected to rapid eutrophication.

(2) **The distribution of the speciation of the phosphorus in the Dianshan Lake**

The distribution characteristics of each speciation of phosphorus in the sediment cores of the Dianshan Lake are shown in Table 3.56, Fig. 3.51.

The soluble phosphorus concentration in the Dianshan Lake sediments ranges from 0.0008 to 0.0032 mg/g, a relatively small proportion of total phosphorus. Because of the soluble phosphorus is easily released within the water and used by aquatic organisms, so this part of P is with strong effects on nutritional status of the water. The changes in physical and chemical conditions, such as temperature, pH, hydrodynamic conditions, and bioturbation factors, can lead to the release of the phosphorus to the overlying water and then affect the nutritional status of the water.

Table 3.57 Concentration of organic phosphorus from the lakes in Yangtze Delta

Lakes	Wuchang Lake	Pohu Lake	East Lake	Poyanghu Lake	Yellow Lake	Daguan Lake	Lu Lake	Dianshan Lake
Or–P(μg/g)	106.2	155.5	202.1	34.5	106.6	122.4	153.6	80.3

The concentration of the iron and aluminum combined phosphorus accounts for 15% of total phosphorus and varies from 0.4344 to 0.0672 mg/g, and its average value is 0.05276 mg/g. As an important part of phosphorus source, Fe–P and Al–P plays an important role in phosphorus cycle. If the redox conditions change, the Fe–P may be dissolved due to the reduction of iron, these parts of P will be released to the lake water again. This process might have important impacts on the lake nutrition conditions and water quality.

The concentration of occluded phosphate is accounted for 33% of the total phosphorus, the highest part accounted for in the total inorganic phosphorus. Its concentration ranges from 0.0973 to 0.1247 mg/g, and the average value is 0.1126 mg/g. Due to tightly packed by minerals such as iron and aluminum in the particles, the occluded phosphate is difficult to release into the water. Therefore, in the sediment, the percentage of occluded phosphate is higher, and the water quality generally is more stable.

The concentration of authigenic calcium phosphorus accounts for 7% of the total phosphorus and varies from 0.0153 to 0.0441 mg/g, and the average value is 0.0238 mg/g. Authigenic calcium phosphorus is generally regarded as permanent phosphorus. However, when the circumstance is in week acid, a certain amount of it could be released to the water from the sediment.

The concentration of clastic calcium phosphorus accounts for 21% of the total phosphorus and varies from 0.0456 to 0.1817 mg/g, and the average value is 0.0717 mg/g. The concentration of organic phosphorus accounted for 23% of the total phosphorus and varies from 0.6452 to 0.9070 mg/g, and the average value is 0.0803 mg/g.

The average value of the organic phosphorus concentration in the Dianshan Lake sediment is 80.27 μg/g. The organic phosphorus concentration in the Dianshan Lake sediment is significantly lower than that of the shallow lakes in the Yangtze River basin, shown as in Table 3.57. It shows that by Table 3.57, the organic phosphorus concentration in some lake sediment accounts for 80% of the total phosphorus, and this part of P is known as an important role often in phosphorus recycle of the lakes.

The organic phosphorus in the lake sediment comes mainly from soil organic matter, such as the metabolites, the residues of aquatic organisms, and from unmineralized organic pollutants. Many factors may affect the concentration of phosphorus in lake sediment, such as the type of sediment, deposition rate, and the clay concentration. The organic phosphorus in the sediment will be released to the water during the degradation of the organic matter. However, changes of external conditions as storm and hydrodynamics can also have an effect on this process.

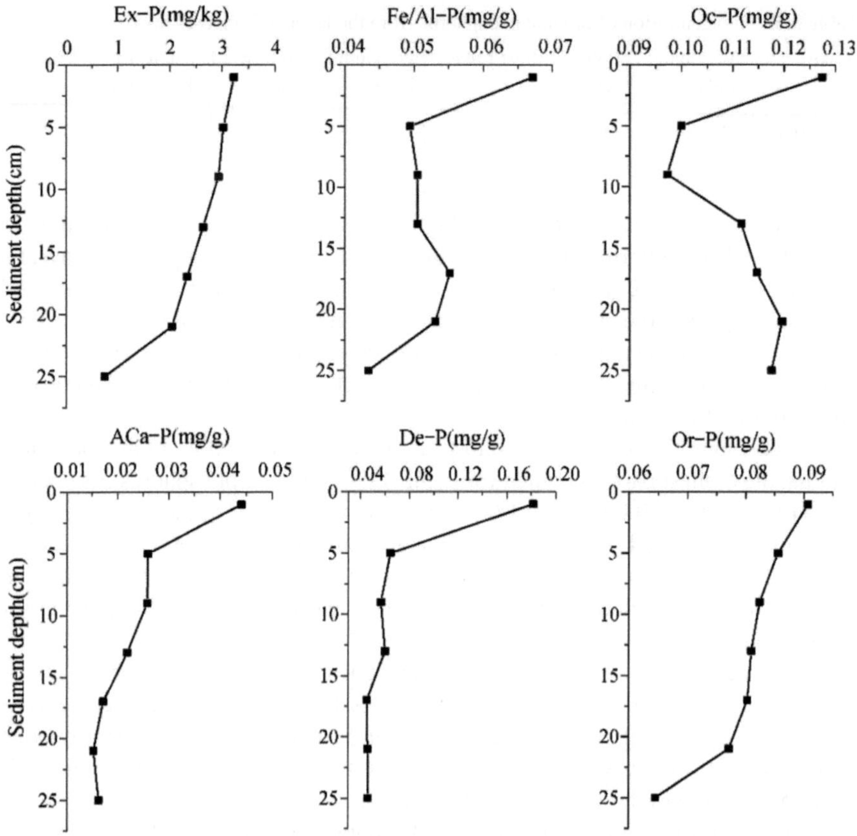

Fig. 3.52 Change feature of phosphorus species from the sediment of the Dianshan Lake

The trends of each speciation of phosphorus in the Dianshan Lake sediment are shown as Fig. 3.52. It indicates by Fig. 3.52 that except for the occluded phosphate concentration decreases at first and then rises with the sediment depth increasing, other phosphorus speciation and total phosphorus show a decreased tendency with the increasing of the sediment depth. Among them, the exchangeable phosphorus and organic phosphorus change slowly. However, iron and aluminum combined phosphorus, authigenic calcium phosphorus, clastic calcium phosphorus show a rapidly decreased tendency with increasing of the sediment depth within the shallow layer (0–5 cm).

Each fraction of the speciation of phosphorus in the sediment only indicates its absolute concentration. Nevertheless, phosphoric percent concentration indicates abundance and it could well show the change of phosphorus with time during its deposition. The percent concentration of each speciation of phosphorus in the Dian-

Table 3.58 Correlation between species ratios in the total of phosphorus in the Dianshan Lake

	Ex–P	Fe/Al–P	Oc–P	ACa–P	De–P	Or–P
Ex–P	1					
Fe/Al–P	0.1939	1				
Oc–P	−0.5253	0.6436	1			
ACa–P	0.4747	−0.7037	−0.9152	1		
De–P	0.0000	−0.8648	−0.8516	0.7246	1	
Or–P	0.5403	0.7658	0.3977	−0.2823	−0.8287	1

shan Lake sediment is shown in Table 3.58. The vertical variation of each speciation of phosphoric percent concentration is shown as Fig. 3.53. The change trend of percent concentration of authigenic calcium phosphorus and clastic calcium phosphorus is consistent and showed a rapidly decreased tendency with increasing depth of the sediment. The exchangeable phosphorus, iron, and aluminum combined phosphorus, and organic phosphorus shows a first increase and then decrease with increasing of the sediment depth. Both the exchangeable phosphorus and iron and aluminum combined phosphorus are reactive phosphorus, and they all have a smaller percent concentration in the lower sediment may be caused by the effects from water disturbance of the lake.

The changing trend of the percentage concentration of the occluded phosphorus is not obvious, and its percentage concentration increased with the growth of the sediment depth. The occluded phosphorus, which is wrapped by Fe_2O_3, is quite stable iron or aluminum phosphate, and it is difficult to release. The changing trends of percentage of this part of P show that the composition of the input of the external P in the Dianshan Lake sediment may be in changing.

3. **The geochemical process and ecological significance marked by the distribution of the phosphorus in the sediment of the Dianshan Lake**

The change of pH values of the sediment of the Dianshan Lake is shown as Table 3.59. The phosphorus is one of the most important factors on lake eutrophication for freshwater lake systems. Many studies indicate that phosphorus in sediment release is as an important factor that affects lake eutrophication. In lacustrine facies, various speciations of phosphorus will convert to the orthophosphate finally through processes of migrations and conversions. The orthophosphate can enter into organic matter or combine with metal ion (Ca, Fe, Al, etc.) to form insoluble phosphate. The metal phosphate may be dissolved, and the phosphate anion can be released when the redox condition or pH change. The phosphorus released from sediment can combine with other particles, or it can be absorbed to develop unconsolidated inorganic compounds or organics, which may be released to overlying water by horizontal diffusion, ion exchange, and bioturbation to change nutritional status of the overlying water. Therefore, the speciation and distribution of phosphorus from lake sediment is a very important role for the phosphorus cycle as well as the mechanism research of phosphorus pollution and control of eutrophication of lake.

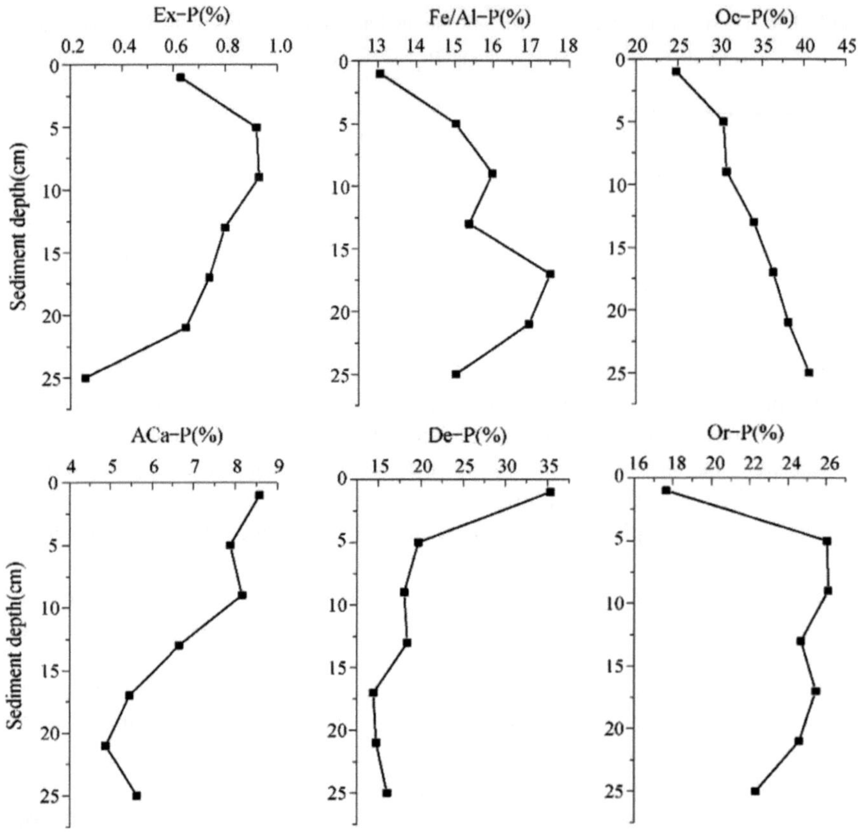

Fig. 3.53 Ratio of phosphorus species in the total phosphorus from the sediment of the Dianshan Lake

Table 3.59 Feature of pH in the depths of the sediment in the Dianshan Lake

Depth (cm)	0–1	4–5	8–9	12–13	16–17	20–21	24–25	Average
pH value	6.68	6.52	6.87	7.21	7.53	7.40	7.47	7.10

According to the continuous extraction of this work, the speciation of phosphorus is divided into organic phosphorus (Or–P) and inorganic phosphorus (IOr–P), while inorganic phosphorus includes readily soluble phosphorus (Ex–P), Fe–bound P (Fe–P), Al–bound P (Al–P), authigenic Ca–bound P (ACa–P), detrital P (De–P), and occluded P (Oc–P). The gross of the phosphorus extracted from the sediment is called the total extracted phosphorus [135].

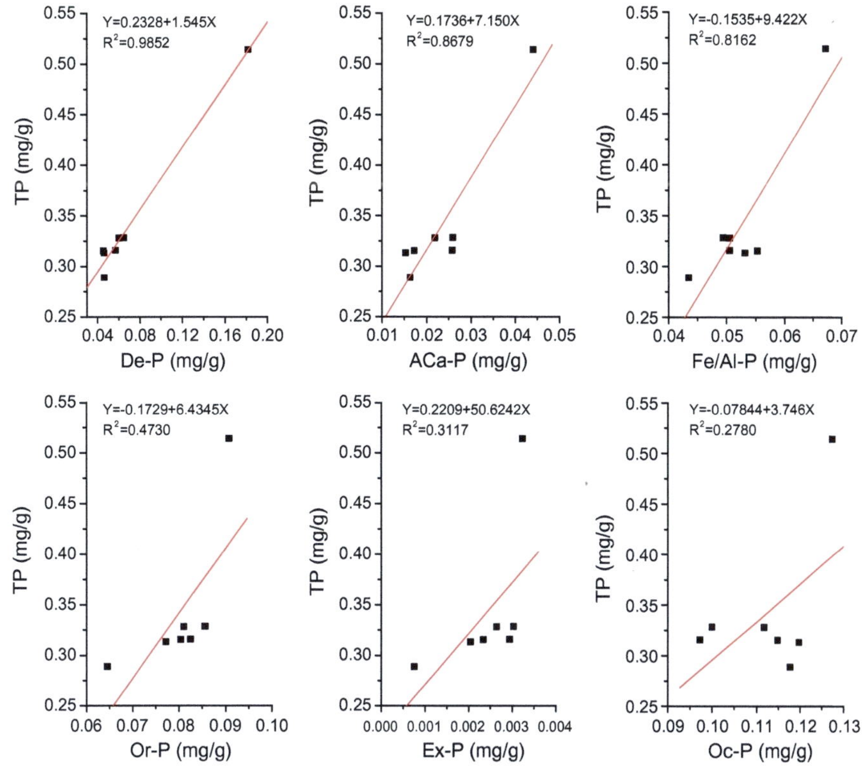

Fig. 3.54 Correlation of the total and the species of phosphorus in the sediment of the Dianshan Lake

(1) The features and correlation among the speciations of the phosphorus in the sediment of the Dianshan Lake

As the characteristics of the speciation of the phosphorus show above in the sediment of the Dianshan Lake, the sequence that is based on the amount relationship between various phosphorus and total P is from more to less orderly as follows: De–P > ACa–P > Fe/Al–P > Or–P > Ex–P > Oc–P shown as Fig. 3.54. The accumulation of the phosphorus is related to the accumulation of inorganic phosphorus in the Dianshan Lake, the biggest part in the phosphorus speciation is Ca–bound P, followed by Fe/Al–bound P. However, the impact of organic phosphorus on accumulate of the sediment phosphorus in the sediment of the Dianshan Lake is also not ignorable.

The total P in the sediment of the Dianshan Lake is with certain relations to readily soluble phosphorus and Al/Ca–bound P, and particularly having an obvious correlation to organic phosphorus shown as Fig. 3.54. Readily soluble phosphorus shall be used directly for aquatic organisms, and it affects the nutritional status of the water. Al–bound P and Fe–bound P are released to water because of reductive dissolution of Fe when the redox conditions change. Oxide or hydroxide absorbed in the sediment and clay mineral particle surfaces may be transferred to readily soluble

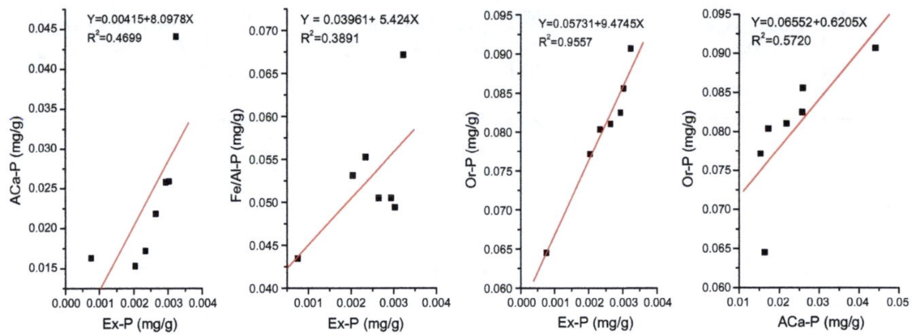

Fig. 3.55 Correlation of the Ex–P and the other species of phosphorus in the sediment of the Dianshan Lake

phosphorus. The relationship between readily exchangeable phosphorus and others is shown as Fig. 3.55.

(2) **The origin and change of the speciation of the phosphorus in the Dianshan Lake**

A. Organic phosphorus

Organic phosphorus is mainly from soil organic matter, metabolites, debris of aquatic organisms, and degradation of organic pollutants. However, organic phosphorus released by mineralization of organic matter is hard to be used by organisms. As mentioned above, the concentration of organic phosphorus in sediment is related to the factors such as sediment type, depositing rate, and clay concentration. Organic phosphorus in lake sediment is in the form of phospholipid or organophosphate [136–138]. The impact of external conditions such as wind wave, hydrodynamic factor on release of organic phosphorus is very important. Besides, bacteria can absorb excess phosphorus to reserve in the form of polyphosphate under aerobic conditions, and use organic phosphorus as an energy carrier for their metabolism in anaerobic condition circumstances [139, 140]. This process leads to the migration and transformation among various speciations of phosphorus in sediments. The concentration of Or–P may share 80% of TP in some lakes. The impact to the lake water quality and release mechanism to overlying water of organic phosphorus in the sediment of lake on lake phosphorus cycle are attracting attention increasingly today. The organic phosphorus of the sediment in the Dianshan Lake is 23% of TP, less than occluded phosphorus (33%) only.

B. Inorganic phosphorus

Inorganic phosphorus is mostly soluble phosphate that is adsorbed on the particles of sediment or combined with metal ions such as Fe, Ca, Al in water, and organic phosphorus can also be transformed into inorganic phosphorus under the influences of microorganisms and mineralization. The cases of this part of phosphorus in the sediment of the Dianshan Lake are as follows:

a. Readily soluble phosphorus

Readily soluble phosphorus is mainly the part of phosphorus adsorbed on the surfaces of oxides or hydroxides, which includes exchangeable and weakly adsorbed phosphorus. Readily soluble phosphorus in the sediment is easily released to water and is then utilized by aquatic organisms, having a significant impact on nutrition to organisms. Physical and chemical conditions such as temperature, pH, hydrodynamic, and bioturbation can make the soluble phosphorus to release and diffuse into overlying water. The concentration of readily soluble phosphorus is 1% of TP in the sediment of the Dianshan Lake.

It is the most common method that NH_4Cl as well as $MgCl_2$ solutions are used as extraction agents to extract this part of phosphorus [141]. However, Froelich introduced a method called infinite dilution and extrapolation (IDE), which could characterize available phosphorus from exchangeable phosphorus [142], and the IDE method is a meaningful way to measure the potential for phosphorus release from the sediment.

b. Iron/Aluminum/Calcium-bound phosphorus

Iron-bound phosphorus is a kind of phosphorus which is easy to combine with iron oxides or hydroxides. Iron-bound phosphorus is considered to be a permanent phosphorus remit under oxide circumstances and to be a temporary phosphorus remit under aerobic condition [143]. Iron-bound phosphorus as an important remit of the phosphorus is critical in phosphorus cycle. Aluminum/Iron-bound phosphorus and readily soluble phosphorus are easy to release relatively in the circumstances and called by a joint name "active phosphorus." Aluminum and iron-bound phosphorus are 15% of the TP of the Dianshan Lake sediment with little increase trend as the depth growth from shallow to deep layer. Calcium-bound phosphorus includes authigenic calcium-bound phosphorus and detrital calcium-bound phosphorus (original phosphorus). The authigenic calcium-bound phosphorus is formed by the deposition, solidification, and coprecipitation with $CaCO_3$ is affected by biological processes, such as hydroxyapatite. This part of phosphorus is difficult to be used by organisms as one of permanent phosphorus remits, but it can release in weak acidic circumstances [144]. Calcium-bound phosphorus is about 28% of the TP of the sediment of the Dianshan Lake, and the concentration decreases from the shallow to deep in the sediment.

c. Occluded phosphorus

Occluded phosphorus is reductive iron phosphate and aluminum phosphate encapsulated in the film ferric oxide. The films such as iron and aluminum tightly pack the occluded phosphorus, and it is generally difficult to release in circumstances. The occluded phosphorus in the Dianshan Lake sediment is 33% of the TP of the sediment, and the concentration of it increases significantly from the shallow to deep in the sediment.

(3) The variation and factors for the phosphorus in the sediment of the Dianshan Lake

The phosphorus in sediments of lake is mainly in the form of inorganic orthophosphate. The P may be released from the sediments when the circumstance conditions were conducive to the dissolution of calcium, aluminum, iron, and other insoluble phosphate. This part P released from sediments always diffuses from the interstitial water of sediments into the sediment–water interface and then diffuses into the overlying water. It is shown that the release of P from sediments is influenced by dissolved oxygen (DO), pH, redox potential, temperature, organisms, water disturbance, etc., represented as follows:

The effect of calcium: The precipitation of $Ca_3(PO_4)_2$ is formed in high calcium concentration water. However, the binding ability of calcium to P is weak, and CO_2 in water plays a key role in the dissolution of Ca–P.

The effect of aluminum: The range of pH between 5.4–6.2, the reaction of orthophosphate and aluminum can generate aluminum phosphate (Al–P), and can be in a state of dynamic equilibrium too. Any physical and chemical variations can change the direction of the reaction. The $Al(OH)_3$ is formed while the pH of the system is increased to 7, it could be with a large specific surface effects and the orthophosphate might be strong adsorbed on. Since the valence state of aluminum is not affected by the redox potential, the solubility of Al–P is not enhanced while the reduction condition occurring.

The effect of iron: The binding of P in the phosphate is interpreted as a process of the ligand exchanges, which hydroxyl group on the surface of iron oxide particles is exchanged by the phosphate groups. When the pH value is near not to 5.5, Fe–P tends to dissolve, and when the pH value is around 5.5 the lake water may be a saturated solution of iron P. The P in the deep sediments will not form into iron P precipitation even orthophosphate concentration is at high level because of the reducing circumstance here. Iron phosphate is dissolved when the pH value is greater than 6:

$$FePO_4 - 2H_2O + OH^- = Fe(OH)_3 \downarrow + H_2PO_4^-$$

The solubility product of $Fe(OH)_3$ is higher than $FePO_4$'s; it plays a hiding role for the interior iron P and reduces the solubility of iron phosphate and the release rates of the P.

The effect of redox potential: The reaction of $Fe^{3+} \rightarrow Fe^{2+}$ occurs with the decrease of redox potential, and it causes the release of iron P. The change of redox potential may cause this reaction, and it always is in decreases with the increasing of the pH value of the system.

The effect of temperature: Increasing temperature can enhance the P absorption by sediments, which is an endothermic reaction that results in increase of the temperatures and microbial activities. As the aerobe increasing and dissolved oxygen decreasing, the system redox potential will be reduced associated with the reaction $Fe^{3+} \rightarrow Fe^{2+}$ occurring, which can induce the P to be released from that bound in

Fe–P. In addition, under these conditions the organophosphorus (Or–P) may transform to the inorganic-P, and the insoluble-P transform to the soluble P caused by the activity of microorganism in the system. Temperature is with significant effect on the containing calcium sediment, of which substantial amount of carbon dioxide may be produced due to the increase of the organic matter mineralization as the temperature increase in the process. Therefore, the phosphorus in the sediments will be easy to release rapidly associated with the dissolution of containing calcium sediments.

The effect of organic matter: The impact of the organic matters on phosphorus is various and tanglesome. It is generally assumed that the P fixed by organic matter is generally weak because of the following reasons: ① The organic matter colloid–humus which is generated from the disintegration of organic matter can form glue film covering the surface of inorganic mineral particles, such as Fe/Al oxide and calcium carbonate, it reduces the contact between inorganic matter and P anion, which can avoid or relieve the degree of the P fixation by inorganic matter; ② organic acid and other chelating agent in the disintegration of organic matter may relieve P fixation to be soluble phosphorus; ③ CO_2 generated from the decomposition of the organic matter reacts with water to form H_2CO_3, which increases the solubility of Ca/Mg–P; ④ there is a competition reaction between the P anion and the polyanion of fulvic acid in the organic matter, and the organic polyanion can promote the release of the phosphorus by special adsorption as it entering into mineral particles [145]. However, some studies also suggest that the organic matter reacts with Fe and Al to form inorganic compounds, which may offer important inorganic phosphorus adsorption points to enhance the adsorption action for the P. Other researchers believe that the plenty of H^+ released by organic matter can also promote the adsorption for the P via making the group's protonation on the surface of minerals.

The effect of pH: The release of the P from sediments obeys the U-shape changes rule with the rise of pH value of the system, it is the minimum when the pH value is at about 7, and both the conditions of acidic and alkaline are beneficial to the release of P. The reduction of the pH value is probably beneficial the P to be dissolved with Al–P firstly release, and the rise of pH value is primarily to generate ion exchange in the system, which causes the competition action between OH^- and the P anions, it also enhances the release of the phosphorus [146].

The effect of DO: The release of endogenous P may be enhanced in anaerobic conditions, and that may be inhibited in aerobic conditions. In anoxic conditions, Fe^{3+}–P may be reduced to Fe^{2+}–P which is easily soluble to lead P to be released from sediments, meanwhile the acid matter, which is caused by the anoxic conditions, to be accumulated in the sediments can lead to the Ca–P release. On the contrary, in the aerobic environment it could inhibit the release of P from the sediment too. The capacity from the aerobic inhibition of P released from sediments is closely related to pH value of the system, in the conditions of acid to medium or week alkali (pH 5.0–8.5), the P in sediment is stable, and in the alkaline environment (pH>8.5), the P in the sediments will be unstable [147].

The effect of water disturbance: In theory, disturbance can increase the dissolved oxygen in the system, which is not conducive for the release of P from the sediments. However, the disturbance also promotes the mixing and exchange between the muds and water; this effects on the release of P from sediments is more important than that of dissolved oxygen. Disturbance is an important factor to the reaction on water–sediment interface, especially for the shallow lake which the layering of temperature and other physical and chemical properties are inconspicuous, the wind wave may cause a great disturb on the water–sediment interface. The disturbance can cause the particles in the sediment to be re-suspended, which accelerates the diffusion and increases the release of P in the interstitial water of the sediment. When over the time of maximum P release disturbance, the effect of the kinetic on the release of P is not obvious even with a decline. This indicates the hydrodynamic effect on the release of P is short term [148, 149].

The effect of microorganisms: The migration and transformation of P in the sediment can be directly or indirectly affected by microorganisms. Microorganisms may release or combine with P through metabolic responses, extracellular release, and cell dissolution. At the same time, the behavior of P can be influenced by changing the physical and chemical conditions of the sediments to stimulate the different physical, chemical, and biological processes. Therefore, the release of endogenous P is a complex process related to physical, chemical, and biological factors. The release of P in sediments with microorganism's activity is significantly higher than that of the sediments without microorganisms. The presence of microorganisms can speed up the consumption of dissolved oxygen. Microorganisms with the ability of degrading organic matters can release the Or–P through mineralization and inorganic phosphorus associated with the acid production. The Or–P in sediments can be transformed and decomposed into inorganic phosphorus. Meanwhile, the insoluble phosphide in the sediment can be transformed into soluble P, and it was reported that the release of P was almost zero without microorganisms while the release of phosphorus in the sediment was improved by 50–100% as compared with the presence of microorganisms [150–153].

The effect of P concentration in overlying water: There is an exchange process of phosphorus between the sediments and overlying water. The P separates oneself from water to the sediments when the concentration of P in water is high to a certain level, and on the contrary, it may turn back to the water from sediments at the certain conditions too [154]. Therefore, the release of P is related to the concentration difference between the sediment and the overlying water. Due to high concentration of P in the water may occur after the strong wind, the rate of release P from the sediment generally may be then decreased, or the source of the sediment of P may be converted to a remit in a short period.

The effect of chemical and mechanical composition of sediment: Non-calcareous sediments contain more iron and organic carbon than calcareous sediments. Under aerobic condition, due to iron oxide adsorbs P may lead to the release of P from non-calcareous sediments for that to be limited in. At the same conditions, it is more for the P release from calcium sediments to the overlying water than that of the non-calcareous sediments. The release P from sediments is also affected by mechanical composition of the sediment, when the clay is in high concentration in the absorbed particles the P is adsorbed in more level because of the large surface area and strong surface energy of the clays [155].

The effect of solar radiation: Algae assimilates the nutrients in the water, and a certain amount of nutrients is always required during the growth of algae. The release of nutrients from sediment to the overlying water becomes the source of nutrients for algae growth while the low concentration of nutrients in overlying water. Therefore, the benthic algae may become a barrier to prevent the release of P from the sediment to the overlying water. According to the biological function of the benthic algae, the benthic algae can indirectly inhibit the release of the P from the sediments. So, the release of P from sediments to overlying water is indirectly inhibited by biological action of the benthic algae [156].

As Fig. 3.55 showing, there is an obviously relation between authigenic calcium phosphate (ACa–P) and organic phosphorus (Or–P) in the Dianshan Lake. ACa–P (authigenic apatite) is mainly composed of particle P, which forms from biological effects such as deposition, consolidation, and coprecipitation by authigenic calcium carbonate. The weathering of shells from underwater zooplankton can form ACa–P. The Or–P mainly comes from drainage basin soil organic matter, metabolites of aquatic organisms, residues of dead bodies, organic pollutants that are not mineralized or degraded. The obvious relationship between ACa–P and Or–P shows that Or–P of the Dianshan Lake sediments mainly comes from the eobiont which lives in the Dianshan Lake, and the Or–P inputted by human activities might not be obvious. In contrast, the concentration of the Or–P in the sediment of Dianshan Lake is significantly lower than that in the other shallow lakes of the Yangtze River basin.

Generally, the Or–P in sediments are difficult to be used by organisms, and when the changes of the external conditions such as pH value, redox potential, and so on occur in the circumstance, the Or–P may release from the sediment. Therefore, the degradation of Or–P also controls at a certain extent on the concentration of other speciation of phosphorus in sediments. The ACa–P (authigenic apatite) is mainly composed by the particle P that forms from biological effect such as deposition, consolidation, and coprecipitation with authigenic calcium carbonate. It is difficult to be released unless the conditions are weak acidic [145]. The study shows that the pH of the sediments in the Dianshan Lake varies from weak acid to weak alkaline while the sediment changes from shallow to deep (Table 3.59). In lower pH value condition, the phosphorus is easy to be released from sediments, and the release rate is fast too. Particularly, the release of Ca–P is more obvious to be effected by pH value, which is related to the high concentration of Ex–P in the surface sediments.

In conclusion, the behavior of P in lake sediments is a very complex problem which is affected by many factors to release or precipitate in lakes and sediments. According to the analysis above on different characteristics of P speciation variation in the sediments of the Dianshan Lake, the following items of it can be demonstrated:

A. The concentration of P in the sediment of the Dianshan Lake is in the range of 0.2776–0.5197 mg/g within the layer of 0–25-cm-depth sediments, and the average concentration is 0.3335 mg/g. The concentration of P in surface sediments has reached 0.5197 mg/g. The Dianshan Lake is closing to the eutrophication level of lakes compared within the total phosphorus (TP) concentration of the Dianshan Lake and that of the hyper-eutrophication of other lakes. The vertical variation characteristics of the TP in the Dianshan Lake show that the process of P pollution from earlier to late sediment is stable increasing slowly, then increasing sharply in the recent ten years. The eutrophication phenomena would be happened in the Dianshan Lake in the next decade according to this situation.

B. The concentration of speciation of the P in the sediments of the Dianshan Lake is as following: the concentration of Ex–P was in 0.001–0.003 mg/g, accounts for 1% in the TP; Fe/Al–P is in 0.434–0.0676 mg/g, accounts about 15% in the TP; Oc–P is in 0.097–0.125 mg/g, accounts about 33% in the TP which is the maximum proportion of inorganic phosphorus; ACa–P is in 0.015–0.044 mg/g, accounts about 7% in TP; De–P is 0.04558–0.1817 mg/g, accounts about 21% in the TP; Or–P is 0.6452–0.907 mg/g, accounts about 23% in the TP which is lower than that of the shallow lake sediments in the Yangtze basin of China.

C. In the sediment core of the Dianshan Lake except the concentration of Oc–P is decreased first and then increased as the sediment depth growth, other speciation of the P shows all decreasing trend with the sediment depth increasing and corresponded with the changing trend of the TP. Among them, the changes of Ex–P and Or–P are slower, while the Fe/Al–P, ACa–P, and De–P in lower sediments (0–5 cm) show a trend of sharp increase with the decreasing of the sediment depth. The percentages of Ex–P, Fe/Al–P, and Or–P are lesser in the shallow sediments, which might be related to the impact of water disturbance on the release of Ex–P from water. The change trend of Oc–P with the change of depth of the sediment is not more obvious, while the percentage in the TP increases with the sediment depth increasing which indicates that Oc–P might be with a trend of conversion to other speciation of the P with time and the depositing processes of the sediment in The Dianshan Lake.

D. The various speciations of the P in the sediments are sorted by the correlation with the TP was De–P > ACa–P > Fe/Al–P > Ex–P > Oc–P, and this indicates that the accumulation of the P in the Dianshan Lake sediments is related mainly to the accumulation of inorganic phosphorus. The proportion of the Ca–P is maximum, followed by the Fe/Al–P. The Or–P is one of the factors to affect the accumulation of the phosphorus in the sediments of the Dianshan Lake. The correlation between the Ex–P and Fe/ Al/Ca–P, and Or–P is, respectively, with an obvious correlationship, and the Oc–P is not with any correlation with other speciation of the phosphorus.

E. The pH value of the sediments in the Dianshan Lake is from weak acid to weak alkaline with the change from shallow to deep (Table 3.59) in the range of 6.68–7.53. The concentration of the TP in the sediments of the Dianshan Lake generally shows an increasing trend with the going of time, the growth of Ca–P (ACa–P and De–P) is significantly positive correlated with time; the situation of Fe/Al–P (Fe–P and Al–P) and Oc–P is seemly special, and the general trend is decreased with time associated with the partially sharp increase on the top layer of the sediments (0–5 cm). This suggests that the input rate and amount of the P of the Dianshan Lake is likely to significantly increasing in recent years, at the same time it also shows that these two speciations of the P are variated obviously as the enhancement of the reductive factor of the system.

F. The concentration of Or–P, Ex–P, and TP in the Dianshan Lake sediments increases with the going of time, but the Or–P and Ex–P proportion of the TP sharply decreases at the top layer of sediments with significant correlation between the changes of their concentrations. Above that change features of the speciation of the P, the Ca–P seems as the main contributor for the phosphorus inputted to the Dianshan Lake.
G. The transformation of the P speciation in the sediment of the Dianshan Lake is with complex mechanisms, and the Ca–P is one of the main contributors in the sediments. There is an obvious effect to the fixation of P by biological process because the Ca–P (ACa–P) is mainly deposited by biological action. This shows the Dianshan Lake is in the process of the strong bio-geochemical actions, on the other hand it also demonstrates that the P nutrition in the Dianshan Lake offered predominantly by the form of Ca–P, which signifies it being in a very fragile situation. When the changes of the physical and chemical conditions, especially with the decrease of the pH value, the P might be released in water that would be in bioavailable forms to in the water. This is a serious environmental problem that needed to be urgent attention.

(4) **The evolution of the total organic carbon in the Dianshan Lake**

The total organic carbon (TOC) in the sediment is investigated by the potassium dichromate method in this research. In the method of the research, the organic materials are oxidized by dichromate in sulfuric acid, and the final amount of dichromate is determined by titration using ferrous ammonium sulfate as the titrant. The ferroin (1,10-phenanthroline ferrous sulfate) is as the indicator in this way, and the concentration of TOC can be calculated by the consumption of the dichromate. The theoretical reaction stoichiometry is shown below:

$$2K_2Cr_2O_7 + 3C + 8H_2SO_4 \rightarrow 2K_2SO_4 + 2Cr_2(SO_4)_3 + 3CO_2 + 8H_2O$$

$$K_2Cr_2O_7 + 6FeSO_4 + 7H_2SO_4 \rightarrow K_2SO_4 + Cr_2(SO_4)_3 + 3Fe_2(SO_4)_3 + 7H_2O$$

The equipment and reagents used were test tubes (18 × 180 mm), funnel, flask (500 ml), acid burettes, etc.

The reagents used in the measurement were prepared as follows:

The potassium dichromate ($K_2Cr_2O_7$) as the standard solution (0.8 mol/L) prepared in the following schedule: Analytical pure grade potassium dichromate was dried at 130 °C for 3–4 h, then 39.225 g of dried potassium dichromate was dissolved in 400 ml water, the dissolution was enhanced by heating if necessary, and then diluting and to be constant with water to 1 L after cool.

Ferrous sulfate ($FeSO_4$) solution (0.2 mol/L): After taking 56 g $FeSO_4 \cdot 7H_2O$ (chemical pure grade) or 78.4 g of $FeSO_4 \cdot 6H_2O$ to dissolve in 30 ml sulfuric acid solution (3 mol/L), the solution was diluted and to be constant with water to 1 L, shaking well before use.

1,10-phenanthroline indicator: After taking 0.695 g $FeSO_4$ and 1.485 g 1,10-phenanthroline to dissolve in 100 ml water, then the process was until the indicator formed red-brown complex $(Fe(C_{12}H_8N_3)_3)^{2+}$).

The experimental steps were as below:

a. 0.100–0.500 g dried soil was taken accurately in test tubes. Adding 5.00 ml of potassium dichromate standard solution (0.8 mol/L) and 5 ml of thick sulfuric acid, shaking well before use was, respectively, operated.
b. Heating the paraffins to about 185 °C, then the samples were, respectively, put in oil bath to boil at 170–180 °C for 5 min.
c. Taking out the test tubes and waiting to be cool, the samples were washed, respectively, to flasks in about 50 ml distilled water, then 3 drops of 1,10-Phenanthroline indicator were added, then the samples were, respectively, titrated using the ferrous sulfate solution (0.2 mol/L). When the color of the solution each turned from yellow to green and became red-brown suddenly, that were the titration to the endpoints.
d. Making two times for the sample blank experiments.

The concentration of the TOC was calculated by the following formula:

$$\mathrm{TOC}(\%) = \frac{\frac{0.800\times 5.00}{V_0} \times (V_0 - V) \times 0.003 \times 1.1}{\text{sediment weight}} \times 100$$

In the formula, V_0 means the volume of $FeSO_4$ when titrated with the blank sample; V means the volume of $FeSO_4$ when titrated with soil samples; 5.00 means the volume of $K_2Cr_2O_7$ solution and 0.008 means its concentration; 0.003 is achieved by dividing 0.12 (the m mol weight of C) by 4 (the total amount of electrons exchanged in the reaction); 1.1 is the correction parameter.

It was revealed that the concentration of the TOC in the Dianshan Lake sediment was below 2% by the pre-experiment, so the soil was taken more than 0.5 g for saving samples and ensuring the accuracy in this study. The boiling time and temperature were strictly controlled to 5 min and 170–180 °C, respectively (as the digestion time and temperature would be with a great effect on the results). Blank and parallel experiments were conducted, and the average values from the experiment were used for the calculations. Analysis accuracy is shown in Table 3.60.

The concentration of organic matter in sediments can be affected by many factors, its main input source can be divided into internal sources and external sources. The internal organic matter is the earliest source of the soil organic matter, it is formed mainly with biological evolution and process of the soil formation process, in which sediments of animal and plant residues, plankton, and microorganisms are formed by the productivity of the water system. The external organic matters are mainly particles and dissolved organic matters to be carried in the external water through replenishment process, including organic fertilizers, industrial and agricultural or domestic wastewater, sludge, microbial products, organic pesticides, and other organic matter. These are outside source of the organic matter inputs after the lake formed and was disturbed by human activities [157]. The lake with less anthropogenic interference

Table 3.60 Parameters of the analysis accuracy for the total organic carbon from the sediment of the Dianshan Lake

Sample from sediment depth (cm)	1	3	10	18	24
Repeating measuring 1	0.6468	0.5891	0.4559	0.3302	0.2668
Repeating measuring 2	0.6719	0.5756	0.5188	0.3553	0.2758
Average value	0.6594	0.5824	0.4873	0.3428	0.2713
Relative mean variation	0.0190	0.0115	0.0645	0.0366	0.0166
Standard deviation (*s*)	0.0177	0.00948	0.0445	0.0177	0.00637
Variable coefficient (*CV*)	0.0269	0.0163	0.0912	0.0518	0.0235

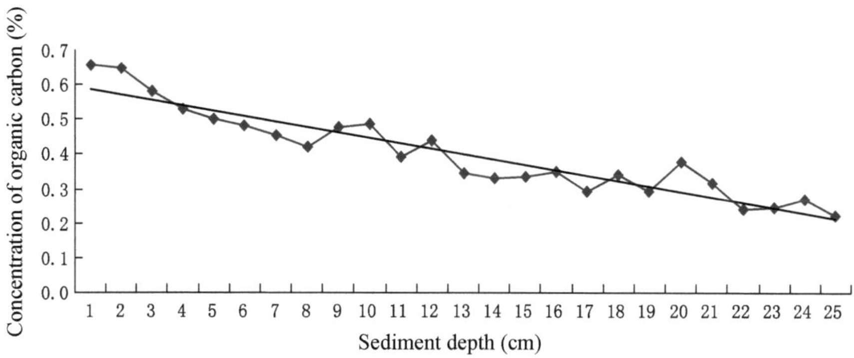

Fig. 3.56 Distribution of the organic carbon from the sediment of the Dianshan Lake

or at the period in which with less anthropogenic interference for the same lake, the organic matter contained the sediment may mainly be from the internal source. Human activity can enhance the organic carbon by a long way input from the external sources.

The changes of organic matter in sediment include mineralization and humification. The mineralization is a process in which the part of easily degradable organic matter is decomposed to CO_2, H_2O, NH_3, and other inorganic components by microorganisms, and other biological macromolecules are re-broken up and compounded by microorganisms to form humus, this is what we called humification process. These two processes coexist and cycle all the time in the sediment [158].

The concentration of the TOC in the sediment of the Dianshan Lake is shown in Fig. 3.56, which interprets the change of the TOC in 0–25-cm sediment of the lake. As Fig. 3.56 showing, the concentration of the TOC in the Dianshan Lake is between 0.221 and 0.659%, and the concentration decreases with the increase of sediment depth with the wide variation within 8–25-cm section. This might be caused by casual rainstorm to sharply wash the vegetation at the surface of earth [159].

The correlation between total organic carbon, total phosphorus, and fractions of the phosphorus speciation in the Dianshan Lake is shown as in matrix Table 3.61. The correlation between the TOC and the fractions of phosphorus speciation is shown in

Table 3.61 Correlation matrix between TOC and phosphorus in the sediment of the Dianshan Lake

	Ex–P	Fe/Al–P	Oc–P	ACa–P	De–P	Or–P	Org–C	TP
Ex–P	1							
Fe/Al–P	0.6238	1						
Oc–P	−0.2785	0.5345	1					
Aca–P	0.6855	0.7661	0.2051	1				
De–P	0.5180	0.8477	0.5001	0.9497	1			
Or–P	0.9776	0.7561	−0.09366	0.7563	0.6414	1		
Org–C	0.834	0.7125	−0.02909	0.9415	0.8279	0.87	1	
TP	0.5883	0.9035	0.5272	0.9316	0.9926	0.6878	0.8664	1

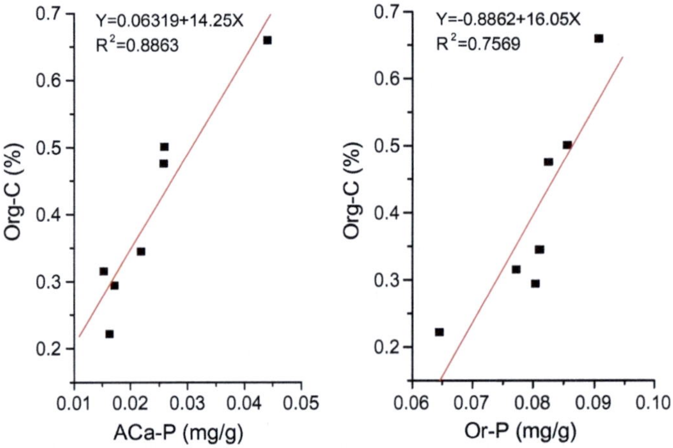

Fig. 3.57 Correlation of the TOC and species of phosphorus in the sediment of the Dianshan Lake

Fig. 3.57. It can be derived from the data that there is an obvious correlation between the TOC and the organic phosphorus, fractions of the phosphorus in the Dianshan Lake.

From Table 3.61, it can be shown that there is also a better correlation between Oc–P and Fe/Al–P in the sediment. Some researchers consider that organic matter can form complex of organic and inorganic to provide adsorption sites for inorganic phosphorus, to increase the sediment adsorption to phosphorus. There is another viewpoint that organic matters improved the adsorption of phosphorus by release of proton to protonate groups in the mineral surface [146]. Therefore, it could infer that the correlation between Oc–P and Fe/Al–P might be relevant in the combination of organic matter and Fe/Al.

There is an obvious positive correlation between the concentration of the TOC and the organic phosphorus, the concentration of the TOC decreases with the sediment depth increase, and so did the organic phosphorus. The correlation between the concentration of the TOC and the organic phosphorus is shown in Fig. 3.58.

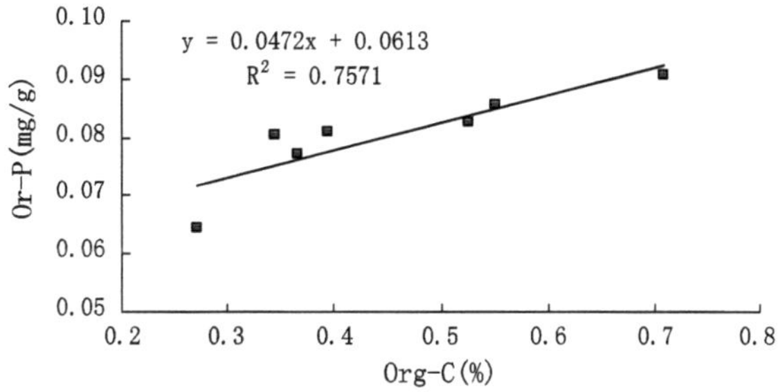

Fig. 3.58 Correlation of the TOC and Or–P in the Dianshan Lake

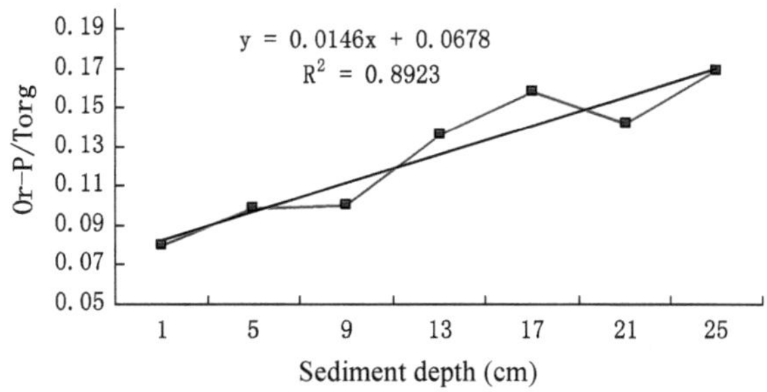

Fig. 3.59 Change of the ratio of Or–P and TOC versus the sediment depth from the Dianshan Lake

The change in ratio of organic phosphorus to organic carbon (Or–P/OrgC) with the increase of sediment depth is shown in Fig. 3.59. It reveals that the ratio of Or–P/OrgC increase with the decrease of organic carbon and the increase with the sediment depth. Combining the changes of organic phosphorus and organic carbon in the sediment, it demonstrates that relatively substantial accumulation of the organic phosphorus occurring when the organic matter degrades in the Dianshan Lake sediment.

The urban lake sediment is an accumulation bank of water pollutant, and it plays an important role in the research of pollution in different periods of the urban development. The urban economic construction can affect water quality and accelerate the eutrophication process. This research based on the changes of the environment matters in the sediments is in order to understand the environmental geochemistry and ecological significance of the pollutant evolutionary trends with the pollutant variations. Based on the distribution of organic carbon in the Dianshan Lake sediment, some items can be concluded as below:

A. The concentration of TOC in the Dianshan Lake sediment is between 0.659 and 0.221%, and it decreases with the increase of the sediment depth. The TOC concentration in the sediments within 8–25 cm fluctuates in wide range, while between 0 and 8 cm is relatively stable. These might be related to the frequent heavy rains before 1980s, in those days the serious soil erosion influenced the input of organic matters in the sediment. In the past 20 years, the water and soil loss had slowed down as a result of the better protection to vegetation around the Dianshan Lake, so as to increase the input of organic matters and it was mainly from the impacts of human activities.

B. There is an obvious correlation between the TOC and the organic phosphorus, the calcium and the phosphorus in the sediment of the Dianshan Lake. The obvious correlation is between the calcium and the phosphorus, and the organic phosphorus reflects that the organic phosphorus in the sediment mainly comes from living matters of the lake. There is a positive correlation between the concentration of the TOC and the organic phosphorus, both of them decreased with the increase of sediment depth. From the correlation fitting line of Org versus Or–P and the curves of change of Or–P/Org with sediment depth, it is concluded that substantial accumulation of the organic phosphorus occurs while the organic matter degraded in the Dianshan Lake sediment.
C. The change trends of the concentration of the TOC and total phosphorus in the Dianshan Lake sediments are responded to the economic development trend in Shanghai. The growth rate of industrial and agricultural output in Shanghai was relatively slow until the China's economy system reform and opening up from 1980s, and Shanghai economy grew rapidly after that. The increase of the TOC and the total phosphorus shows also a sharply high increase trend. It is revealed that there is a definite relation between the pollution of the Dianshan Lake, the accumulation of C and P elements in the sediment of the lake, the lake eutrophication, and the development of Shanghai industry and agriculture.

3.2.2.3 The Features of Organochlorine Pesticide Residues in the Sediment of the Dianshan Lake

As discussed above, as one of the main water sources of Shanghai, the water quality of the Dianshan Lake directly affects the health and life quality of 24 million Shanghai residents. There is a long history of the agriculture development around the Dianshan Lake, and a long-term use of pesticides must definitely lead the water quality of the lake. Pesticides in water eventually may settle in the sediment through bioaccumulation and adsorption occurred in the sediment. It will be important that to know the adding and residual concentrations for the pesticides in different periods of the sediments in the Dianshan Lake. In this study, the investigation for the changes of organochlorine pesticides concentrations in the Dianshan Lake sediments is designed to explore the coupling relationship between organochlorine pesticides and trace metals in the lake in order to provide a reference for the water environment protection of the lake.

About 70–80% pesticides people used are endocrine-disrupting chemicals. The endocrine-disrupting chemicals, also known as environmental hormones, are exogenous, and it means their presence in the environment can interfere with human or animal endocrine system in all links of substance. It is accepted that these chemicals can cause abnormal effect, through animal intake and accumulation in variety of ways but not directly as toxic to organisms, and these substances are as a role similar to estrogen hormones in the organism to affect human health. These chemicals, even though few in account, also make the organism endocrine imbalance and lead the emergence of various anomalies. The organochlorine pesticides can cause animal and human genital system disorders, behavioral abnormalities, fertility decline, larvae died, and even extinction. The endocrine disruptors are mostly organic pollutants and trace metals.

In recent years, the studies on endocrine-disrupting chemicals attract widespread attention in the scientific community and have become a new hot spot in the field of environment science and engineering. In a study about dioxins (PCDDs, PCDFs) and polychlorinated biphenyls (PCBs), human intake in Japan showed that the greater intake of contaminants was seaside residents, followed by the inhabitants working on planting, the least intake of that for urban resident. The seaside residents get this contamination through the intake from eating of fish and shellfish. The situation of the enrichment of the organochlorine pesticides in sediments can directly reflect the situation of their variation in different periods of the water. Mainly human in the agricultural activities causes the spread of organochlorine pesticides. Although since the 1980s, China Government has promulgated a relevant policy documents to forbid the use of organochlorine pesticides, but the organochlorine pesticides in large areas of farmland have remained in the soil or sediments because of the use before that. These residues would spread into the circulating water by the circulation of hydrosphere, biosphere, and soil systems, and they are enriched by the biota or transferred in the kinds of them each other, and all of these changes can be recorded in water sediments.

The organochlorine pesticides are used for controlling plant diseases and pests. As the raw material in making their components, it is generally divided into benzene and cyclopentadiene two categories. The former as the earliest and most widely used pesticide were dichloro-diphenyl-tricgloroethane (DDT) and hexachlorocyclohexane soprocide (HCH), acaricide tetradifon, dicofol, fungicide as PCNB, chlorothalonil, etc. In addition, insecticides chlordane, heptachlor, aldrin, etc., also are widely used. Turpentine as raw material in making camphene alkenes pesticides, the toxaphene and terpene bornyl chloride as raw materials in making pesticides are also organochlorine pesticides category. The main features of the organochlorine pesticides are as follows:

A. These chemicals are all with low vapor pressure, small volatile, slowly disappear in the circumstances after use.
B. They are with strong liposoluble property, and the water solubility of most of them are less than 1 μg/ml.
C. The chlorobenzene is with structural stability in the component of it, and it is hard to be degraded by the enzymes in organisms, and it is with very slowly disappears rate in the organism.
D. The result products from them affected by soil microbial can also have residual toxicity to organisms, such as DDT generated by the reduction DDD, DDE generate the same biological toxic after dehydrochlorination.
E. Some organochlorine pesticides, such as DDT can be suspended in the water and evaporated along together with the water molecules. In environment, the

Fig. 3.60 Molecular frame of HCHs

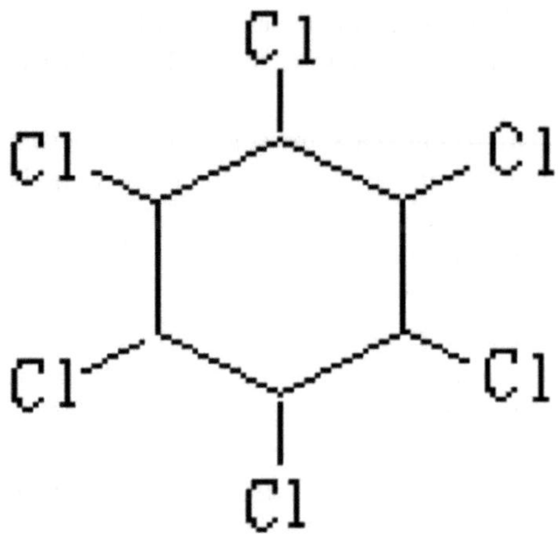

organochlorine pesticides can also expand its harmful impacts to organisms through the food chain and bioaccumulation.

Organochlorine pesticides can affect the central nervous system leading to human acute toxicity, chronic poisoning manifested as loss of appetite, weight loss, cerebellar disorders, and hematopoietic organ disorders. Some organochlorine pesticides are carcinogenic in animals tested. The poisoning from the organochlorine pesticides is with strong irritation symptoms like headache, dizziness, red eye hyperemia, and so on such as tearing, photophobia, cough, sore throat, fatigue, sweating, salivation, nausea, loss of appetite, insomnia, and head and facial sensory abnormalities.

As the main kind of the organochlorine pesticides, BHC and DDT are discussed for their main features as follows:

A. BHC—hexachlorocyclohexane (commonly known as BHC, HCHs)

The hexachlorocyclohexane molecular structure is shown in Fig. 3.60.

Hexachlorocyclohexane (BHC) is a saturated compound which is together with six chlorine atoms on benzene. Molecular formula is $C_6H_6Cl_6$, and it is the product of six chlorine atoms added to benzene. As white crystals, there are eight kinds of isomers, called α, β, γ, δ, ε, η, θ, and ξ. Their major features are as follows: The α isomer is a crystal with monoclinic arris, and with melting point 159–160 °C, boiling point 228 °C. It is soluble in chloroform and benzene, and volatile with water vapor. It is with persistent spicy odor, and with vapor pressure 0.06 mm Hg (40 °C). It can decompose to be 1,2,4-trichlorobenzene (three molecules of hydrogen chloride molecule removed) at boiling state. The β isomer is with a melting point 314–315 °C, density 1.98 g/cm^3 (19 °C). It can sublimate when is melted. It is slightly soluble in chloroform and benzene and no volatile together with water. Its vapor pressure is 0.17 mm Hg (40 °C) and can form 1,3,5-trichlorobenzene under the reaction with alcohol and potassium hydroxide solution. The γ isomers are needle-like crystals with a melting point 112–113 °C, a boiling point 323.4 °C. It can dissolve in acetone and benzene, and with very soluble feature in chloroform and ethanol. It is generally with mildew odor and with volatile property.

HCHs can kill insects by the action of stomach poisoning, smelling, and directly contacting, etc. Its γ isomer is with a highest insecticidal potency, and α isomer second, followed by δ isomer. The β isomer poisoning is the lowest efficiency in the isomers of it. HCHs are stable in acids, and in alkaline solution or together with zinc, iron, tin, etc., can easily be decomposed. It may lose the efficacy under long-term moisture or sunlight. HCHs can be compounded from benzene and chlorine under UV irradiation exposure. As it has toxicity toward human and animals, that is why China has stopped its production and its usage in early 1980s.

It is known that HCHs are with less acute toxicity, and the γ isomer is with the maximum toxicity in HCHs. HCHs are into the animal body mainly to accumulate in adipose tissue and the central nervous system. They will stimulating animal brain and cerebellum activity and affect the autonomic nervous system through the skin and peripheral nerves. They can affect cell phosphorylation in animal organs to make organ in malnutrition and degeneration or necrosis. It can induce hepatocyte microsomal oxidase endocrine activity and inhibit ATP liveness. HCHs can also cause animal and human acute toxicity, chronic toxicity, cancer, and mutations in the body. In the body of plants, insects, microorganisms, and animals, it can generate a variety of metabolic product and conjugates of sulfur and glucuronide, and then excreted out. In almost all cases, the initial product of HCHs metabolism is pentachloroethylene, which is isolated in the form of several allogeneic molecules. Phenols generated in the body of the warm-blooded animals are excreted out in forms of acid sulfate or glucuronide with urine and feces. Under the influence of microorganisms, they can also generate phenols, but they are also further decomposed in soil and destroyed all of the molecules. In the body of animals (rats), they can generate dichloro-, trichloro-, and tetrachloro-phenol isomers. The reaction for HCHs and pentachlorophenol cyclohexene to sulfur hydrogen of amino acid first occurs in insects and generates cyclohexane-based, cyclohexene-based, and aromatic-based derivatives. The thiophenol and its derivatives thereof may be the final product for these reactions

The organochlorine pesticides can be decomposed in the environment biologically, chemically, or biochemically. The biochemical decomposition of pesticides is an important cause for the disappearance of the pesticides. The degradation of HCHs in the environment occurs by the activities of microorganisms, generally considered under anaerobic conditions faster than that under aerobic conditions for biodegradable HCHs. The HCHs can be decomposed by many microbes, such as the Clostridium, the pseudomonas bacteria. The chemical decomposition of organochlorine pesticides can be conducted under various physical and chemical factors, and these factors include sunlight, alkaline environment, air, humidity, etc., where the sunlight is always as an important role for the decomposition of organochlorine pesticides. Under normal conditions of circumstances, the HCHs are completely degraded naturally in 6–7 years.

The HCHs can be accumulated biologically through the food chain in the environment; the rice and aquatic plants generally are with the potential to enrich BHCs in

Fig. 3.61 Molecular frame of DDT

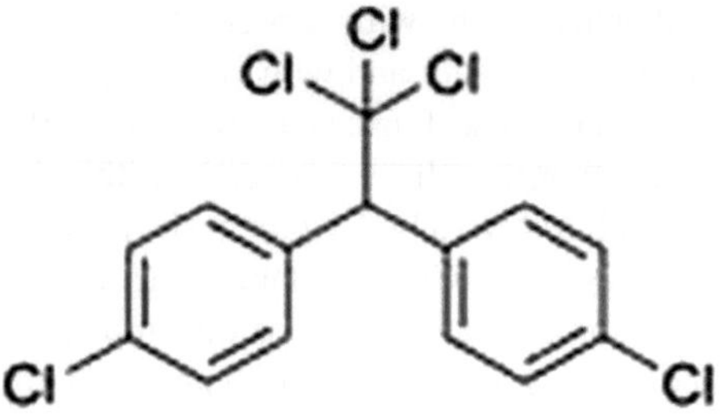

their bodies. BHCs as other organochlorine pesticides enter the environment through physical, chemical, and biological means and then eventually disappear in it gradually. Its disappearance in the environment can be conducted by the actions of diffusion, decomposition, and bioaccumulation. There are several forms for BHCs to be diffusion in the environment such as dissolved, suspended, volatilization, sedimentation, and penetration. Studies show that at 25 °C, the solubility of α-BHC in water is 1630 μg/L, β-BHC is 700 μg/L, γ-BHC is 7900 μg/L, and δ-BHC is 21,300 μg/L. Suspended substances (including sediment, plankton, and organic particles, etc.) may also adsorb these pesticides into the aqueous environment. The organochlorine pesticides in the water and in the soil surface can also be released to the atmosphere through volatilization. The particles of the organochlorine pesticides in the air as gaseous form also can be transported with the air dust drifting and are carried away to distance places, eventually settling in soil, water, or sediments. The pesticides in soil can also penetrate from the upper to the lower layer of the soil with water, thereby causing the pollution of surface water or groundwater.

B. DDT 1,1,1-trichloro-2,2 bis (p-chlorophenyl) ethane (DDTs)

1,1,1-trichloro-2,2 bis (p-chlorophenyl) ethane molecular structure is shown in Fig. 3.61.

DDT 1,1,1-trichloro-2,2 bis (p-chlorophenyl) ethane formula is $C_{14}H_9Cl_5$. Generally, it is used as emulsions, wettable powders, dusts, and aerosols. Its melting point is 107–109 °C, and boiling point is 260 °C. Relative density (water = 1) is 1.55 (25 °C), and the saturated vapor pressure is 2.53×10^{-8} kPa (20 °C). It is insoluble in water, and is soluble in acetone, benzene, dichloroethane. It is mainly used as agricultural pesticides. DDTs cause acute toxicity, chronic toxicity, cancer, aberrations, and mutations in the body of animals and humans. Human symptoms of acute poisoning by the DDT include headache, dizziness, nausea, vomiting, limb paresthesia or ataxia, severe fever, tachycardia, dyspnea, coma, and even death. It also has a stimulating effect on the skin. This pesticide's half-life is up to 2.5–5 years because of a strong fat soluble, and it is easy to accumulate in the body of animals from pollution to harm human health by food and potential threats. It has been known the DDT residues are with a positive correlation with the fat concentration of the organ in human organs. From the early 1970s, DDT has been banned in many countries; China stopped the production of it and use in agriculture in 1983.

There are two approaches to let the DDT to be degraded in the human body; one is by dehydrochlorination to generate DDE, and three years are generally required for DDT converting to DDE with less than 20% DDT of the human body. Another approach for DDT change in human body is to be converting DDD from the reduction, and the DDD can be converted DDA which is more soluble in water than DDT so that to be eliminated with its biological half-life only about 1 year. A survey of USA from 1964 for American national suggested that the total amount of DDT stored in body fat was at an average of 10 μg/g, of which about 70% was DDE, and the DDE discharge from the body was especially slow, biological half-life was in about 8 years [160].

DDTs can change in series actions with complex biological and chemical degradations in environment, and the main reaction is dehydrochlorination to generate DDE. The DDE is relatively less toxic for insects and higher animals, and it is almost no role in the biological and chemical degradation. Thus, DDE is the main residuum stored in the biological tissues [160]. DDT in biological systems can also be reduced by dechlorination to generate DDD. The DDD is formed at the first step in the actions of degradation in the body of animals and the environment, and it is less stable than that of DDT or DDE. DDD can generate DDMU (2,2-bis-(chlorophenyl)-1-chloride) by dehydrochlorination, then can be reduced to DDMS (2,2-bis-(chlorophenyl)-1-chloroethane), then can be dehydrochlorinated to generate DDNU(2,2-bis-(chlorophenyl)-1-ethane), eventually can be oxidized to DDA (bis-(chlorophenyl) acetic acid). The DDA in water is better soluble than that of DDT, and it is the final product excreted from higher animals and human for that stored and taken. DDT can also be degraded by oxidase particulars to form dicofol by the reaction at its α-H site. In the environment, DDT residues can be converted into pairs—dichlorobenzophenone. Recently, a new anaerobic degradation phenomena have been found, especially DDT in the sludge can be converted into a bacterial DDCN (bis-(chlorophenyl) acetonitrile).

The DDT disappears slowly in soil environment; under normal circumstances it may be in about 10 years. A recent study shows that DDT can degrade into carbon dioxide and hydrochloric acid under the conditions similar as the upper atmosphere. Because DDTs are with high stability, they can be detected after use in farmland for six months. Nowadays, DDT pollution in environment has been a problem around the world. It can be detected from the dust drifted 2000 km and Antarctic melting snow water. In general, the concentration range of DDT in non-agricultural area air is less than $1.00–2.36 \times 10^{-6}$ ng/m^3, in agricultural area with residents its concentrations range is $1.00–22.00 \times 10^{-6}$ ng/m^3. In the area of inhabitants indoor with DDT used for mosquito spray, the concentration of DDT can be up to 8.50×10.00^{-3} mg/m^3. In the area of agriculture and non-agricultural outlying areas, DDT concentration in rainwater is often in the same range ($1.8–6.6 \times 10^{-5}$ mg/L), which indicates that the distribution of this compound in the air is fairly same. DDT concentration in surface water is related to rainwater and soil concentration of DDT levels. The highest concentration detected in US drinking water in 1960 was amounted to

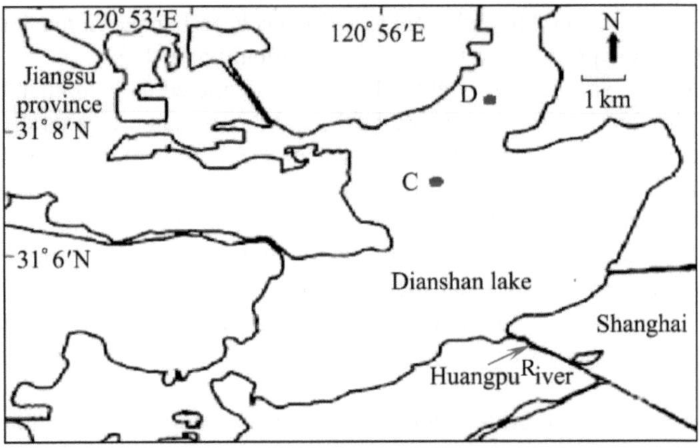

Fig. 3.62 Sampling sites for the research on the OCPs from the Dianshan Lake

0.02 mg/L. The DDT concentration in no-sprinkle soil is founded in 0.10–0.90 μg/g, only slightly lower than that in cultivated soil with sprinkle DDT for more than 10 years (0.75–2.03 μg/g). Most of DDT is present in surface layer soil within the 2.5 m depth.

The DDT can translate into other forms in the environment by photolysis, biotransformation, soil conversions, etc. In addition the biotransformation metabolic in mammals, there are also different approaches of for birds, insects, microorganisms, and higher plants, and the conversion resultants (including mammal metabolites) include about 20 species. Many other compounds are still unknown in the chemical structure of the conversion resultants of the DDT. Excepting the main products of the conversion resultants of DDT such as DDE and DDD, the toxicological properties of these transformation products for people are still not at all known. There is still a considerable research issue for DDT and its homologue in the whole environment cycle.

1. **Samples and the measurement for the organochlorine pesticides**

(1) **Sampling**

The samples were taken from sites C and D shown as Fig. 3.62. The site C was located in the center of the lake where sediments were less interrupted. The site D was located besides an aquatic sport plot of the Dianshan Lake (Fig. 3.62) [161].

The sediment was, respectively, sampled at above-mentioned sampling sites using a PVC pipe with a diameter of 8 cm and a length of 4 m. The PVC pipe was inserted with pressure directly into the sediment. After the pipe being taken out, the empty part was sawn and section with the sediment core was immediately transported to the laboratory after sealed with kraft paper twined with adhesive tape. In laboratory, the sample cores were taken out from the pipes with the sediments integrally and segmented with a bamboo chip with 1-cm interval. Segmented samples were dried naturally covered with kraft paper, grinded, sieved through the sieve of 200# and stored in kraft paper bag under 4 °C temperature condition.

(2) **The treatment of extraction and concentrating for the organochlorine pesticides**

The most common methods for the extraction of the organochlorine pesticides are with means of microwave, ultrasonic, accelerated solvent, and Soxhlet extraction, reviewed summarily as follows for understanding the means in this work:

The microwave extraction uses microwave digestion system (MARS-Xpress, CEM, USA). 5 g of sample is placed into the extracting tank, and 25 ml mixed solvent of n-hexane and acetone (1:1, V/V) is added into the extraction machine with microwave power of 1200 W, at 110 °C temperature for 10 min [115]. This method is employed the program to raise the temperature, and due to with a capillary chromatographic column and μECD it is with advantages such as shorter analytical time, sound separation effect, and high sensitivity. The all process are with the features such as simple operation, homogeneous heating, high selectivity of extractant, little dosage of organic solvent, low energy consumption, high recovery, nice reproducibility, it is a good method as a fine pretreatment technology for qualitative and quantitative measurement of the semivolatile organic compounds.

The ultrasonic extraction method uses strong vibration, high accelerated velocity, strong cavitation effect, and the stirring action is generated from ultrasonic wave to accelerate the effective constituents in the samples into solvent. That is, the mechanical effect of ultrasonic wave attaches greater seepage force of the solvent for the extracted components as well as their interaction. The optimal conditions for the analysis of organochlorine pesticides in the sediment are with the chromatographic column DB-1701, extraction solvent n-hexane/CH2 Cl2 (1:1, V/V), the extraction time 60 min in each extraction [150]. This method is with advantages such as short time, simple operation in fast process, and applications in broad cases. It is enable to be used for mass samples to be extracted.

Accelerated solvent extraction (ASE) is a new automatic pretreatment method for organic solvent extraction under high temperature and pressure. Compared to above methods, ASE is with some prominent advantages such as little dosage of organic solvent requiring, fast process, and high recovery. It has been recommended by US EPA as a standard method for the extraction of organochlorine pesticides.

The Soxhlet extraction as a kind of liquid–solid extraction it uses condensate extracted solvent through boiling to extract samples repeatedly, thus reducing the solvent dosage used greatly. This extraction can prevent the solvent loss when the sample was heated to boiling, and also is recommended by US EPA as a standard method for the extraction of organochlorine pesticides, such as EPA3540 and EPA3541. Although Soxhlet extraction method saves solvent compared with traditional methods, but there are difficulties in practice because of complex operations and longer time-consuming [162]. In addition, the back-flow velocity in the Soxhlet extraction also influences the extraction effect although the cost is relatively lower.

In this work, the Soxhlet extraction was used for sample pretreatment. The procedure in details was briefly as follows (Fig. 3.63).

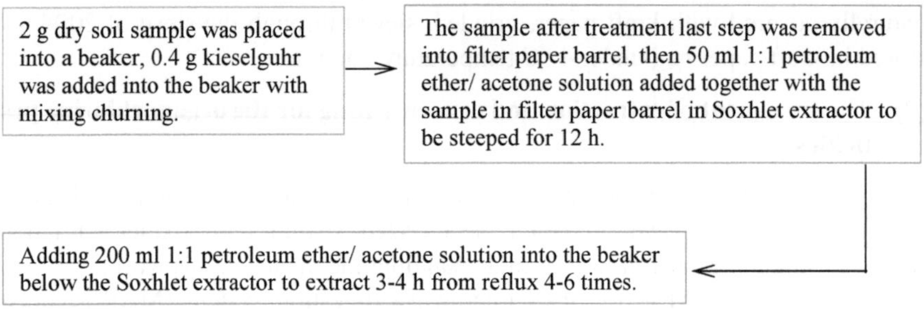

Fig. 3.63 Extracting steps for HCHs and DDTs

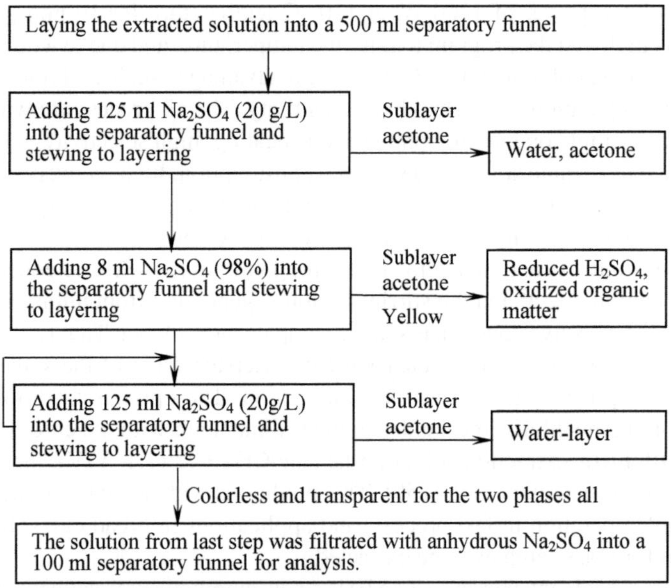

Fig. 3.64 Purified steps for HCHs and DDTs analysis

In the sample treatment processes, no any loss of the sample was required and controlled in the operations theoretically. The sediment sample was completely soaked in the solvent for over 12 h, and the back-flow time should be kept same for every extraction. The back-flow in five times for each extraction was considered to be extract completely for the target substances. The purification process after the extraction is shown in Fig. 3.64.

In the purification procedure, the concentrated sulfuric acid was used to oxidize other organic matters in the system of the liquor as much as possible. The process involved the two operations, transferring of the extracted liquor and discarding the water layer, and these might cause at a certain degree some losses for the objective substance.

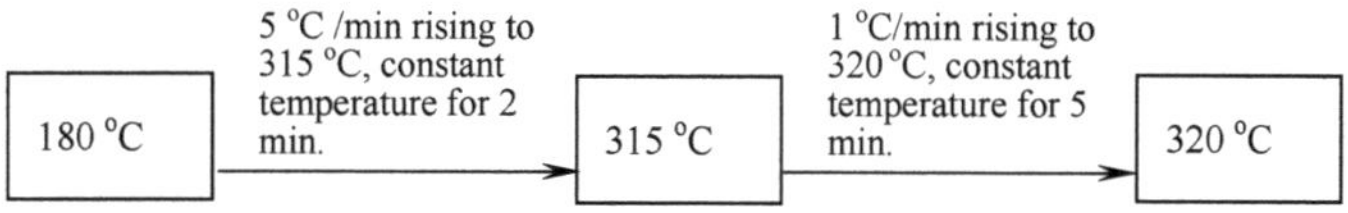

Fig. 3.65 Temperature rising schedule of the chromatographic instrument

After the extraction, the target substances were also purified and concentrated. A special-purpose equipment with homemade nitrogen-blowing concentrator (nitrogen cylinder, reducing valve, rubber tube, suction head) was used in the work to blow slowly the liquid at 1 cm above the surface of the liquid, and let it, respectively, be only 1 ml left, before transferring into the graduated test tube.

(3) **The measurement for the organochlorine pesticides**

The HCHs and DDTs are both with poor solubility in water and good that in fat. The azeotrope of petroleum ether mixed with acetone (1:1, V:V) was used to extract these two substances repeatedly in Soxhlet extractor, and then acetone layer was extracted out with Na_2SO_4 to dissolve HCHs and DDTs in petroleum ether. After the extraction, concentrated sulfuric acid was used to oxide the organic matter in extract liquid. Because too low concentrations of HCHs and DDTs might be in the extract liquid in the case, the two substances should always be concentrated to 1 ml and then detected qualitatively and quantitatively using gas chromatography.

Necessary reagents in the extraction above were as follows:

There were, respectively, including petroleum ether, acetone, Na_2SO_4 solution (20 g/L), 98% concentrated sulfuric acid, isooctane, anhydrous Na_2SO_4, industrial nitrogen, and 100 mg/L standard solution for α-HCH, β-HCH, γ-HCH, δ-HCH, p,p'-DDE, p,p'-DDD, o,p'-DDT and p,p'-DDT standard solutions, purchased from the Institute for Environmental Reference Materials of China Ministry of Environmental Protection.

The instruments used in the extraction experiments were as follows:

There were, respectively, including gas chromatography (Agilent Technologies-6890), electrothermal isothermal concussion flume, nitrogen cylinder, volumetric flask (10, 100, 250 ml), beaker (100 ml), Soxhlet extractor (100 ml), graduated test tube (10 ml), separating funnel (500 ml), triangle funnel, glass rod, rubber burette, pipette (10 ml), microsyringe (10 μL), and filter paper.

All reagents were analytically pure grade and water used was deionized. The operating parameters for the instruments used in the extraction were as follows:

Water bath concussion groove used was the water bath designed heating at 80 °C, and the temperature of gas chromatography was set as shown in Fig. 3.65. The temperature at the injection port and detector were 250 and 315 °C, respectively. The flow rates of nitrogen and tail-blowing were 1.3 and 21.4 ml/min, respectively.

Each 1 μL of samples was injected by a microsyringe for measurement. The standard samples were detected for measuring the appearing time of each component (Fig. 3.66).

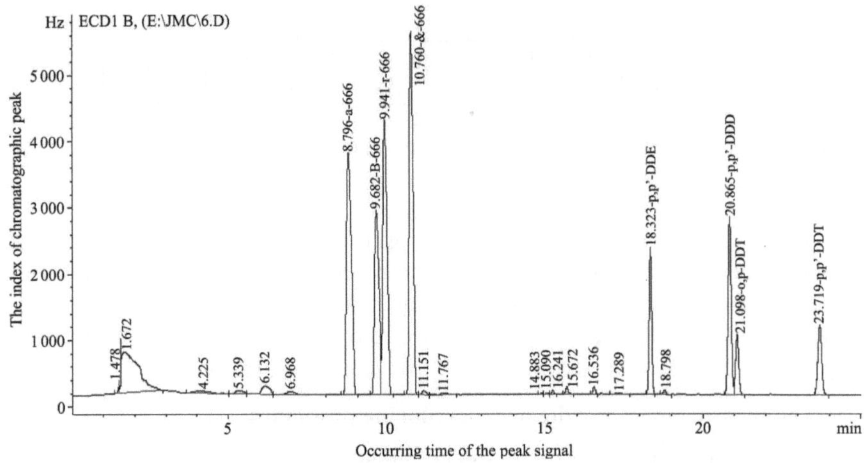

Fig. 3.66 Holding time of components of the reference

The standard curve used in calculating the amount of the target component was as below:

$$\alpha - \text{HCH} : Y = 742.42666X \; R^2 = 0.9960$$
$$\beta - \text{HCH} : Y = 524.96214X \; R^2 = 0.9903$$
$$\gamma - \text{HCH} : Y = 714.55348X \; R^2 = 0.9952$$
$$\delta - \text{HCH} : Y = 879.39368X \; R^2 = 0.9871$$
$$p,\ p' - \text{DDE} : Y = 237.77718X \; R^2 = 0.9934$$
$$p,\ p' - \text{DDD} : Y = 347.7421X \; R^2 = 0.9855$$
$$o,\ p'\text{DDT} : Y = 123.65115X \; R^2 = 0.9774$$
$$p, p' - \text{DDT} : Y = 170.64819X \; R^2 = 0.9688$$

The calculation formula used was as follows:

$$X = \frac{C_{is} \times V_{is} \times H_i(S_i) \times V}{V_i \times H_{is}(S_{is}) \times m}$$

In the formula above, X is the concentration of the pesticide residues in the samples (mg/kg), and C_{is} is pesticide concentration of the component in the standard solution (μg/ml), V_{is} is the injection volume of standard solution (μl), V is the constant volume of the sample solution (ml), V_{i} is the injection volume of sample solution (μl); $\mathrm{H_{is}}$ (or $\mathrm{S_{is}}$) is the peak height (or peak area) of the component in standard solution (mm or $\mathrm{mm^2}$), $\mathrm{H_i}$(or $\mathrm{S_i}$) is the peak height (or peak area) of the component in sample solution (mm or $\mathrm{mm^2}$), m is the quality of sample (g).

Table 3.62 Detecting limits and recovery ratio in the experiment

Pesticides	Detecting limits (ng/g)	Recovery ratio (%)
α-HCH	0.64×10^{-3}	62.37
β-HCH	0.58×10^{-3}	68.53
γ-HCH	0.68×10^{-3}	59.22
δ-HCH	0.75×10^{-3}	53.51
p, p'-DDE	0.62×10^{-3}	64.71
p, p'-DDD	0.58×10^{-3}	71.76
o, p'-DDT	0.61×10^{-3}	65.62
p, p'-DDT	0.55×10^{-3}	72.42

Five levels of mixed standard solution (2.75, 5.5, 11, 27.5, 55 μg/L) are prepared and the standard curves obtained from these concentrations, respectively, are used to calculate the concentration of the target substances quantificationally according to the formulas mentioned above, shown as Fig. 3.67.

The standard curve of each component is shown as Fig. 3.68. In this study, GC with ECD was used to measure the standard solution independently, nitrogen as carrier gas. The recovery rates of the target substances is not too good in the measurement than that expected, which may be due to nitrogen-blowing leading to some losses of the target substance. It can be understood from this fact inspiring that if the samples are concentrated with a rotary evaporator to 3 ml and then nitrogen-blowing it to 1 ml, the recovery rate may be improved at a certain extent. Another reason for the recovery question may be the multiple dilutions to be extracted in one process, which is caused by too much extracted solvent and little sample amount. It is found by the study that the different kinds of chromatographic columns can lead to be completely different retention time and even an order inversion for the retention time of the samples. When chromatographic column used is same, the temperature raising procedure commonly cannot invert the sequences of the retention time for different components. The faster the temperature raises, the quicker the peak becomes visible.

From above phenomena, it is indicated that chromatographic condition changed without changing the chromatographic column can enable the sample measurement more accurate. Under the same chromatographic condition with the same column, compounds of different properties, especially for the samples with complicated matrix as vegetables and fruits, are more likely to retain for the same length of the peak appearing time. In this case, it may cause inaccurate qualitative, while it is likely with very less probability to retain for the same length of the peak appearing time when two or more columns are used.

Therefore, two or more chromatographic columns can be applied to analyze pesticide residues qualitatively more accurately. In conclusion, the fused silica capillary column (DB-1) is able to analyze 12 kinds of organochlorine pesticides in 20 min, and the baseline separation for most kinds can be obtained [163]. The analyses' accuracies are shown in Table 3.62.

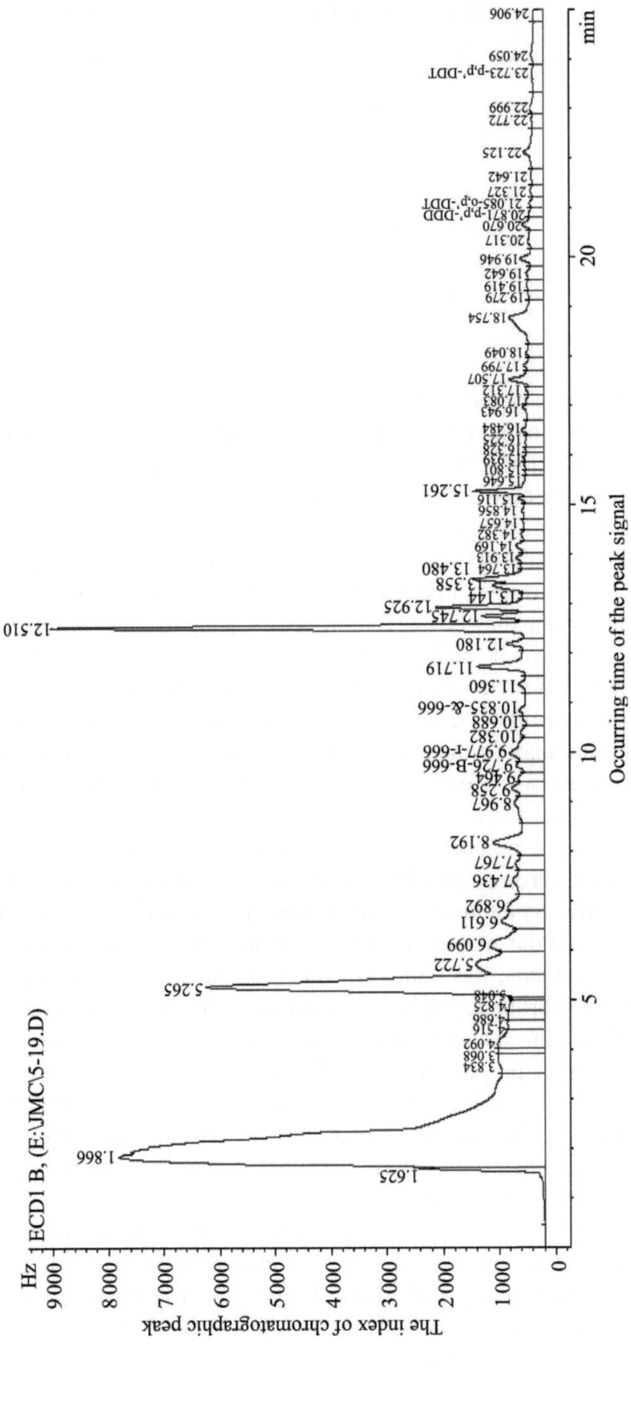

Fig. 3.67 Peaks of the components in the mixed references

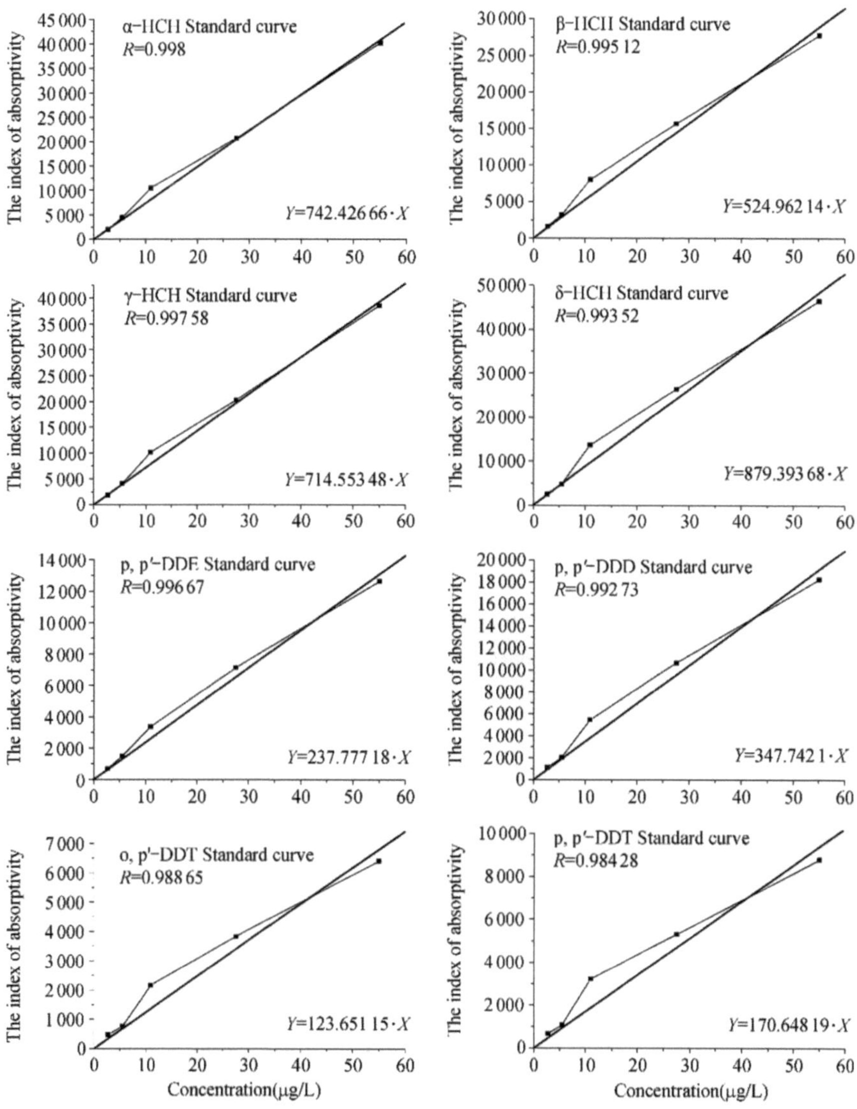

Fig. 3.68 Standard curves of the objective matter for α-HCH, β-HCH, γ-HCH, δ-HCH, *p*, *p*′-DDE, *p*, *p*′-DDD, *o*, *p*′-DDT, and *p*, *p*′-DDT

Table 3.63 Concentrations of the HCHs and DDTs in the sediment at site *C* (ng/g)

Pesticide component	Sediment core C depth (cm)									
	19	17	15	13	11	9	7	5	3	Average
α-HCH	14.33	–[a]	–	–	–	–	–	–	–	–
β-HCH	21.95	–	32.37	49.06	–	–	–	–	–	–
γ-HCH	–	–	–	–	–	–	–	–	–	–
δ-HCH	–	–	–	–	–	–	–	–	–	–
p, *p*′-DDE	–	–	1.84	0.58	–	0.59	0.39	0.38	0.51	–
p, *p*′-DDD	0.30	0.21	3.16	1.45	–	0.60	0.24	0.52	4.07	–
o, *p*′-DDT	–	–	4.34	9.11	–	1.60	–	–	–	–
p, *p*′-DDT	–	2.00	19.49	1.92	1.04	1.87	1.60	2.15	3.43	–
总HCH	36.28	0.00	32.37	49.06	0.00	0.00	0.00	0.00	0.00	13.08
总DDT	0.30	2.21	28.83	13.05	1.04	4.65	2.22	3.05	8.01	7.04
HCHs + DDTs	36.58	2.21	61.19	62.11	1.04	4.65	2.22	3.05	8.01	20.12

[a]Means no detecting out

Table 3.64 Concentrations of the HCHs and DDTs in the sediment at site D (ng/g)

Pesticide components	Sediment core D depth (cm)								
	23	19	16	13	10	7	4	2	Average
α-HCH	27.21	–[a]	–	–	–	–	–	–	–
β-HCH	32.79	7.91	–	–	–	–	–	–	–
γ-HCH	–	16.00	17.99	–	–	–	–	–	–
δ-HCH	–	–	18.01	–	–	–	–	–	–
p, *p*′-DDE	25.02	–	–	–	–	0.63	0.34	–	–
p, *p*′-DDD	5.73	5.44	–	22.25	6.68	0.20	–	–	–
o, *p*′-DDT	50.50	18.76	15.80	–	–	–	–	–	–
p, *p*′-DDT	41.06	26.77	4.38	61.08	0.86	–	0.56	–	–
Total HCH	59.99	23.90	36.00	0.00	0.00	0.00	0.00	0.00	14.99
Total DDT	122.32	50.97	20.18	83.33	7.54	0.83	0.90	0.00	35.76
HCHs + DDTs	182.31	74.87	56.18	83.33	7.54	0.83	0.90	0.00	50.75

[a]Means no detecting out

2. **Results and discussion**

The organochlorine pesticides measurement results were shown in Tables 3.63 and 3.64.

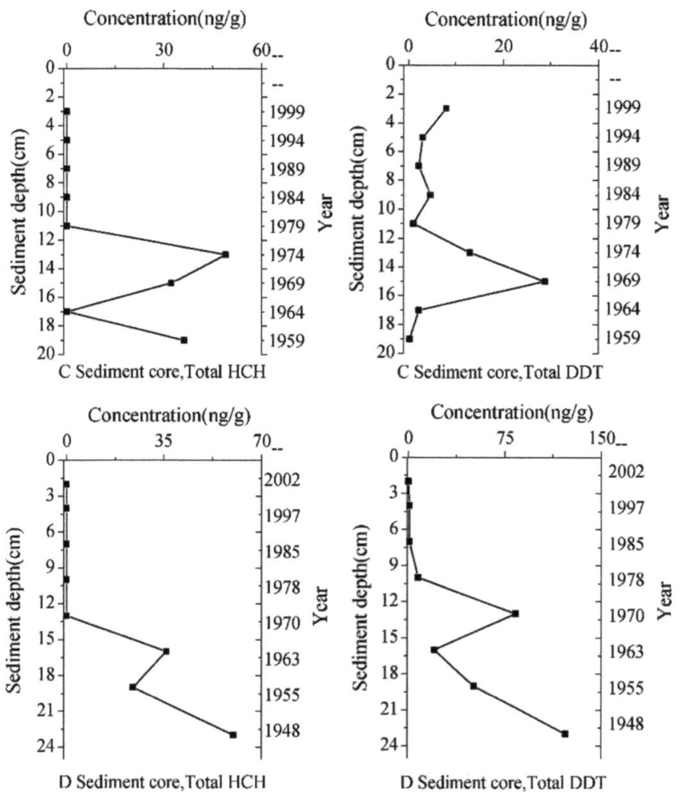

Fig. 3.69 Vertical distribution of HCHs and DDTs in the sediment at sites C and D from the Dianshan Lake

The isotope dating research data show that the deposition rate of the Dianshan Lake is about 0.4 cm/a [104, 116]. According to this research, the concentration and vertical distribution characteristics of HCHs and DDTs based on the reckoning are shown as in Fig. 3.69.

Figure 3.69 shows that the residual amount of the organochlorine pesticides in the sediment of site C and site D has started decreasing from the 1980, and after 1980s the concentration of the organochlorine pesticides is, respectively, below the detection limits of the instrument (Table 3.62). The residues of the HCHs in the core sample column C in the period before 1980 was generally in a higher concentration trend, which was with a peak value at 1974. The HCHs residues in the core column sample at the site D were with a long increased trend period before 1980. The trend of DDTs residuals was similar to HCHs, which began to decline after 1980. The residues of DDTs in core column sample at site C was generally with higher level before 1980, which was with a peak concentration around 1969, while DDTs residues in the core sample column at the site D were increased with the length of the time from 1980.

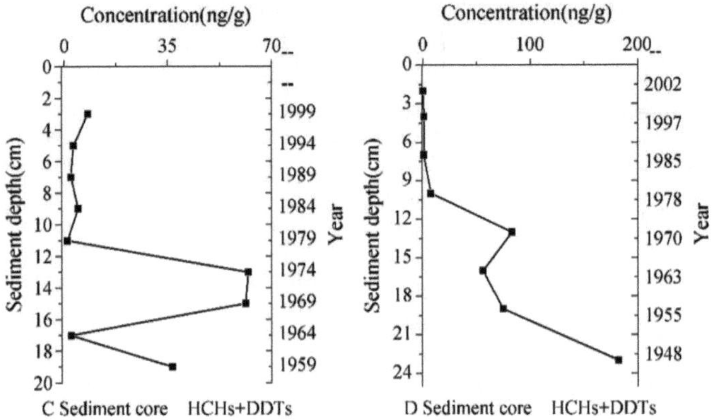

Fig. 3.70 Distribution feature of HCHs + DDTs in the sediment at site C and site D from the Dianshan Lake

The concentration and vertical distribution of HCHs + DDTs versus time based on the calculating is shown as Fig. 3.70.

From Fig. 3.70, it is shown that the residual amount of HCHs + DDTs in the sediment remained stable and is with relatively lower values after 1980. The level of the residues in the core sample column site C is little higher than those of the core sample column site D. The residual amount of HCHs + DDTs was higher in the column C before 1980, with a peak level around 1974. However, the residual amount of HCHs + DDTs in the column D has gradually increased as the length of the time from middle of 1980s.

Each monomer concentration of DDTs and HCHs in the sediment and their vertical distribution characteristics are, respectively, shown in Fig. 3.71.

According to Fig. 3.71, the β-HCH in HCHs in the sediments from the sites C and D is all relatively with higher concentration, and the residual amount of the α-HCH increased with increasing depth of the sediment while the residual amount of the δ-HCH declined with increasing depth of the sediment. The β-HCH is the most stable monomer in all of HCHs monomers, with the lower saturated vapor pressure and good resistance to degradation, and it might be the reason that the residues of β-HCH in sediment was with larger proportion of the whole HCHs.

The monomer concentration of DDTs in sediments and the vertical distribution are shown as Fig. 3.72. Figure 3.72 shows that the residues of p, p'-DDT in sediments are higher than that of the other monomer. The DDTs are the basic forms of this chemical into the system initially, while DDE and DDD are the metabolites of DDTs. The DDT residues in sediments are mainly pesticides, which have been applied in the fields for agricultural purpose through air or water, and the sediments finally become its container. From the 1980s to 2000, the parameter (DDD + DDE)/DDT was less than 1 in the sediments of the Dianshan Lake, while after 2000 (DDD + DDE)/DDT more than 1 in the sediments. Based on the estimated rate of decay of DDT, it would completely disappear in soil system after 10 years, the situation illustrated

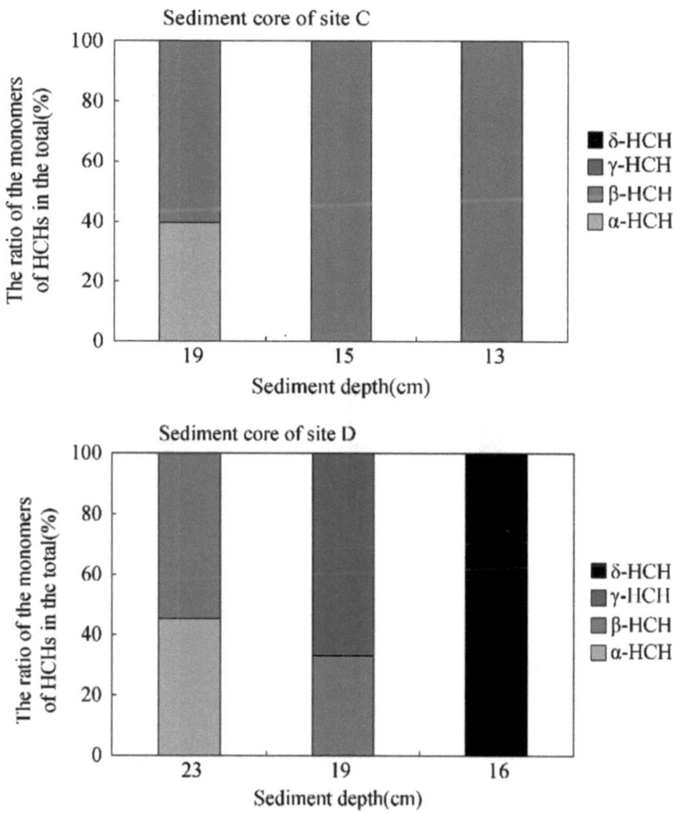

Fig. 3.71 Distribution of the monomers of HCHs versus sediment depth at the site C and site D

that after the ban of pesticide issued by China Government in 1983, there still might be in illegal use for DDT in the farmland around the Dianshan Lake, and it was in gradually reducing over time.

The coupling relations between the concentrations of HCHs, DDTs, HCHs + DDTs in the sediments of the Dianshan Lake are shown as Fig. 3.73.

Figure 3.37 shows that it seems not with any correlation between the concentrations of HCHs and DDTs in the sediment of the Dianshan Lake ($R^2 = 0.26712$). Combined with the distribution of the pesticide residues (Fig. 3.69), this phenomenon can be roughly interpreted as there might be significant differences in the degradations for HCHs and DDTs. The correlation coefficient of DDTs and HCHs + DDTs is equal to 1 after 1980s, while before 1980s it is equal to 0.5133, which is obviously lower than that of HCHs and HCHs + DDTs at the same time, 0.8087. Aforementioned, before the 1980s the residual amount of HCHs + DDTs was mainly related to the residual amount of HCHs, and in the 1980s the residual amount of organochlorine pesticides in sediments of the Dianshan Lake mainly related to DDTs residues.

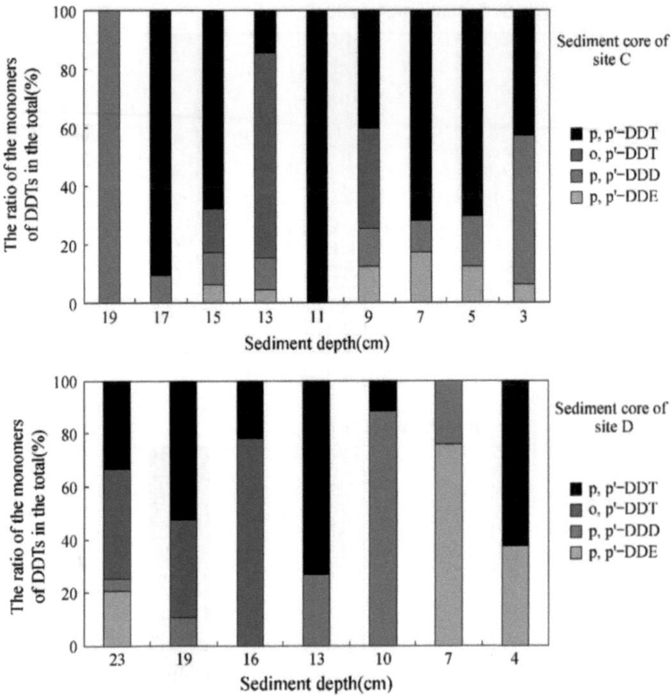

Fig. 3.72 Distribution of the monomers of DDTs versus sediment depth at the site C and site D

When organ chlorine pesticides enter the water, they can enter the sediment through cycling, bioaccumulation, settling, and other processes. The residual amounts of organochlorine pesticides in the sediments may represent the contaminations of the water by organochlorine pesticides during a certain period. In general, from the distribution of two main organochlorine pesticide residues, DDTs and HCHs, in the sediment of the Dianshan Lake, the points of the environmental meanings can be concluded as below:

A. The concentrations of residues of HCHs, DDTs in the Dianshan Lake are, respectively, in range of 0–59.99395, 0–122.3471 ng/g. The residual concentrations of HCHs after the 1980s are below the detection limit of the instrument, and the concentrations of DDTs after the 1980s are all with relatively lower values. This might be related to the water solubility and the residual capacity of the particulate phase of HCHs less than that of DDTs; moreover, it might be related the HCHs with stronger biodegradability also than that of DDTs.
B. The residual amounts of HCHs and DDTs in the sediments of the Dianshan Lake have a similar distribution trends, which are toward two trends divided by the early 1980s. Before the 1980s, the distribution was higher with increase trend as the time, peaked at the middle 1970s. After 1980s, there is a sharp reduction in the concentration showing with less and steady values as time.

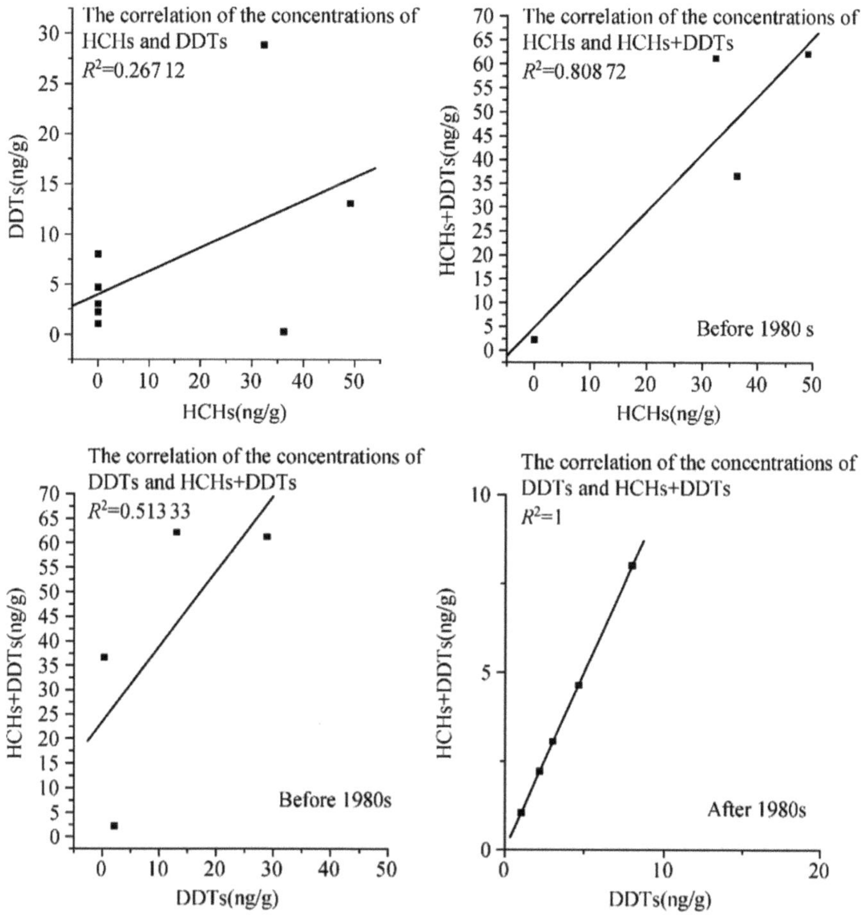

Fig. 3.73 Correlation of the concentrations of HCHs, DDTs, and HCHs + DDTs in the sediment from the Dianshan Lake

C. The β-HCHs are with the maximum proportion in the total concentrations of HCHs in the sediments of the Dianshan Lake, which is related to that the β-HCHs are as the most stable and the most hard to degrade monomer in the monomers of the HCHs. In the era of the legitimate use for the organochlorine pesticides in China, the δ-HCH residual amount in the sediment decreased with the increase of the sediment depth, which might be caused by the good degradation ability among all the HCHs monomers.
D. After promulgation of the ban for use of the organochlorine pesticides (1983), the index (DDD + DDE)/DDT was less than 1 in the sediments of the Dianshan Lake before 2000; however, after 2000 the index (DDD + DDE)/DDT has been

more than 1. It can be inferred that there is still the illegal use for organochlorine pesticides around the farmland around the Dianshan Lake at that time.

E. The coupling relationship among organochlorine pesticides in the sediments of the Dianshan Lake is divided toward two trends around 1980. Before 1980s, the R value of HCHs versus HCHs + DDTs ($R^2 = 0.8087$) is higher than that of DDTs versus HCHs + DDTs ($R^2 = 0.5133$), while the R value of DDTs and HCHs + DDTs is equal to 1 after 1980s. It could illustrate that the residual distribution of HCHS + DDTs in the sediments of the Dianshan Lake is mainly related to the concentration of HCHs before 1980s, while the residual distribution of HCHS + DDTs in the sediments of the Dianshan Lake is mainly related to the concentration of DDTs after 1980s. It also can explain that the distribution of HCHs residual amount in the sediments of the Dianshan Lake is literally no correlation with the residual amount of DDTs, which could be related to the difference of degradation ability for these pesticides in the system.

3.2.3 *The Implication from the Coupling Features of the Trace Metals, N, P, C, and Pesticides in the Sediment of the Dianshan Lake*

3.2.3.1 The Traits of the Dianshan Lake Sediment

In general, there is no obvious stratification in the sediment of the Dianshan Lake; its color represents different degrees of dark gray to light gradually along with the increasing depth. At the bottom, there are some mixed biological debris such as shells. It is with a gray-black flocculation within the top 1–3 cm and mainly is consisted of clay and fine silty sand particles. The following section below 3 cm is in priority with argillaceous sediment, and the texture gradually becomes dense as the depth increasing. The pH value distributes in the Dianshan Lake sediments (measuring method introduced at 3. 2. 2. 2. 1(2)) for the depth, respectively, at 0–1, 4–5, 8–9, 12–13, 16–17, 20–21, 24–25 cm, shown as in Fig. 3.74. The results show that the pH of the Dianshan Lake sediment is between 6.68 and 7.53, belonging basically neutral state. From shallow to deep depth, the pH value is in the change from weak acid to weak alkaline, generally in 0–10 cm in weak acid and 10–25 cm in weak alkaline.

The moisture concentration of the Dianshan Lake sediment is shown in Table 3.65 (measuring method introduced in Section 3. 2. 2-(2)-1-B). From Table 3.65 data, the moisture change range for the dry sample of the Dianshan Lake sediment is very less, and it is between 4.16 and 5.43%. The moisture is with the average value 4.75% and is no obvious change with its sedimentary depth.

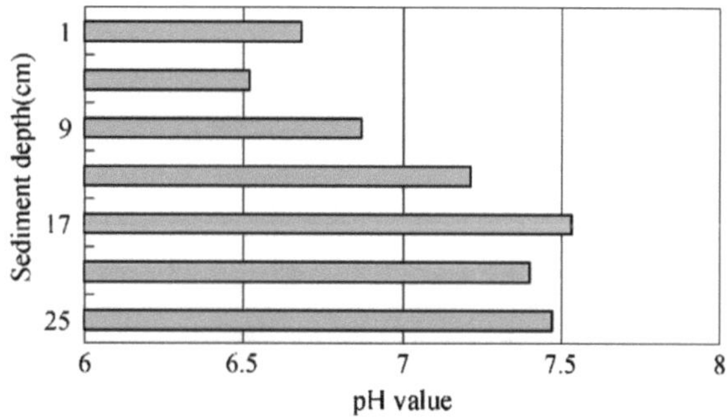

Fig. 3.74 Feature of pH versus the depth of the sediment in the Dianshan Lake

Table 3.65 Distribution of water in the sediment of the Dianshan Lake

Depth (cm)	Water contained	Depth (cm)	Water contained
1	0.0416	14	0.0509
2	0.0448	15	0.0484
3	0.0497	16	0.0517
4	0.0501	17	0.0474
5	0.0489	18	0.0452
6	0.0476	19	0.0459
7	0.0450	20	0.0484
8	0.0492	21	0.0468
9	0.0453	22	0.0466
10	0.0496	23	0.0445
11	0.0543	24	0.0444
12	0.0512	25	0.0424
13	0.0484	Average value	0.0475

3.2.3.2 The Change of Trace Metals and Coupling with N, P, TOC, and Pesticides in the Sediment

The foregoing results suggest that the pH and moisture of the Dianshan Lake sediments belongs to the normal situation of recent sediments, but the substance composition change of the sediment such as trace metals, N, P, TOC, and pesticides is, respectively, with obvious characteristics as discussed above. These substances or pollutants all are the discharged matters to the natural system associated with human's activities. The changes in different material concentration recorded in the sediments signify the effects of human activities on the natural system. The relationship of their coupling each other will be very important to understand the cause of formation of the pollution in the Dianshan Lake.

In the past few decades, with the rapid development of economy and urbanization, Shanghai has become China's financial, business, shipping, and trade hub. However, with the development of urban economy, the environment pollution problem is also increasing sharply. To keep in view the rapidly changing characteristics of trace metal elements, N, P, TOC, and pesticides in the Dianshan Lake sediments pay attention

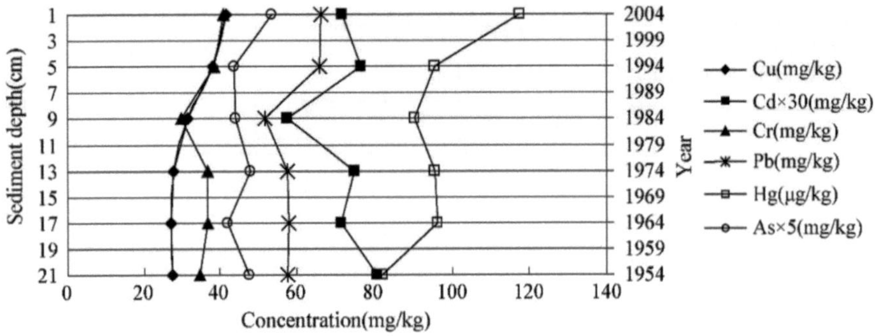

Fig. 3.75 Distribution of the trace metals versus the sediment depth from the Dianshan Lake

close with Shanghai urban development history and present situation in this study for well understanding the pollution history and causes. From the changes and coupling relationships of these substances in the sediment, which are closely with correlations in the production, it can be inverted out what is the contribution of industry, agriculture, and human daily life in the urban development to the water environment. In addition, it can well understand the actual process of environmental pollution, the pollution reduction from the environmental protection means of Shanghai in the development of the urban.

1. **The changes of trace metals in the sediment of the Dianshan Lake**

The average concentrations of the trace metals from five sampling sites are, respectively, used to show the concentration distributions for the metals in the sediment of the Dianshan Lake. The change features of the trace metals versus the depth of the sediment are shown as Fig. 3.75.

In several major lakes in the middle and lower areas of the Yangtze River basin, the average deposition rate of Chaohu Lake is 0.25 cm/a from ^{210}Pb CIC mode dating system and 0.27 cm/a from ^{137}Cs mode system [104]. The average deposition rate of Honghu Lake is 0.14–0.17 cm/a from ^{137}Cs mode dating system. The average deposition rate of the Taihu Lake is 0.41 cm/a from ^{210}Pb CIC mode dating system, and 0.34 cm/a from ^{137}Cs mode dating system [104]. The water of the Dianshan Lake mainly comes from the upstream of Taihu Lake, and the deposition rate of the Taihu Lake was similar to the cases of the Dianshan Lake. Therefore, the Dianshan Lake's deposition rate could be about 0.4 cm/a as that of the Taihu Lake [104], and this value is generally response to the data of the depositing rate of eastern China's lake cases, 0.1–0.5 cm/a [164]. So, according to these data, the sediment core used in this study can reflect about 60 years depositing history for the Dianshan Lake. The change of trace metals in the Dianshan Lake has two sections with obvious variation trends. The first stage was from the early 80's last century with sharply increasing, which was related to the high-speed development of Shanghai as well as the Yangtze River Delta area in the period. The second stage was from the New China establishing (1949) to the early years after the major reforms of China's economy system with the increasing and stable change developing trend, responded to the beginning of

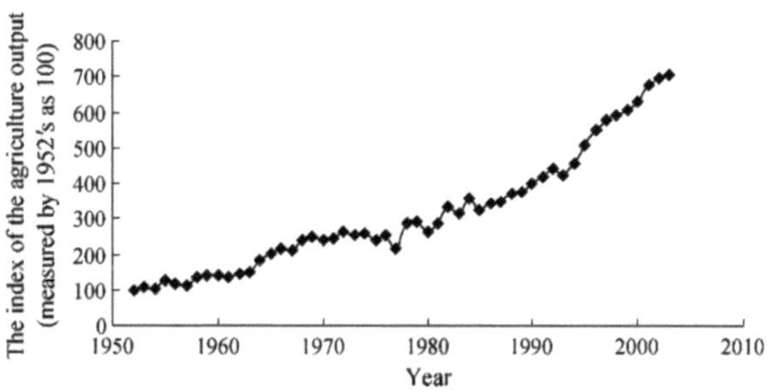

Fig. 3.76 Index of agriculture output value of Shanghai (measured by that of 1952 as 100)

production resuscitating. These results show that the trace metals in the Dianshan Lake are with great correlation with human activities around the lake, and the metals in the sediment are the adding results of them from the human production and daily activities.

2. **The change of the nitrogen in the sediment of the Dianshan Lake**

In this study, the nitrogen evolution in the sediment of the Dianshan Lake and human activities is investigated to discuss the relationship of human activity with the pollution by means of its concentration change in the two core samples took from the lake.

The nitrogen in the Dianshan Lake sediments is primarily from farmland's surface runoff and the surface runoff from upstreams of the lake coast. The nitrogen concentration is well associated with nitrogen fertilizer's application for agriculture around the lake. The feature of agriculture development in Shanghai is shown in Fig. 3.76 [165]. Before China economy system reform at the early of 1980s, the agriculture development in Shanghai is generally with a low increase speed. The total value of output of the agriculture in Shanghai is 2.1 times to that of 1952, and after the early of 1980s the agriculture development in Shanghai is going in a relative fast speed period. At the 2003, the total value of output of the agriculture reached the value as seven times to that of 1952. The requirement for the nitrogen fertilizer must be more and more under this situation, and the high-speed development of the agriculture would lead the N fast increase in water systems through every roads under the efficiency of N use no visibly improvement. The nitrogen change in the Dianshan Lake sediments versus time is shown in Fig. 3.77.

These data show that the N in the sediment of the Dianshan Lake is increasing relatively at a higher speed after 1960, and it is in good response with the agriculture feature at this period. Around 1970, the total value of output of the agriculture in Shanghai was with a little peak (Fig. 3.76), and the N in the Dianshan Lake sediment was with an obvious higher value around 1970 too (Fig. 3.76). From the late of 1970s to the early of 1980s, as the agriculture develops trend in the Shanghai the N in the sediment of the Dianshan Lake was with a decline trend, both with a very good correlations shown as Figs. 3.76 and 3.77. After the late of 1990s, the N in the sediment of the Dianshan Lake against the feature of agriculture trend is in

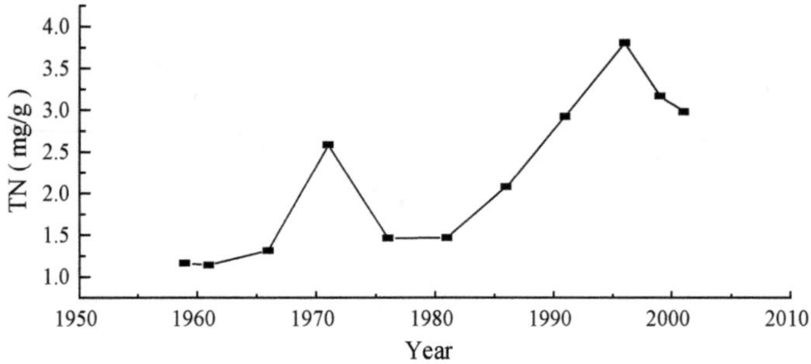

Fig. 3.77 Trend of TN variation as time in the sediment from the Dianshan Lake

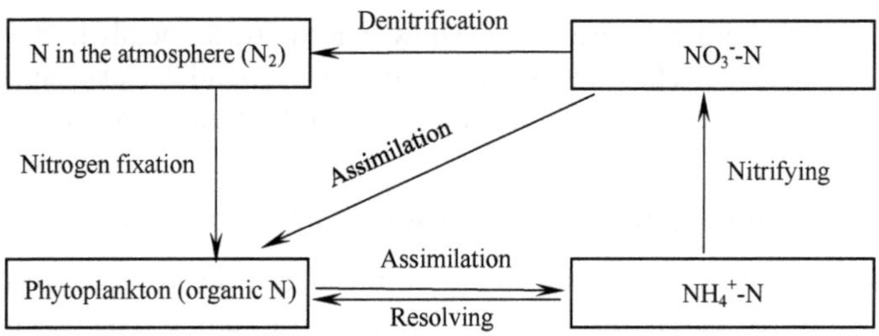

Fig. 3.78 Circle of nitrogen in lake water

an obvious variation of decrease, and it might be associated with the impacts of environmental protection efforts of Shanghai; the N in the Dianshan Lake is to be controlled efficiently.

In the sediment of the Dianshan Lake, the nitrogen is usually found in forms of N_2, NH_4^+, NO_3^-, NO_2^-, and organic nitrogen. Under aerobic conditions, the nitrogen in water undergoes nitration reaction, which is conducted by two clustering chemoautotrophic bacteria. Firstly, ammonium is oxidized to nitrite by nitrosomonas, and then the nitrite is oxidized to nitric acid by nitrobacter. In the anoxic conditions, the nitration reaction generally cannot happen. From microbial activities NO_3^-/NO_2^-, the denitrification occurs, and it restores the nitric acid to N_2. This process can reduce the nitrogen in the lake environment and is good for to limit the eutrophication of the lake water. This nitrogen's cycle in a lake is summarily shown in Fig. 3.78.

In the water area, the nitrogen is mainly found in the forms of nitric nitrogen and ammonium. Some phytoplankton, especially some special blue-green algae, can directly absorb the nitrogen from air and made it to solidify. After the algae death, the organic nitrogen in the algae would be decomposed to ammonia by bacteria presented in water, or broken down to ammonia nitrogen by phytoplankton, and then would be discharged into the environment [166].

Many studies have reported the transformation process about nitrogen under the effects of particulate matter in water. Zhang et al.'s work found that the nitration reaction on the interface of sediment and water was faster than that in water [167, 168].

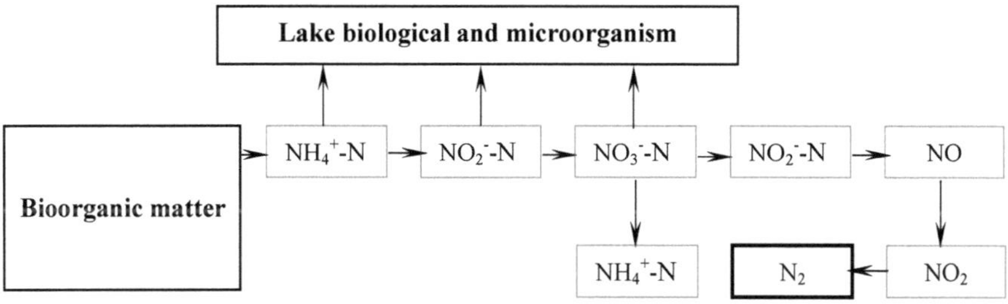

Fig. 3.79 Migration and transfer of nitrogen in the lake sediment

He et al. have studied the transformation process of organic nitrogen in the biofilm on sand, activated carbon, and zeolite, and finds that the reaction rate on the sand is minimum and zeolite maximum [169]. Xu et al. studied the characteristics about organic nitrogen mineralization in soil; their findings showed that the conversion speed of organic nitrogen on solid particles was more slower than that in water [170], and the absorption from the sediment to organic nitrogen increased its staying time in the environment [170]. However, Yu et al. discovered that the particulate matter in water could promote the nitrification of the ammonia nitrogen [171]. In Wang's research, it was reported that the nitrogen could trigger severe bio-geochemical reactions on interfaces of sediments and water [172]. The lake sediments are very important for the nitrogen in a lake as not only main source but also a confluence site. The sediments can accept nitrogen of depositing from the overlying water and the release from the particulate matter transported by water. The nitrogen in lake can also release under certain conditions to take part in its recycle [173]. Exchangeable nitrogen is an important source of nitrogen in biological activities in a lake, which can be directly absorbed by some primary producers in the water ecological systems. The fixed ammonium is generally as a buried form in mineral crystal, and it can only be released under some special conditions. The bio-geochemistry feature for the cycle of the nitrogen in the sediment of lake is schematic as in Fig. 3.79.

The Dianshan Lake is an important water source of Shanghai, and its environmental quality is becoming poor as more and more disturbing from human activities in recent years. The eutrophication sign in the Dianshan Lake had happened in 1985. The transparency of the Dianshan Lake is only 36.6 cm more less than that in 1994 (0.9–1.1 m) and 1951 (1.3–2.0 m), and according to the water quality monitoring data (GB3838-2002) the Dianshan Lake's total nitrogen (TN) level is inferior to the grade V, and the total phosphorus (TP) is between grade IV–V. The eutrophication will lead to the plankton blooms, some organisms, which are with resistant pollution ability such as snails and water earthworms will increase [174].

3. **The features of P and TOC in the sediment of the Dianshan Lake**

At low pH condition, the phosphorus in sediments would be released easily and rapidly, especially for the Ca–P, and it would be difficult to be release at high pH condition. In the sediments of a deep lake, the accumulation of acidic substances may be led due to being lack of oxygen here and will promote the release of Ca–P.

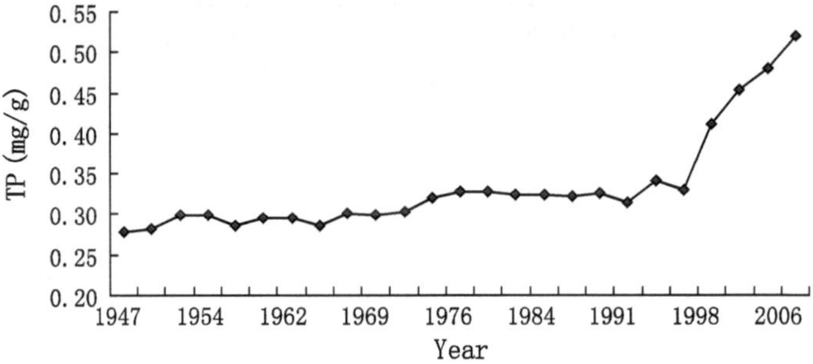

Fig. 3.80 Total agriculture production value of Shanghai; the rate of sediment calculated as 0.44 cm per year

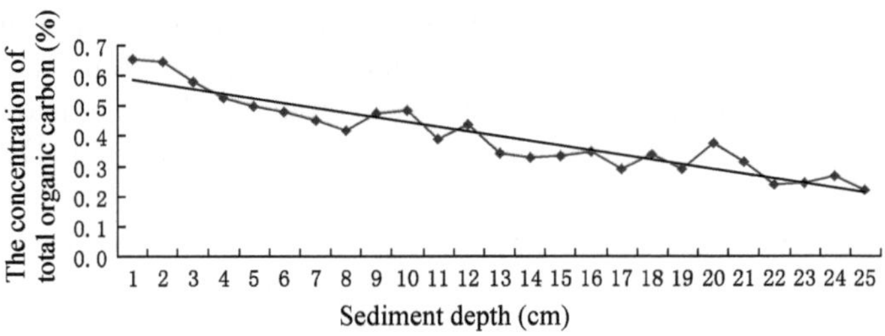

Fig. 3.81 Variation of the total organic carbon in the sediment of the Dianshan Lake, Shanghai

However, from the distribution characteristics of pH of the Dianshan Lake sediments, this accumulation of acidic substances is not obvious. The stability of the phosphorus in sediments is well when the pH value of the system was between 5.5 and 8.5 [147]. Therefore, the stability of the sediment phosphorus in the Dianshan Lake seems not to be affected by the pH in the 0–25-cm sediment in the Dianshan Lake.

The phosphorus sources of lake are commonly from some aspects, such as the wastes that come from overland runoff, farmland drainage, the lake's tourism activities, the baits placed by human, industrial sewage, municipal wastewater [175, 176]. The lake water phosphorus overload problem caused by anthropogenic pollutants is mainly as a focused goal in this study. The Dianshan Lake is greatly influenced by human activities, and the addition of phosphorus by artificial drainage is far more than that by the natural process, which has greatly affected the ecosystem of the lake. The concentration changes of P and TOC in the Dianshan Lake sediments are shown, respectively, in Figs. 3.80 and 3.81.

Many researchers have believed that extensive use of industrial and household detergents is also one of the biggest phosphorus pollution sources in lakes. In the agricultural production, the use of chemical fertilizers and pesticides and plastic mulches is one of the main sources of phosphorus pollution.

From the early New China established (1950s) to the China system's reform (1980s), the growth rate of Shanghai industrial total value output was relatively slow, and the industrial GDP in 1978 was 12 times to that of 1952. After the China

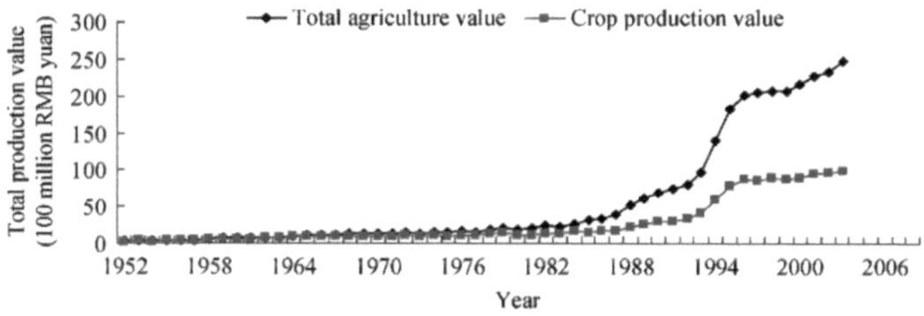

Fig. 3.82 Total agriculture production value versus time in Shanghai

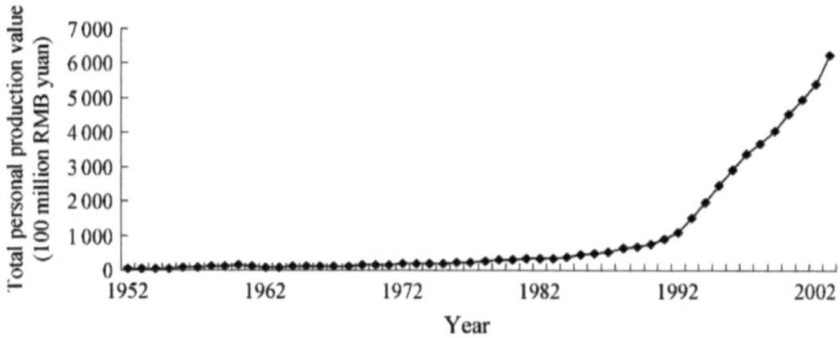

Fig. 3.83 Personal production value versus time in Shanghai

economy system reform and industrial revolution, the industrial GDP growth had been in relatively rapid speed. The GDP in 2003 was 14 times to that of 1978.

Before the China economy system reforms, the growth rate of agricultural GDP was relatively slow, and the agricultural GDP in 1977 was 2.1 times to that of 1952. After economy system reform, the agricultural GDP in 2003 was 7.0 times to that of 1952 as mentioned in front. The agricultural output and per capita output of Shanghai are shown, respectively, in Figs. 3.82 and 3.83 [165].

In contrast, the data of Figs. 3.80, 3.81, 3.82, and 3.83, the change trend of P and TOC in the Dianshan Lake sediments as time, are similar to the variation tendency of Shanghai industrial GDP, agricultural GDP, and Per capita output. These obvious coupling phenomena suggest that there is a clear correlation between phosphorus pollution of the Dianshan Lake, sediment accumulation of phosphorus element, eutrophication process, and the development of industry and agriculture in Shanghai [165].

4. **The changes of organochlorine pesticides in the sediment of the Dianshan Lake**

In 1950s, the organochlorine pesticides began to be used in Shanghai, and the use amount went on increasing until 1970s. The ban on organochlorine pesticide use issued in 1980s by China Government, and after the ban on the organochlorine pesticide sale and usage, they in the lake sediment and water came mainly from the residual in the sediment. Therefore, the change trend of HCHs and DDTs in the sediment of the Dianshan Lake with time is generally similar. From 1950s to 1980s,

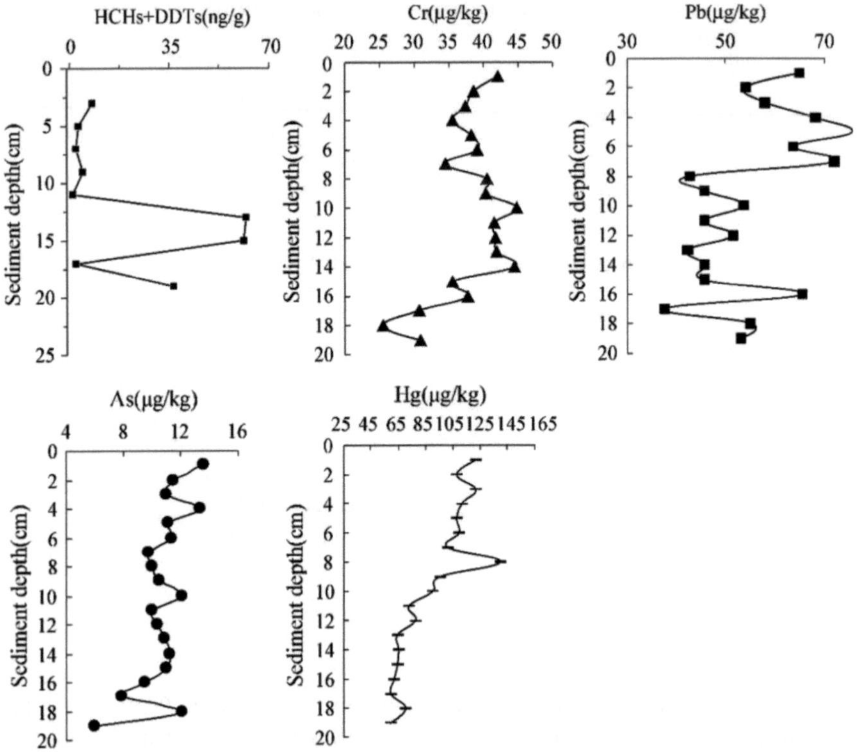

Fig. 3.84 Distribution of HCHs + DDTs, Cu, Cd, Cr, Pb, Hg, and as in the sediment of the Dianshan Lake

the residuals of the HCHs and DDTs in the sediments were very high. Due to its hysteresis effect in the sediment, the reflection of higher organochlorine pesticide values was shown obviously in middle 1970s, when was after using in great amount in early 1970s. The residual in the sediment began to decline after 1980s. So far, the residual of HCHs and DDTs in the sediment of the Dianshan Lake was very less.

For HCHs, the residual of α-HCHs are increasing with the increase of the sediment depth, which is contrary to δ-HCHs's feature. This may be related to property of the δ-HCHs that is easy degradable than that of α-HCHs in environment. For all of HCHs, the concentration of β-HCHs is relatively higher, which is corresponding to the β-HCH properties stable in nature, low-saturated vapor pressure, and difficult to be degraded.

The coupling correlations between the organochlorine pesticides and the changes of the trace metals are shown in Fig. 3.84.

The relationship between residual trend of HCHs + DDT and change trend of N, P is shown in Figs. 3.85 and 3.86. As shown in Figs. 3.85 and 3.86, the trend of HCHs + DDT residual in the sediment below 12 cm is similar with TN. As mentioned above, the developments of agriculture in 1970s were fast in Shanghai, and the

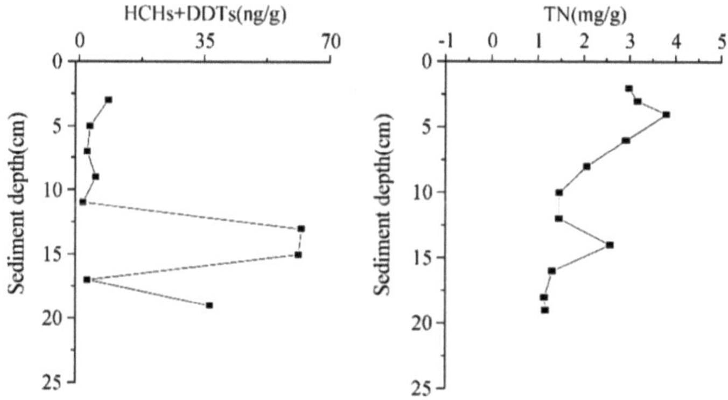

Fig. 3.85 Distribution of HCHs + DDTs and TN in the sediment of the Dianshan Lake

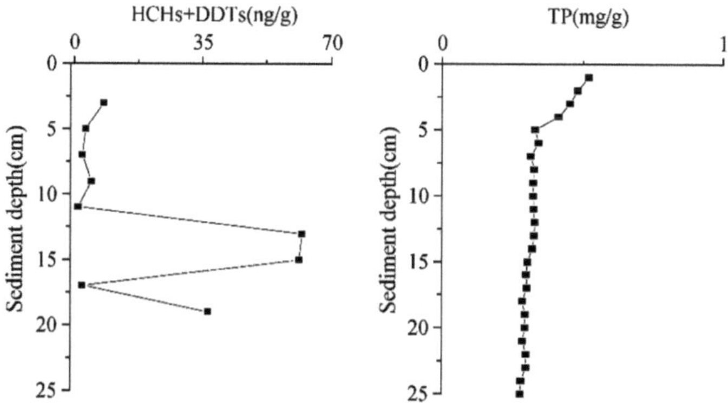

Fig. 3.86 Distribution of HCHs + DDTs and TP in the sediment of the Dianshan Lake

concentration of organochlorine pesticide, TN, TP in the sediments of the Dianshan Lake were higher than that above 12 cm on the sediment due to the usage of pesticides and fertilizers in the period. Corresponding the China ban for the organochlorine pesticide use at 1980s, the residual of HCHs + DDT in the sediment began to decline; however, the TN and TP continued to increase in this period. As shown in Fig. 3.86, the concentration of the P in the sediment is with a turning point at the depth of 5 cm and begins sharply to increase with reversing stably increasing trend before. These might be caused by the scale extension, as time, of the breeding industry in the lake and the wash care products from human daily life around the Dianshan Lake.

The coupling relationship of the pollutants in the sediment reveals, at a certain extent, the pollution process and possible cause of formation of the Dianshan Lake. The 1970s was the beginning period of the economic system reform in China. At this period, the GDP began to increase, and the development of agriculture along with the slathering use of pesticides and fertilizers could cause an increase of organochlorine pesticide, TN, TP in the sediments of the Dianshan Lake. The development of industry must cause heavy metals release to water, which was the main cause of formation for the continual increasing of trace metals in the sediment (above 12 cm) of the Dianshan Lake. After 1980s, with the rapid development in economy in Shanghai, the concentration of TN and trace metals in Lake and sediment increases rapidly, and corresponding the China ban for the organochlorine pesticide use the concentration of organochlorine pesticides in the sediment is contrary to increase in decline change trend.

5. **Discussion on the significance of the coupling correlations among trace metals, N, P, TOC, and the pesticides in the sediment of the Dianshan Lake**

The pesticides human used have no natural source and all of them in the natural systems come from human's activities, especially from agricultural activities. The sources of nitrogen and phosphorus can be varied in natural or artificial processes, respectively, and at the present, their rapid increase might be related mainly to the human's agricultural activities. The extra trace metals in natural system are accepted mainly from industrial production, and the pollution of them in lake is mainly from human's daily life activities as well as the industry activities around the lake. From the data above, the concentrations of the trace metal, the nitrogen, and the phosphorus in the Dianshan Lake increase along with the time, primarily because of human's activities. Therefore, the trace metals, nitrogen, phosphorus, TOC, and pesticides stored in the sediment of the Dianshan Lake are related to its discharge and dissemination characteristics in the social and economic development, the amount of pollutant discharge and the means and potency dimension of environmental protection. In the factors of the pollutant increasing in the Dianshan Lake, environmental protection means are different from the other factors; they inhibit the increase of pollution concentration in the sediments. From the interrelations between pollutant concentration changes in the Dianshan Lake and the factors of human's activities, the understanding and conclusion about the pollution formation and some revelations for the processes in the lake may be summarized out with discussion.

During the period before People's Republic of China establishing and the period after the establishment of PRC from 1949 with planned economy system, the industrial development level of Shanghai generally took the leading role in China. As the biggest city of industry, commerce, and the economic center, Shanghai was at a unique position then in China not only in industry and physical distribution, but also in domestic and foreign trades. Since the establishment of PRC, the economy in Shanghai has developed comprehensively and rapidly. To 2006, the gross domestic product had increased to 500 times of that in 1950, from 2.028 billion RMB (1950) to 1036.637 billion RMB (2006) (Fig. 3.87) [165]. Especially, since reform of China economic system from 1980s, the GDP has increased drastically. Comparison within the changes of trace metals, nitrogen, phosphorus, and TOC concentrations in the sediments of the Dianshan Lake and the GDP growth situation, it can be inferred that the trace metals, the nitrogen, the phosphorus, and the TOC concentrations in the sediments of the Dianshan Lake, are closely related to the GDP growth. From early days of foundation of PRC in 1949 to the economy system reform in 1980s, the concentrations of the trace metals, the nitrogen, the phosphorus, and the TOC

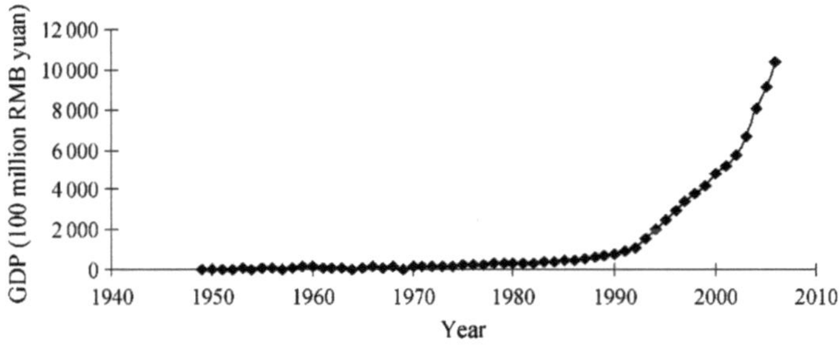

Fig. 3.87 GDP increasing versus time in Shanghai

in the sediment increased slowly. In contrast, since the economy started booming in 1980s, the corresponding concentrations of the pollutants, trace metal, nitrogen, phosphorus, and TOC all have increased rapidly with time.

The industrial production capacity and product output in Shanghai accounted for a large proportion of China's output, the industrial structure of old Shanghai before 1949 spontaneously formed the sequence of light industry, textile industry, and heavy industry. After the establishment of PRC, especially after 1953, China Government, proposed heavy industry as a priority developing category in China, and Shanghai developed the heavy industry simultaneously along with the precedence of development of light textile industry. From 1953 to 1957, the steel industry, mechanical and electronic industry, shipbuilding industry, chemical industry, and building material industry all in Shanghai developed rapidly. Compared to that of 1952, the light industry's total output in Shanghai in 1957 increased 105.5%. From 1958 to 1965, Shanghai total industrial output value transferred from large growth during the great leap forward to a relatively stable growth state. Between 1966 and 1976, Shanghai industry suffered huge losses due to the disturbances of the political movement then, and it still made progress generally in a lower degree. Calculated on the current prices then, the output value of Shanghai industry reached 25.959 billion RMB in 1966, and 42.345 billion RMB in 1976. In the ten years, it had increased 16.386 billion RMB with the growth rate 63.12%. Calculated on the industrial exponent, the growth rate in the ten years was 103.0%. During the economic system reform in the early 1980s, because of the central government policy, the southeastern China developed rapidly, which led to a marginalization crisis of Shanghai industry.

Since the mid-1990s, with the development of Pudong and the reduction of the proportion of financial transfer payment, Shanghai's industry has been revitalized, and Shanghai's industrial output value accounts for one tenth of the country's total, mainly contributing from the fields of light textile, heavy industry, metallurgy, petrochemical, mechanical and electronic industries, as well as automotive, aviation, aerospace, etc. In the 1990s, due to the increasing cost of land and labor, coupled with economic policy orientation, low-end manufacturing and labor-intensive industries such as textile and heavy pollution industries began to move to the suburbs or to be closed. Automobile manufacturing, telecommunication equipment manufacturing, power plant

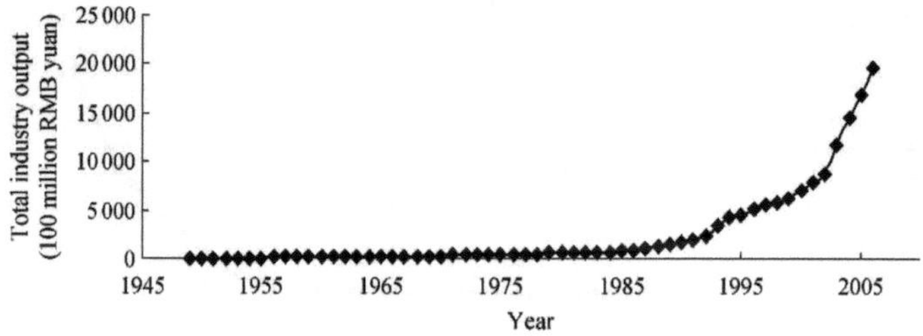

Fig. 3.88 Total industrial output versus time in Shanghai

equipment manufacturing, petrochemical and fine chemical industry, steel industry, and household appliances manufacturing had became the pillar industries of Shanghai's rapid development in the 1990s. Since then, the overall industrial structure of Shanghai had made remarkable progress and has formed and gradually consolidated its core competitiveness in China and worldwide. The above development history has been recorded in the sediments of the Dianshan Lake from the perspective of pollutants.

The gross output of Shanghai industry in recent years is shown in Fig. 3.88. Corresponding to the above-mentioned Shanghai social and economic development features, the value of gross output of Shanghai industry increased slowly from the early 1950s to economy system reform of China (1980s). The total industrial output value in 1978 was 16 times higher than in 1949. After the economy system reform, the total industrial output value of Shanghai increased quickly, and the value in 2006 was 36 times higher than in 1978 [165], which was in accordance well with the increasing trend of the pollutant (trace metals, nitrogen, phosphorus, and TOC) concentrations in the sediments of the Dianshan Lake. The trace metals, nitrogen, phosphorus, and TOC concentrations in the sediment core under 9 cm (corresponding to the period before the mid of 1980s) increase slowly, which are related to the low speed of the industry development before economy system reform and the discharge of the industrial pollutants. Nevertheless, the trace metals, nitrogen, phosphorus, and TOC concentrations in the sediment core between 0 and 9 cm (corresponding to the period after the mid of 1980s) increase quickly, which are related to the rapid increase in industry development after the economy system reform and the discharge of industrial pollutants [177, 178]. The anthropogenic sources of trace metals in the environment are mainly industry. In terms of heavy metals, the types of them emit by various industries in Shanghai, which are almost associated with the every process in the industry, are summarized in Table 3.66. The change of the pollutants in the sediments of the Dianshan Lake is related obviously to the industrial gross output values, which signifies the fact case in point to that the pollution in the Dianshan Lake is resulted from the causes of the human activities.

With the rapid development of industry, the discharge of industrial wastes accordingly increases. The discharge situation of the industrial wastewater, waste gas, and solid wastes in 1990 s is, respectively, presented as in Figs. 3.89, 3.90, and 3.91[165].

Table 3.66 Trace metals discharged from the main branches of the industry [9]

Branches of industry	Trace metals						
	As	Cd	Cr	Cu	Hg	Pb	Zn
Mining and beneficiation	√	√			√	√	
Metallurgy, electroplate	√	√	√	√	√	√	√
Chemical engineering	√	√	√	√	√	√	√
Tincturing	√	√		√		√	
Ink manufacture				√	√		
Ceramics	√		√				
Coating			√			√	√
Photography		√	√			√	
Glass	√						
Paper manufacture			√	√	√	√	
Tanning	√		√	√	√		√
Pharmacy				√	√		
Spinning	√	√		√	√	√	
Fertilizer	√	√	√	√	√	√	√
Chlor-alkali manufacture	√	√	√	√	√	√	√
Petroleum refining	√	√	√	√		√	√

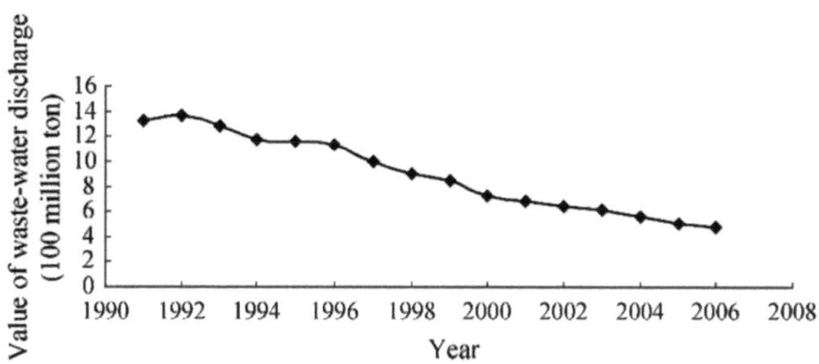

Fig. 3.89 Industrial wastewater discharge versus time in Shanghai

In 1991, the discharged amount of industrial waste gas reached 400 billion m^3, and the value of it was with increased trend constantly until recent years. By 2006, it had reached 942.8 billion m^3/a. The discharge amount of the industrial solid waste also increased constantly, and from 2001 to 2006, it increased from 13.5474–20.6319 million t/a. The discharge amount of the industrial wastewater started to decrease in 1990s, from 1.325 billion t/a (1991) to 0.483 billion t/a (2006), which was obviously related to the readjustment of industrial structure and the enhanced potency on the environmental protection in Shanghai [165].

From the fact that the trace metal concentrations in the sediments of the Dianshan Lake increased in last 20 years (Fig. 3.75), it can be inferred that the trace met-

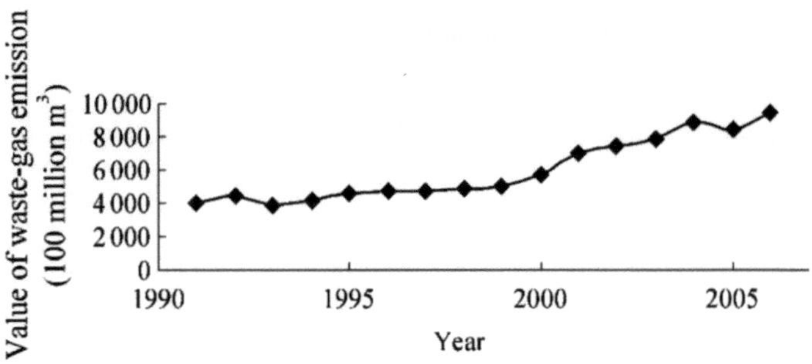

Fig. 3.90 Waste gas emission from industry versus time in Shanghai

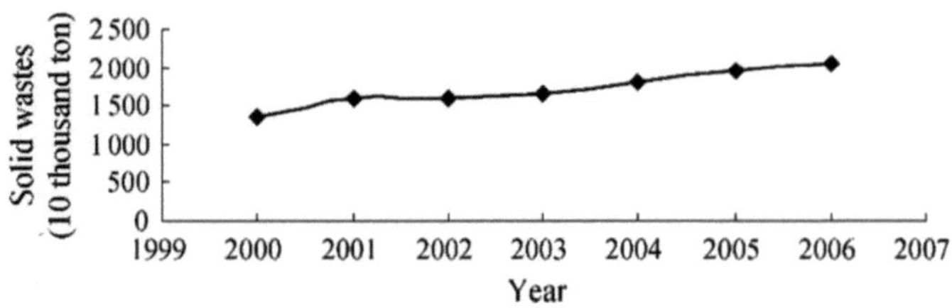

Fig. 3.91 Solid wastes from industry versus time in Shanghai

als, nitrogen, phosphorus, and TOC concentrations entering the environment of the Dianshan Lake may be still in an increasing trend. Though the discharge amount of industrial wastewater decreasing continuously, the total amount of pollutants entering environment is still on the rise, which is a very serious concern.

The urbanization degree of Shanghai is in fast speed today. The proportion of primary industry is comparatively lower, but the long-term use of fertilizer, pesticides, and the mulching film can be in a result in the increase in the pollution of trace metals, nitrogen, phosphorus, and TOC in soil. The large dose of chemical fertilizers is the main reason of nitrogen and phosphorus pollution in Shanghai. In addition, organic fertilizer is the main source of trace metals, and the trace metals may mainly come from the fodder additives. With the development of modern graziery, fodder additives are used widely, which may contain a certain amount of trace metals. The researcher has reported that during 16 years of continuous use of organic fertilizer, the copper and chromium concentrations in the local soil vary from 43.7 to 253.4 μg/g for Cu, and from 2.7 to 104.0 μg/g for Cr [179]. In addition, some pesticide includes trace metals, such as Hg, As, Cu, and so on, which is an important source of trace metals into the ecosystem. For example, the fungal pesticides, which are largely used for fruiter and greenhouse crops often, contain Cu, which may lead Cu to be accumulated in soil with long time use. When mulching films are produced, the heat stabilizer containing Cd and Pb can add these elements into the films, and from large-scale use of these films together with pesticides and fertilizers, the trace metals, nitrogen, phosphorus, and TOC will increase in soil [179]. Moreover, as previously mentioned, Shanghai's total agricultural output value increases as time; however,

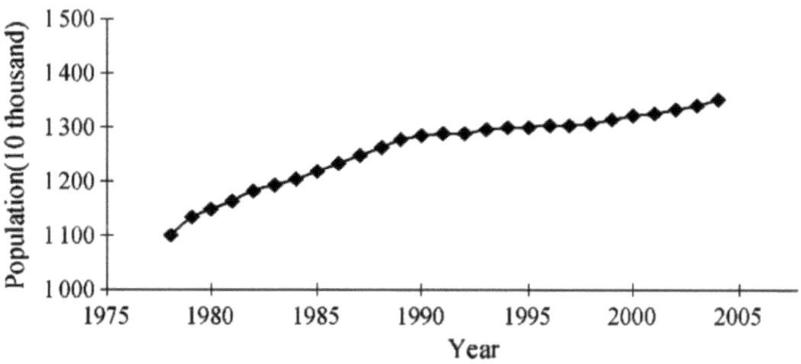

Fig. 3.92 Growth of the population versus time in Shanghai

from 1949 PRC establishing to the national economy system reform in China, the increasing speed was slow and the value in 1977 was only 2.1 times of it in 1952. After the economy system reform, the economy increasing speed has been very rapid, and the total agricultural output value in 2007 was 7.0 times to that of in 1952. This change in increased trend of the total agricultural output value is obviously related to the increase in the concentrations of trace metals, nitrogen, phosphorus, and TOC in the sediments of the Dianshan Lake.

With the constant economic development and increasing, population has entered to Shanghai from all parts of the country. The population growth situation in Shanghai in recent 20 years is shown as Fig. 3.92. In 1978, Shanghai population registered was 10.9828 million, and permanent residents were 11.04 million. In 2004, the population registered increased by 13.5239 million, adding close to 23%; in 2013, the permanent residents were 13.5239 million, increasing 0.1311 million people in 35 years; the growth rate was 118.8%. The continuous increase in population inevitably leads to the wastes from life and production increasing due to large consumption from the pollution increase. To comparing the change in trace metals, nitrogen, phosphorus, and TOC concentrations with the increasing population, it is found that the increasing trend of population is related well to the change s of the concentrations of trace metal, nitrogen, phosphorus, and TOC in the sediments of the Dianshan Lake.

To improve Shanghai environmental quality, Shanghai has formulated and implemented "The plan of fifteen years for Shanghai environmental protection," which is aimed to manage the water and atmospheric environment and control the total discharge amounts of some main pollutants. Water environment protection is focused on improving water quality and along with improving the drainage. Atmospheric environment protection is focused on clean energy replacement, vehicle pollution control, power plant desulfurization, and comprehensive improvement of key industrial regions, as well as to improve the ability for urban environmental management. According to national policies and environmental protection situations, Shanghai manages to control the total discharge of main pollutants, such as pesticides, SO_2, soot and dust, COD, ammonia nitrogen, and industrial solid wastes.

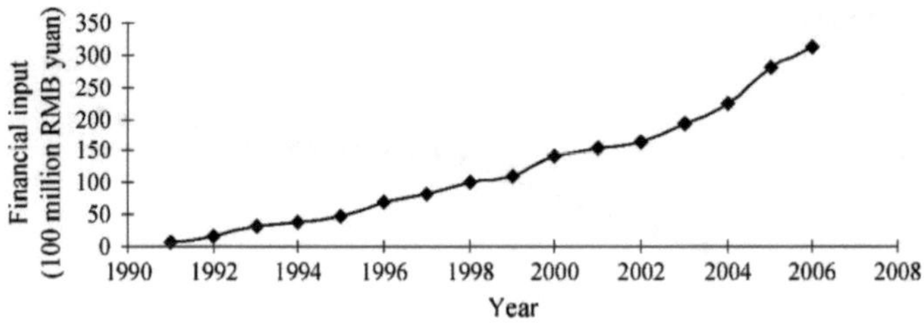

Fig. 3.93 Financial input for environmental protection versus time in Shanghai

Fig. 3.94 Compliance rate for the industrial wastewater discharging versus time in Shanghai

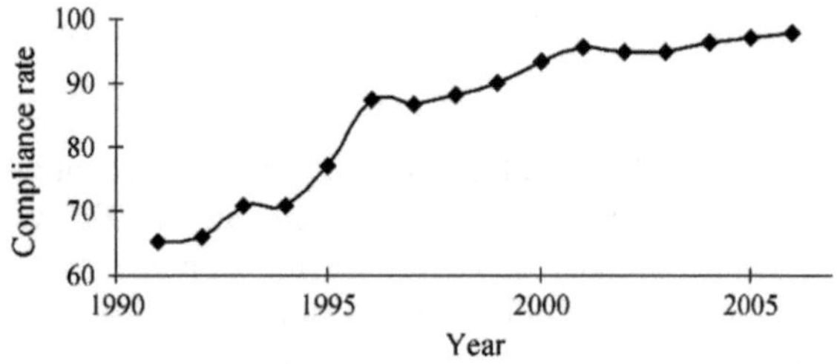

The devoted potency for the environmental protection in Shanghai from 1990 to 2006 is shown as in Fig. 3.93 [165].

In 1991, Shanghai invested 0.76 billion RMB in environment protection affairs, while in 2006, it reached to 31.085 billion RMB. Environmental protection investment accounted for 0.90% of the GDP in 1991 and increased to 3.00% of that in 2006. With the environment protection regulation strengthened constantly, the discharge amount of industrial wastewater obviously decreases on yearly basis (Fig. 3.89). The control potency of industrial wastewater discharge also increases rapidly, and by 2006 the all discharged wastewater in Shanghai to conform to the relative state criterion had reached to 97.61%, Fig. 3.94 [165], which is the evident too as the pollutants decreased in the Dianshan Lake. Nevertheless, the important aspect is that the trace metals, nitrogen, phosphorus, and TOC concentrations in the sediments of the Dianshan Lake between 0 and 5 cm increase still fast.

The situation of the pollutants in the Dianshan Lake as mentioned above may be related to the following two factors, firstly, the trace metals, nitrogen, phosphorus, and TOC concentrations in the sediments from wastewater discharge probably decrease, but the pollutants from atmospheric transmission, direct solid waste emissions, and sporadic point-source emission do not downright decrease. Secondly, because of the hysteresis effect of records retained in the sediment, even though the pollution emission reduction may be decreased from the inputs, the trace metals, nitrogen, phosphorus, and TOC concentrations in water and sediment still cannot decrease synchronously. There are many precedents in the world, for example, the Hg concentration in the sediments in Yatsushiro in Japan did not decrease substantially after the pollution sources of mercury emission were closed 30 years [180]. Because of the pollution of the mercury mine in Trieste Italy, which ranged number second in the world, the serious environmental impacts as well as the industrial and domestic wastewater led to the water quality to be serious deterioration. After 10 years when the mercury mines were closed there, the Hg concentration in the sediments and river water was still high and did not decrease as expected [181]. In addition, the complex physical and chemical factors between sediment and water trigger may cause the trace metals, nitrogen, phosphorus, and TOC pollutants in the sediment in migration or transformation. Even with the release of the pollutants existed in stable states in the sediment, it may arise that the pollutants enter the water again. Based on these, the trace metals, nitrogen, phosphorus, and TOC in the Dianshan Lake may not decrease significantly in a short term.

In conclusion, the trends of industrial, agricultural, and socio-economic development in Shanghai and the increasing population are directly influencing the increasing trend of the trace metals, nitrogen, phosphorus, and TOC concentrations in the sediments of the Dianshan Lake. After the economy system reform of China, the industry, agriculture, and social economic development increases quickly and the population increases fast, and during this period, the trace metals, nitrogen, phosphorus, and TOC concentrations in the sediment also increase rapidly.

These facts indicate that the main source of pollutants is human activities. The change in the concentrations in the sediments of the trace metals, nitrogen, phosphorus, and TOC in the Dianshan Lake is obviously affected by human's activities, and the abnormal contribution of these pollutants has disturbed the normal material exchange between water, sediment, and organisms in the nature system. The massive use of metals, organic matter, and artificial chemicals for industrial development and also huge amount of the pesticide and fertilizer use for increasing the crop yield will all finally lead to the pollutants accumulation in ecosystem, which can cause serious damage to humans and the environment later.

In recent years, the eligible discharge rate of industrial wastewater increases, and the direct discharge of the pollutants decreases with the continual increase of the investment in environmental protection. Nevertheless, a serious attention is the concentrations of the trace metals, nitrogen, phosphorus, and TOC in water of the Dianshan Lake no decreasing obviously. This may be affected by the complicated exchanges between water and sediments, and increasing total emission of various pollutants and the hysteresis effects in the process of depositing in water system may affect the feature of pollutant concentration in the system too. From above-mentioned situation, it is reasonable to believe that the trace metals, nitrogen, phosphorus, and TOC pollutants in the Dianshan Lake will not decrease significantly in a short term, and the ecological risk still needed to be evaluated on regular basis and should be timely supplemented and perfected with efficient environmental protection strategies.

References

1. Chen Y. The contamination characteristics of heavy metals and polychlorinated biphenyl in urban rivers: a case study on Suzhou Creek Shanghai (in Chinese with english abstract). Shanghai Jiaotong University. Shanghai Jiaotong University: Theses of School of Environmental Science and Engineering; 2007.
2. Tessier A, Campbell PGC, Bisson M. Sequential Extraction Procedure for the Speciation of Particulate Trace Metals. Anal Chem. 1979;51(7):843–50.
3. Zhang LC. Geochemistry on aquatic environment of the source region of the Yangtze River (in Chinese). Beijing: China Environmental Science Press; 1992.
4. Ni Q, Jiang JY, Ma ZD, et al. The heavy metal pollution status of the lake from the urban industry zone: a case study on Moshuihu lake Wuhan (in Chinese). Secur Environ Eng. 2005;12(1):13–6.
5. Liu C, Wang ZY, He Y, et al. Analysis of water quality of the estuary around Bohai Sea Bay. Environ Pollut Prev. 2003;25(4):222–5.
6. Qiu ZN, Zhang ZY, Qi DY. Distribution characteristics and correlation of pollutants in the bottom mud of the Suzhou Creek. Shanghai Environ Sci Chin. 1998;17(2):10–4.
7. Zhu SQ, Zang XP. Heavy metal pollution and its characteristics in main cities on the banks of the Yangtze River (in Chinese). People Yangtze River. 2001;7:23–6.
8. Bevtley GE. Analytical methods and problems Boca Raton. USA: CRC Press; 1989.
9. Chen JS, Deng BS, Cheng CQ, et al. Environ Geochem. Beijing: Ocean Press; 1990.
10. Ji WN, Zhang SB, Deng YE, et al. Background values and characteristics of 19 elements in main grain crops in Taihu Lake Basin. China Bull Environ Sci Chin. 1987;1:86–92.
11. Urban and rural construction and environmental protection ministry of people's Republic of China. Control standard of pollutants in the sludge used in agriculture (GB4284-84) 1984.
12. Zhang CL, Zhang S, Zhang LC, et al. Calculation method of heavy metal elements in river sediments of the water system of the Yangtze River (in Chinese). China Bull Environ Sci. 1995;15(3):257–64.
13. He J, Wang XW, Li CS, et al. Pollution characteristics of heavy metals in water sediment system of Baotou section of the Yellow River (in Chinese). China Bull Environ Sci. 2003;23(1):53–7.
14. Zhu GW, Chen YX, Zhou GD, et al. Distribution and variation of heavy metals in sediments of Hangzhou canal. China Environ Sci Chin. 2001;21(1):65–9.
15. Qu WC, Kelderman P. Heavy metal contents in the Delft canal sediments and suspended solids of the River Rhine: multivariate analysis for source tracing. Chemosphere. 2001;45:919–25.
16. Jia ZB, Liang T, Lin JZ. Heavy metal pollution and potential ecological risk in rivers of Hong Kong. Bull Beijing Univ (Nat Sci Ed Chin). 1997;33(4):485–92.
17. Ramesh R, Ramanathan AL, Ramesh S, et al. Distribution of rare earth elements and heavy metals in the surficial sediments of the Himalayan river system. Geochem J. 2000;34:295–319.
18. Awadallah RM, Soltan ME. Relationship between heavy metals in mud sediments and beach soil of the River Nile. Environ Int. 1996;22(2):253–8.
19. Osman AGM, Kloas W. Water quality and heavy metal monitoring in water, sediments, and tissues of the African catfish clarias gariepinus (Burchell, 1822) from the River Nile. Egypt J Environ Prot. 2010;1:389–400.
20. Daifullah AAM, Elewa AA, Shehata MB. Evaluation of some heavy metals in water, sediment and fish samples from River Nile (Kafr El-Zyat city), Egypt: a treatment approach. Afr J Environ Assess Manage. 2003;6:16–31.
21. Xu SY, Chen ZL, Yu SZ, et al. The bottom mud pollution of the Suzhou Creek and politics (in Chinese). Beijing: China Science Press; 2003.
22. Zhang BC. An estimation of transportation of some important pollution elements through the IRO River to the Yellow River (in Chinese). China Environ Sci. 1985;6(4):34–9.
23. Xue HB. Chemical stability and adsorption mode of heavy metals in water environment (in Chinese). China Environ Sci. 1985;4(3):9–20.
24. Sims J T, Kline J S. Chemical fractionation and plant uptake of heavy metals in soils amended with co-composted sewage sludge. Journal of Environmental Quality, 1991, 20: 387–395.

25. Lamy I, Bourgeviset S, Bermond A. Soil cadmium mobility as a consequence of sewage sludge disposal. J Environ Qual. 1993;22:731–7.
26. Allen HE, Hansen DJ. The importance of trace metal speciation to water quality criteria. Water Environ Res. 1996;68(1):42–54.
27. ASTM. Annual book of ASTM. 11(1), 1983.
28. Zhao J, Zheng XM, Bi CJ, et al. Characteristics of heavy metal pollution and its effect on river dredging in the outskirts of the Suzhou Creek (in Chinese). Agric Environ Prot. 2001;20(1):27–30.
29. Environmental Protection Bureau of Shanghai. Shanghai Environmental Bulletin 2005, 2006.
30. Muo BL. Element geochemistry. Beijing: Beijing University Press; 1999.
31. Chen RS, Huang YK, Gao XZ, et al. Heavy metal pollution in rivers. Beijing: China Environmental Science Press; 1988.
32. Balistrirei LS, Murry JW, Paul B. The cycling of Fe–Mn in the water column of Lake Sammamish, Washington. Limnol Oceanogr. 1992;37(3):510–28.
33. Balistrirei LS, Murry JW, Paul B. The biogeochemical cycling of trace metals in the water column of Lake Sammamish, Washingtion: response to seasonally anoxic condition. Limnol Oceanogr. 1992;37(3):529–48.
34. Davison W. Fe–Mn in lakes. Earth Sci Rev. 1993;34(2):119–63.
35. Muller G. Index of geo-accumulation in sediments of the Rhine River. Geojournal. 1969;2(3):108–18.
36. Forstner U, Muller G. Concentrations of heavy metals and polycyclic aromatic hydrocarbons in river sediments: geochemical background, man's influence and environmental impact. GeoJournal. 1981;5:417–32.
37. Forstner U, Ahlf W, Calmano W, et al. Sediment criteria development-contributions from environmental geochemistry to water quality management. In: Heling D, Rothe P, Forstner U, et al., editors. Sediments and environmental geochemistry: selected aspects and case histories. Berlin Heidelberg: Springer; 1990. p. 311–38.
38. Hakanson LARS. An ecological risk index for aquatic pollution control-A sedimentological approach. Water Res. 1980;14:975–1001.
39. Qiu YW, Wang ZD. Assessment of potential ecological risk of heavy metals in Dayawan sea area. Trop Ocean (Chin). 1997;16(4):49–53.
40. Chen HL, Chen YX. Progress in remediation technology of contaminated sediments. Agric Environ Prot (Chin). 2002;21(2):179–82.
41. He BG, Zhou NS, Yuan XM. Analysis on the second pollution of river confluence from sediment. Environ Poll Prot (Chin). 1999;21(3):41–3.
42. Liu ZQ. Study on the influence of the confluence (stage I) project on the water quality of the Suzhou Creek. Shanghai Municipal Eng (Chin). 1996;39(3):34–8.
43. Chen JS, Zhou JY. Study on heavy metals in water environment in China (Chin). Beijing: China Environmental Science Press; 1992.
44. Long ER, Macdonald DD, Smith SL, et al. Incidence of adverse biological effects within ranges of chemical concentration in marine and estuarine sediment. Environ Manage. 1995;19(1):81–97.
45. Macdonald DD, Dipinto LM, Field J, et al. Development and evaluation of consensus-based sediment effect concentrations for polychlorinated biphenyls. Environ Toxicol Chem. 2000;19(5):1403–13.
46. Zhou CG, Xu HZ, Ma Y-A, et al. Determination of PCBs in environmental samples by capillary GC-ECD. Ocean Environ Sci (Chin). 2000;19(4):57–61.
47. Kang YH, Sheng GY, Fu JM. Preliminary study on polychlorinated biphenyls in surface sediments of the Pearl River Delta. China Environ Sci (Chin). 2000;19(3):262–9.
48. Hong H, Xu LL, et al. Environmental fate and chemistry of organic pollutants in the sediment of Xiamen and Victoria Harbours. Mar Pollut Bull. 1995;31(4–12):229–36.
49. Camacho-Ibar VF, McEvoy J. Total PCBs in Liverpool bay sediments. Mar Environ Res. 1996;41:241–63.

50. Iwata H, Tanabe S, Sakai N, et al. Geographical distribution of persistent organo-chlorines in air, water and sediments from Asia and Ocean and their implication for global redistribution from lower latitudes. Environ Pollut. 1994;85:15–33.
51. Derek CGM, Alex O, Norbert PG, et al. Spatial trends and historical deposition of polychlorinated biphenyls in Canadian Midlatitude and Arctic lake sediments. Environ Sci Technol. 1996;30(12):3609–17.
52. Kang YH, Lin Z, Zhang G, et al. Distribution characteristics of chlorinated organic pollutants in sediments from estuaries of the Pearl River Delta and adjacent coastal areas. China Environ Sci (Chin). 2000;20(3):245–9.
53. Liu J, Wang WH, Wang ZJ. Types and contents of refractory pollutants in sediments of Second Songhua River. China Environ Sci (Chin). 1998;18(6):518–20.
54. Wu Y, Zhang J, Zhou Q. Persistent organochlorine residues in sediments from Chinese river estuary systems. Environ Pollut. 1999;105:143–50.
55. Chen YS, Wu GH, Huang XT. Analysis of environmental pollution in Suzhou Creek. Shanghai Environ Sci. 1997;16(1):1–14.
56. IVL Swedish Environmental Research Institute. HCB, PCB, PCDD and PCDF emissions from ships. Atmos Environ. 2005;39:4901–12.
57. Yu G, Niu JF, Huang J. Persistent organic pollutants. Beijing: China Science Press; 2005.
58. Office Shanghai Local Records. Shanghai yearbook 2006. Shanghai: Shanghai Yearbook Editorial Board; 2006.
59. Xu ZX, Yin HL. An analysis on the water environment simulation of boundary condition of the Huangpu River (in Chinese). Bull Tongji Univ. 2006;34(1):74–9.
60. Ding ZH, Wang WH. Distribution characteristics and ecological effects of mercury in several typical environments (in Chinese with English abstract). Theses of School of Environmental Science and Engineering: Shanghai Jiaotong University; 2003. p. 5.
61. Zhu Y. Distribution of Hg and its environmental effects in the Dianshan Lake and Huangpu River of Shanghai (in Chinese with English abstract). Theses of School of Environmental Science and Engineering: Shanghai Jiaotong University; 2007.
62. Shanghai Flood Control Information Center. Prediction of flood control level in Shanghai (http://service.shanghai.gov.cn/Weibo/WeiboIndex_1776324.html), 2006.
63. Environmental protection ministry of China. Standards for surface water environmental quality GB 3838–2002, 2002.
64. US EPA. Current National Recommended Water Quality Criteria. (http://www.epa.gov/waterscience/criteria/wqcriteria.html), 2006.
65. Tay KL, The SJ, Doe K, Lee K, Jackman P. Histopathological and histochemical biomarker responses of Baltic clam, Macoma balthica, to contaminated Sydney Harbor, Nova Scotia, Canada. Environ Health Perspect. 2003;111(3):273–80.
66. Ollivon D, Garban B, Blanchard M, Teil M, Carru JAM, Chesterikoff C, Chevreuil M. Vertical distribution and fate of trace metals and persistent organic pollutants in sediments of the Seine and Marne Rivers (France). Water Air Soil Pollut. 2002;134:57–79.
67. Shang YN, Ni SJ, Zhang CJ, Huang Y, Yin G. Heavy metal pollution and potential ecological risk assessment of river surface sediments in Chengdu. Ecol Environ (Chin). 2005;14(6):827–9.
68. Meng Y, Liu C, Cheng J. Geochemical characteristics of heavy metal elements in the surface sediments in the Yangtze River estuarine area and evaluations of the bed materials. Marine Geol Quat Geol. 2003;23(3):37–43.
69. US Geological Surve. Sediment Quality Chemical Criteria. (http://www.ecy.wa.gov/programs/tcp/smu/sed_chem.htm), 2006.
70. Xiu GL, Jin QX, Zhang DN, Shi SY, Huang XJ, Zhang WY, Bao L, Gao PT, Chen B. Characterization of size-fractionated particulate mercury in Shanghai ambient air. Atmos Environ. 2005;39(3):419–27.
71. Jaffe D, Prestbo E, Swartzendruber P, Weiss-Penzias P, Kato S, Takami A, Hatakeyama S, Kajii Y. Export of atmospheric mercury from Asia. Atmos Environ. 2005;39(17):3029–38.

72. Environmental Protection Bureau of Shanghai. Shanghai Environmental Bulletin 2003 (in Chinese), 2004.
73. Stuer-Lauridsen F. Review of passive accumulation devices for monitoring organic micropollutants in the aquatic environment. Environ Pollut. 2005;136(3):503–24.
74. Van Metre PC, Mahler BJ. Trends in hydrophobic organic contaminants in urban and reference lake sediments across the United States, 1970–2001. Environ Sci Technol. 2005;39(15):5567–74.
75. White JC, Triplett T. Polycyclic aromatic hydrocarbons (PAHs) in the sediments and fish of the Mill River, New Haven, Connecticut, USA. Bull Environ Contam Toxicol. 2002;68:104–10.
76. Fung CN, Zheng GJ, Connell DW, Zhang X, Wong HL, Giesy JP, Fang Z, Lam PKS. Risks posed by trace organic contaminants in coastal sediments in the Pearl River Delta, China. Mar Pollut Bull. 2005;50:1036–49.
77. Luo XJ, Chen SJ, Mai BX, Zeng YP, Sheng GY, Fu JM. Distribution and sources of polycyclic aromatic hydrocarbons in surface sediments of the Pearl River and Northern South China Sea. China Environ Sci. 2005;26(4):129–34.
78. Zhang H. The orientation of water quality variation from the Metropolis River—Huangpu river, Shanghai. Environ Monit Assess. 2007;127(1–3):429–34.
79. Tang SL. Research method for the natural background value on some elements in environment: The probability distribution of heavy metals in Nanjing region. In: The Proceeding of the workshop of Chinese Academy of Science on environmental background value (in Chinese), 1982. p. 9–15.
80. Yang XY. Research method for the natural background value on some elements in environment: the relationship between soil background value and host substances in Nanjing region. In: The Proceeding of the workshop of Chinese Academy of Science on environmental background value (in Chinese), 1982. p. 16–20.
81. Wang LJ, Zhang S, Zhang CS. The geochemical characteristics of rare earth elements in aquatic environment from middle reaches of the Yangtze River (in Chinese). China Environ Sci Bull. 1995;15(1):57–65.
82. Zhong LN. The depositing reports and significance of the trace metals in the sediment of the Dianshan Lake (in Chinese with English abstract). Shanghai Jiaotong University. Shanghai Jiaotong University: Theses of School of Environmental Science and Engineering; 2007.
83. Yuan XY, Wang AH, Xu NZ. The geochemical speciation and characteristics of heavy metals in the sediment of the Taihu Lake. China Geochem (Chin). 2004;33(11):611–9.
84. Yao ZG, Bao ZY, Gao P. The geochemistry of heavy metals in the sediment of the Dongting Lake. China Geochem (Chin). 2006;35(6):629–38.
85. Chen J, Li SF. The analysis for heavy metal total and speciation concentrations in surface sediment of the Chaohu Lake. Henan Sci (Chin). 2007;25(4):303–8.
86. The Environmental Protection Ministry of China. Soil Environmental Criterion-GB15618-1995, 1995.
87. Weisz M, Polyak K, Hlavay J. Fractionation of elements in sediment samples collected in rivers and harbors at Lake Balaton and its catchment area. Microchem J. 2000;67:207–17.
88. Liu XD, Xu Q, Ge XL. The research on the speciation of metal elements in the sediment of the Miyun reservoir. China Science D (Geosci). 2005;35:288–95.
89. Hlavay J, Polyak K. Chemical speciation of elements in sediment samples collected at lake balaton. Microchem J. 1998;58:281–90.
90. Polyak K, Hlavay J. Environmental mobility of trace metals in sediments collected in the Lake Balaton. Fresenius J Anal Chem. 1999;363:587–93.
91. Bilai LE, Rasmussen PE, Hall CE, et al. Role of sediment composition in trace metal distribution in lake sediments. Appl Geochem. 2002;17:1171–81.
92. Zhang L, Yuan XY, Deng X. The speciation and environmental significance of heavy metals in the sediment of the Xuanwu Lake, Nanjing. J Lake Sci (Chin). 2007;19(1):63–9.
93. Luo JF, Meng WQ, Xia ZL. The geographical differentiation of the speciation of Cd, Pb, and Cu in soil. China Geogr Res (Chin). 1998;17(3):265–72.

94. Morillo J, Usero J, Gracia I. Heavy metal distribution in marine sediments from the southwest coast of Spain. Chemosphere. 2005;55:431–42.
95. Tokalioğlu S, Kartal S, Elçi L. Determination of heavy metals and their speciation in lake sediments by flame atomic absorption spectrometry after a four-stage sequential extraction procedure. Anal Chim Acta. 2000;413:33–40.
96. Álvarez-Iglesias P, Rubio B, Vilas F. Pollution in intertidal sediments of San Simon Bay (Inner Ria de Vigo, NW of Spain): total heavy metal concentrations and speciation. Mar Pollut Bull. 2003;46:491–521.
97. Tessenow U, Baynes Y. Redox-dependent accumulation of Fe and Mn in a littoral sediment supporting Isoetes Iacustris. Naturwissenchaften. 1975;62:342–56.
98. Wang H, Wang CX, Wang ZJ. The speciation of heavy metals in the surface sediment of the Taihu Lake. China Environ Chem (Chin). 2002;21(5):430–5.
99. Mao MZ. Speciation of metals in sediments along the Le An River. CERP Final Report, France. Imprimerie Jouve Mayenne, 1996, p. 55–571.
100. He CM. The depositing symbol of lake eutrophication evolution. Theses of School of Environmental Science and Engineering, Shanghai Jiaotong University (in Chinese with English abstract). Shanghai Jiaotong University, 2009.
101. Soil Institute of Chinese Academy of Science. Physical and Chemical Analysis for Soil. Shanghai: Shanghai Science and Technology Press; 1978. p. 81–2.
102. The China General Bureau of Environmental Protection. Water quality determination of total nitrogen—alkaline potassium persulfate digestion UV spectrophotometry. China National Criteria-GB 11894–89, 1990.
103. The Editorial Board of Water and Wastewater Analysis Method, The China General Bureau of Environmental Protection. Water and Wastewater Monitoring and Analysis Method (fourth edition). Beijing: China Environmental Science Press, 2002. P. 266–267.
104. Yao SC, Xue B, Li SJ. The determination and significance of the depositing rate for the lake sediment from the middle and lower reaches of the Yangtze River: Case study on the Honghu Lake, Chaohu Lake, and Taihu Lake. Resour Environ Yangtze River Drainage Basin (Chin). 2006;15(9):569–73.
105. Sun YC, Wang Y, Qi RH. A research on the eutrophication and prophylaxis and treatment for the Diahshan lake (in Chinese). Shanghai: Eastern China Normal University Press; 1992. p. 23–156.
106. Schelske CL, Hodell DA. Recent changes in productivity and climate of Lake Ontario detected by isotopic analysis of sediments. Limnol Oceanogr. 1991;36:961–75.
107. Bernasconi SM, Barbieri A, Simona M. Carbon and nitrogen isotope variations in sedimenting organic matter in Lake Lugano. Limnol Oceanogr. 1997;42:1755–65.
108. Kaushal S, Binford MW. Relationship between C/N ratios of lake sediments, organic matter sources, and historical deforestation of Lake Pleasant, Massachusetts, USA. J Paleolimnol. 1999;22:439–42.
109. Meyers PA, Teranes JL. Sediment organic matter. In: Last WM, Smol JL, editors. Tracking environmental changes using lake sediments. Physical and geochemical methods, vol. 2. Dordrecht, The Netherlands: Kluwer-Academic, 2001, p. 239–269.
110. Meyers PA. Applications of organic geochemistry to paleolimnological reconstructions: a summary of examples from the Laurentian Great Lakes. Org Geochem. 2003;34:261–89.
111. Routh J, Meyers A, Gustafsson O, Baskaran M, Hallberg R, Scholdstrom A. Sedimentary geochemical record of human-induced environmental changes in the Lake Brunnsviken watershed, Sweden. Limnol Oceanogr. 2004;49:1560–9.
112. Wu JL, Gagan MK, Jiang X, Xia W, Wang S. Sedimentary geochemical evidence for recent eutrophication of Lake Chenghai, Yunnan, China. J Paleolimnol. 2004;32:85–94.
113. Schettler G, Qiang L, Mingram J, Negendank JFW. Palaeovariations in the East-Asian Monsoon regime geochemically recorded in varved sediments of Lake Sihailongwan (NE-China Jilin province), Part 1: hydrological condition and dust flux. J Paleolimnol. 2006;35:239–70.
114. Li JY. The concentration evolution of the eutrophication factor P in the Dianshan Lake Shanghai (in Chinese with English abstract). Shanghai Jiaotong University: Theses of School of Environmental Science and Engineering; 2008.

115. Sondergaard M, Windolf J, Jeppesen E. Phosphorus fraction and profiles in the sediment of shallow Danish lakes as related to phosphorus load, sediment composition and lake chemistry. Water Res. 1996;30(4):992–1002.
116. Reddy KR, Fisher MM, Ivanoff D. Re-suspension and diffusive flux of nitrogen and phosphorus in a hypereutrophic lake. J Environ Qual. 1996;25:363–71.
117. Petterson K, Fstvonavics V. Sediment phosphorus in Lake Balaton: forms and mobility. Arch Hydrobiol Beilh Ergenbn Limnol. 1988;30:25–41.
118. Lin ZX. Analysis for aquatic environment variation and cause of formation in Taihu Lake Basin. J Lake Sci (Chin). 2002;14(2):97–103.
119. Pettersson K, Amiard J, Kouzic B, et al. Mechanisms for internal loading of phosphorus in lakes. Hydrobiologia. 1998;373–374:21–5.
120. Ostrom PH, Ostrom NE, Henry J. Change in the tropic state of Lake Eric: discordance between molecular δ13C and bulk δ13C sedimentary records. Chem Geol. 1998;152(1–2):162–79.
121. Shu JH, Huang WY, Wu TG. Classification of trophic types of lakes in China (in Chinese). J Lake Sci. 1996;8(3):193–200.
122. Jin XC, Tu QY. Investigation specification for lake eutrophication(Second edition, in Chinese)). Beijing: China Environmental Science Press; 1990. p. 10–4.
123. National Environmental Protection Bureau of China. The Plan of Water Pollution Control and Project for "Three Rivers" and "Three Lakes". Beijing: China Environmental Science Press; 2000.
124. Liu JK. Ecological Studies in East Lake. Beijing: China Science Press; 1990.
125. Gu XH, Fan CX, Yang LY, et al. Status of biological resources in winter and evaluation of environmental quality and resource benefit in Gucheng Lake. J Lake Sci (Chin). 2002;14(3):285–8.
126. Weng HX. Phosphorus binding status in river sediments and its environmental geochemical significance (in Chinese). Bull Hydraulics. 1993;38(3):1219–22.
127. Psenner R, Pucskso M Sager. Fraktionierung organoishcher and anorganischer Phosphorverbindungen von Sedimenten. Versucheiner Definition Okologschwichtiger Fractionen, Arch Hydrobiol/Suppl, 1985, 70: 111–115.
128. Ruttenburg K. Development of a sequential extraction method for different forms of P in marine sediments. Limnol Oceanogr. 1992;37:1460–82.
129. Baldwin DS. The phosphorus composition of a diverse of Australian sediments. Hydrobiologia. 1996;335:63–73.
130. Li Y, Wu DN, Xue YX. Improvement of extraction methods for different speciation of phosphorus in sediments and their environmental geochemical significance (in Chinese). China Ocean Environ Sci. 1998;17(1):15–20.
131. Golterman HL. Fractionation of sediment phosphate with chelating compounds: a simplication and comparison with other methods. Hydrobiologia. 1996;335:87–95.
132. Golterman HL. Sediments as a source of phosphorus for algal growth. In: Golterman HL, editor. Interaction between sediments and fresh water. The Hague: Dr W Junk; 1997. p. 286–93.
133. Golterman HL, Booman A. The sequential extraction of Ca–and Fe–bound phosphates. Verh lnt Ver Limnol. 1988;23:904–9.
134. De GC, Golterman HL. Sequential fractionation of sediment phosphate. Hydrobiologia. 1990;192:143–8.
135. Pan CR, Li L, Ye LL, Zhang ZY. Analysis of occurrence forms of nitrogen and phosphorus in sediments of Wabu Lake. China Water Res Prot (Chin). 2007;23(4):10–4.
136. Rutenberg KCR, Berner A. Authigenic apatite formation and burial in sediments from non-upwelling, continental margin environments. Geochim Cosmochim Acta. 1993;57:991–1007.
137. Berner RA, Rao JL. Phosphorus in sediments of the Amazon River and estuary: implication for the global flux of phosphorus to the sea. Geochim Cosmochim Acta. 1994;58:2333–430.
138. Berner RA, Ruttenberg KC, Ingall ED, Rao JL. The nature of phosphorus burial in modern marine sediments: Interactions of C, N, P and S biogeochemical cycles and global change. In: Wollast R, Mackenzie FT, Chou L editors. New York: Springer, 1993. p. 14–521.
139. Khoshmanesh A, Hart BT, Duncan A. Luxury uptake of phosphorus by sediment bacteria. Water Res. 2002;36:774–8.

140. Han SH, Zhuang YH, Zhang HX, et al. Phosphine and methane generation by the addition of organic compounds containing carbon-phosphorus bonds into incubated soil. Chemosphere. 2002;49:651–7.
141. Fu YQ, Zhou YY. Fractionation of phosphorus forms in sediments and its ecological significance. J Lake Sci (Chin). 1999;11(4):376–3811.
142. Aminot A, Andrieux F. Concept and determination of exchangeable phosphate in aquatic sediment. Water Res. 1996;30(11):2805–11.
143. Jensen HS, Kristensen P, Jeppesen E, et al. Iron: phosphorus ratio in surface sediment as an indicator of phosphate release from aerobic sediments in shallow lakes. Hydrobiology. 1992;235:731–43.
144. Li J, Liu CQ, Wang SL, Zhu ZZ, Zhou ZH, Xiao HY. Distribution characteristics of different forms of phosphorus in surface sediments of Wulihu Lake and Taihu Lake. Mineral Bull. 2004;24(4):405–11.
145. Kastelan-macan M, Petrovic M. The role of fulvic acids in phosphorus sorption and release from mineral particles. Water Sci Technol. 1996;34:259–65.
146. Xu YQ. Research progress on phosphorus adsorption and release in sediment. Chonqing Science (Chin). 2003;25:11.
147. Li H. Release and stability of endogenous phosphorus in waters of Zhouzhuang, Shanghai. Shanghai Environ Sci. 2003;22(12):943–7.
148. Wang JQ. Experimental study on simulating phosphorus release from sediment in Chaohu Lake (in Chinese). China Environ Sci Bull. 2002;22:6.
149. Zhu GW. Important factors for phosphorus release from shallow lake sediments, iron and hydrodynamics. China Agric Environ Sci. 2003;22(8):762–4.
150. Yin DQ. Effects of environmental factors on phosphorus release from sediments of Wulihu Lake. J Lake Sci (Chin). 1994;6(3):241–4.
151. Weng S. Environmental Microbiology. Beijing: China Science Press; 1985. p. 64–8.
152. Macia H, Bates N, Neafus JE. Phosohorus release from sediments from Lake Carl Black well, Oklahoma. Water Res. 1980;14:1477–81.
153. Pomeroy LR, Smith EE, Grant CM. The exchange of phosphorus between estuarine water and sediments. Limnol Oceanogr. 1965;10(7):167–72.
154. Liao WG. Phosphorus load analysis for the waters of the Taihu Lake. China J Hydraul Eng. 1994;11:77–81 (in Chinese).
155. Zhou XD. Effects of environmental factors on phosphorus adsorption in sediments of DianchiLake. China J Hydraul Eng, 1998, 12–17 (in Chinese).
156. Yao Y. Effects of light on phosphorus release and phosphorus forms in lake sediments. Environ Sci Res, 2004, 17 (Supplement, in Chinese).
157. James LP. The Role of Nutrient Loading and Eutrophication in Estuarine Ecology. Environ Health Perspect. 2001;109:699–706.
158. Schlesinger WH. Evidence from chronosequence studies for a low carbon-storage potential of soils. Nature. 1990;348:232–4.
159. Fan CX. Estimation and distribution characteristics of sediment accumulation in Taihu Lake. Shanghai Environ Sci. 2000;19(2):72–5.
160. Zhang H. Pollution Ecology (in Chinese). Huhhot: Inner Mongolia University Press; 2000.
161. Jiang MC. Pollutant coupling characteristics and mechanism of lake pollution. Theses of School of Environmental Science and Engineering (in Chinese with English abstract), Shanghai Jiaotong University. Shanghai Jiaotong University, 2009.
162. Li WC. Multi steady state theory of shallow lake ecosystem and its application (in Chinese). J Lake Sci. 1997;9(2):97–104.
163. Qin BQ. Eutrophication mechanism and control approaches of shallow lakes in the middle and lower reaches of the Yangtze River (in Chinese). J Lake Sci. 2002;14(3):194–7.
164. Jin XC. Sediment Pollution Chemistry. Beijing: China Environmental Science Press; 1992.
165. Shanghai Statistics Bureau. Shanghai General Investigation Team of National Bureau of Statistics. Shanghai Statistical Yearbook. Beijing: China Statistics Press; 2006.

166. Jilianglongfu. Protection of Water Resources, Environmental Problems in Biwa Lake, In: Liu H, Chao FZ editors Translated in Chinese from Japanese. Beijing: China Environmental Science Press, 1989: 130–133.
167. Zhang XQ, Xia XH, Yang ZF. Effects of particulate matter on transformation of organic nitrogen (in Chinese). Environ Sci. 2007;28(9):1954–9.
168. Pauer JJ, Auer MT. Nitrification in the water column and sediment of a hypereutrophic lake and adjoining river system. Water Res. 2000;34(4):1247–54.
169. He J, Liu CF, Wu Y. Kinetics of nitrification in biofilms on three carriers. J Appl Environ Biol. 2003;9(5):546–8.
170. Xu XK, Zhou LK, Oswald VC. Effects of urease inhibitor/nitrification inhibitor on urea nitrogen transformation and speciation distribution in soil. Soil Bulletin. 2000;37(3):339–45.
171. Yu H, Zhang XQ, Xia XH, et al. Effects of particulate matter on nitrification process in the Yellow River (in Chinese). China Environ Sci Bull. 2004;24(4):601–6.
172. Wang SR, Jiao LX, Jin XC, Liu JH. Occurrence characteristics of total nitrogen, exchangeable nitrogen, and fixed ammonium in shallow lakes of the middle and lower reaches of the Yangtze River (in Chinese). China Environ Sci Bull. 2008;28(1):37–43.
173. Lu XX, Sung JM. The regional geochemical characteristics of N in the surficial sediment of Southern Yellow Sea China. Progress Ocean Sci. 2003;21(2):174–80 (in Chinese).
174. Ryding SO, Rast W. The Control of Eutrophication of Lakes and Reservoirs. Paris: UNESCO & Hie Parttienon Publishing Group; 1989.
175. Meng F, Yuan XS. The speciation analysis for P in natural water. China Environ Chem. 1987;6(3):29–32 (in Chinese).
176. Wang CY. The geochemistry and environmental effects of P in modern sediments. Chongqing Environ Sci. 2000;22(4):39–41 (in Chinese).
177. Zhang H. A 1955–2004 record of Hg contamination in Dianshan Lake sediments, Shanghai. Environ Chem Lett. 2011;9(4):479–84.
178. Zhang H, Zhang N, Yang Q, Zhong L, et al. The impact of anthropogenic activities recorded by mercury in the sediment from Dianshan Lake. Am J Environ Sci. 2012;8(4):473–8.
179. Wang JY. Study on heavy metal elements in the sediment of lakes of Chengdu. Dissertation of Chengdu polytechnic university (in Chinese), Chengdu Polytechnic University, 2005.
180. Tomiyasu T, Nagano A, Yonehara N, et al. Mercury contamination in the Yat Sushiro Sea, Southwestern Japan: spatial variations of mercury in sediment. Sci Total Environ. 2000;257:121–32.
181. Horvat M, Covelli S, Faganeli J, et al. Mercury in contaminated coastal environments; a case study: the Gulf of Trieste. Sci Total Environ. 1999;237(238):43–56.

Chapter 4
The Experimental Research on the Behaviors of Trace Metals

4.1 The Objectives, Research Idea, and the Approach Design for the Experiments

4.1.1 The Objectives of the Research

To understand the release and containing characteristics of active part of relevant elements from environmental medium and to reveal these impacts as the change of time and other environmental conditions, the desorption and adsorption experiments are designed to simulate the natural process for trace metals being released and detained from the environmental medium under the local natural conditions. So the migration and transform restricting factors of trace metals in the earth surface are found out, which can further understand and evaluate the environmental effects of trace metal contamination.

This study conducts simulation experiments about trace metal desorption from soil in the Hetao Area, sorption from soil in Hetao Area, and its release mechanism from some surface particles (rock and mineral), respectively.

4.1.2 The Approach Design for the Experiments

Based on the basic state of trace metal contamination in nature systems, the nature process for soil and rock on metal release and adsorption was simulated in the laboratory, which were combined with the main issues and specific experimental conditions of the laboratory. The main principles are as follows:

A. The objects used in the experiments were the natural medium (soil, rock, and airborne dust) from the cases of study areas.
B. Based on the specific conditions of the possible or the main occurring in the media of study area, experimental conditions were set in order to explore the

H. Zhang, *Behaviors of Trace Metals in Environment*,
https://doi.org/10.1007/978-981-13-3612-6_4

behavior rule of trace metals when the related condition was changed within certain range.

C. As an exploratory research, the analysis, interpretation, and fitting treatment to the mathematical model about this experimental data focused mainly on the revealing law reproducibility and high fitting degree (correlation), but would not be fettered by relevant theoretical models.
D. Experiments are tried to understand the most probable transformation of trace metals in the medium under natural conditions. The main research variables were the most basic, the most common, and the most direct impact factors for the research system to trace metals' environmental behavior, such as pH, time, and action temperature. For other parameters, such as the medium compositions, the influence on their respective behavior among different elements, temperature conditions of soil were simplified as follows:

   a. It was assumed that medium compositions of system was consistent with the sample used in the experiment.
   b. It was assumed that the temperature of regional soil kept changing between 13 and 16 °C.
   c. It was assumed that the contribution to mutual affecting among trace elements was consistent with the sample used in the experiment.

In addition, according to problems to be tried solving and the experimental conditions, the selected research elements were different from different experiments.

The above assumptions were based on the most common conditions in the study issues, and their coverage contents were limited to the experimental exploration for natural laws. However, it generally was closer to the actual situation for the most likely status of the research system. It was the hope that the experimental results or the contribution to the results from different factor would be meaningful to reveal the mechanism of environment behaviors of trace metals in natural medium system at the earth's surface.

Under the general principles and appoints above, the following three simulation experiments were designed:

A. Simulation experiments on the sorption and desorption behaviors of trace metals in the soil (including urban and regional soil);
B. Simulation experiments on the releasing mechanism of trace metals from some particles from the earth's surface (rock and mineral).

The experiments on the migration and transformation behavior of trace metals in the main solid mediums on the earth's surface such as soil and rocks (mineral) were carried out to explore the rule of behavior of trace metals in the natural systems on the earth's surface. The research emphasizes on the importance of the measured actual data and tries to get inspiration in the cause of formation and the evolvement trends from understanding and evaluating of the possible behaviors, factors, and ecological impacts of trace metals in natural mediums.

4.2 The Simulated Experiments on the Desorption and Adsorption Behaviors of the Trace Metals in the Soil

The desorption and adsorption are the very important and common behaviors of trace metals in environment, and these two actions always arise in the system under certain conditions. Because of the necessary connection in the environmental effects and the adsorption and desorption behaviors of trace metals, it is significant to understand the mechanism and process for governance and prevention of the trace metal pollution.

4.2.1 The Simulated Experiments on the Desorption and Adsorption Behaviors of Trace Metals in the Regional Pollution Soil

Soil as a kind of important medium in nature, it is important to understand the adsorption and desorption behaviors of trace metals in soil, which are associated with the environmental impacts and human health.

4.2.1.1 The Simulated Experiments on the Desorption for Trace Metals in Soil System

The experimental solution was simulating the compositions of the local precipitation, and the samples were collected on time to measure its trace metal element concentration after the solution flowing through the sample. When element concentration reached the original concentration of the experimental solution and kept stable, it indicated that the sample reached to the dissolving equilibrium; meanwhile, desorption value of the sample used in the experiment also reached its limit under the experimental conditions. Because the temperature at the earth's surface changed little in a certain period, its effect on dissolved substances in the soil is much smaller than the acid and alkali conditions (pH) on that. Therefore, this research only simulated the change of the pH value at a constant temperature.

1. **The simulation conditions, experiments, and results**

By simulating local precipitation components [1], the solution was prepared as follows with component concentrations: 0.1 μg/ml Cl^-, 10.0 μg/ml $SO_4{}^{2-}$, 2.0 μg/ml $NO_3{}^-$, 1.0 μg/ml $HCO_3{}^-$. Experimental temperature was 13–16 °C, and experiment was conducted at ordinary pressure (1 atm).

Experimental sample was No. 11 of soil samples from work-line A in the Hetao Area. The amount of experimental sample was 2.0 g for every one of all respectively with pH 7.0. The pH value of experimental solution was adjusted to 4, 5.6, and 7, respectively, with acetic acid and ammonia. The sample was leached with the solution continually, and the leached solution was sampled on time (Tables 4.1, 4.2 and 4.3) according to above rules of sampling. The conducting pipe used in the experiment was PTFE tube ($\Phi = 0.7$ cm), glassware all was immersed 24 h with 10% HCl before using, and the reagents were all guaranteed reagents. Anion concentration measuring instrument was Diorex-300 ion chromatograph (made by Diorex, USA), and pH measuring instrument was pHs-3C precision pH meter (RaMag pHS, Shanghai). Trace metal concentration measuring instrument was JY380 ICP-AES (JY380, France). Element detection limits (mg/kg) were as follows: V – 8×10^{-4}, Cr – 5.7×10^{-4}, Mn – 2×10^{-4}, Co – 4.2×10^{-4}, Ni – 1×10^{-2}, Cu – 2×10^{-3}, Pb – 3.4×10^{-3}, Fe – 3×10^{-3}. All experiments were completed at the Elemental Geochemical Laboratory of the Department of Earth Sciences at Nanjing University. The analysis for As and Cd was completed in the Modern Analytical Center of the Nanjing University. The analysis of other components was completed in the State Key Laboratory for Mineral Deposit Research in Nanjing University. Experimental results were shown in Tables 4.1, 4.2 and 4.3.

2. **Experimental results and discussion**

Simulation conditions (pH = 4, pH = 5.6, and pH = 7) are the most common situations or within the pH range (4–7) of natural precipitation or surface runoff in supergene environment. Results show that the effect of pH value is very small on the of rate element desorption (desorption as time), but significant on the initial desorption rate (in the amount of element desorption at first 2 h of the experiment), and the influence degree is different from different elements as shown in Figs. 4.1, 4.2 and 4.3. The element desorption change trend (the quantity change of element desorption as time) is closed with the different pH values; only it has obvious differences on the desorption intercepts at *Y* axel of the curve, and they mean the initial desorption rates of the elements respectively, and the differences for different elements are different obviously.

Based on these results, desorption curves of the elements can be obtained from using the action time of desorption after being standardized treatment versus the element concentration dissolved (desorption value) from the soil to fit the desorption reactions, and the standard desorption curves of elements can be drawn as shown in Fig. 4.4, respectively. Hereby the desorption fitting equation of soil to release trace metals at the experimental conditions, rule of metal element release from soil in common environment with time, can be obtained as follows:

$$\boldsymbol{Y} = \boldsymbol{C}_{\mathrm{pH}}\mathbf{e}^{-ax}$$

In the equation, $\boldsymbol{C}_{\mathrm{pH}}$ is the element concentration in the desorption solution at the beginning of the desorption reaction, which means the element nature in desorption. $\boldsymbol{C}_{\mathrm{pH}}$ is the function of pH versus element and can be measured by experiment. The $\boldsymbol{a}$ is the attribute of the element desorption with time, which means the intrinsic characteristics of metal elements (cation) to free migration in the experimental solution at certain pH value. It represents element desorption properties and can be obtained by fitting desorption curve from experiment data (Fig. 4.4). $\boldsymbol{Y}$ is the element con-

Table 4.1 Data of the experiment simulated natural conditions at the earth surface for trace metals desorption from the soil in the Hetao Area (pH = 4)

Element	Concentration	Desorption time (h)										
		2	4	6	10	14	22	30	46	62	94	126
V	In solution	0.0226	0.0204	0.0201	0.0157	0.0161	0.0146	0.0128	0.0142	0.0089	0.0058	0.0035
	In blank	0.0000	0.0000	0.0000	0.0000	0.0000	0.0000	0.0000	0.0000	0.0000	0.0000	0.0000
Cr	In solution	0.0385	0.0377	0.0328	0.0299	0.0294	0.0244	0.0217	0.0202	0.0159	0.0100	0.0070
	In blank	0.0056	0.0056	0.0056	0.0056	0.0056	0.0056	0.0056	0.0056	0.0056	0.0056	0.0056
Mn	In solution	0.1489	0.1932	0.1670	0.1542	0.1381	0.1316	0.1128	0.1358	0.0680	0.0547	0.0278
	In blank	0.0000	0.0000	0.0000	0.0000	0.0000	0.0000	0.0000	0.0000	0.0000	0.0000	0.0000
Co	In solution	0.0280	0.0264	0.0249	0.0193	0.0198	0.0180	0.0137	0.0129	0.0091	0.0069	0.0031
	In blank	0.0000	0.0000	0.0000	0.0000	0.0000	0.0000	0.0000	0.0000	0.0000	0.0000	0.0000
Ni	In solution	0.0411	0.0389	0.0346	0.0355	0.0346	0.0282	0.0239	0.0277	0.0113	0.0085	0.0032
	In blank	0.0035	0.0035	0.0035	0.0035	0.0035	0.0035	0.0035	0.0035	0.0035	0.0035	0.0035

Note Sample weight was 2.0 g; the concentration of anion in the solution of the simulated experiment was respectively (mg/kg): $Cl^- = 0.1$, $SO_4^{2-} = 10.0$, $NO_3^- = 2.0$, and $HCO_3^- = 1.0$; the pH of the solution = 4; the temperature of working circumstance was 13–16 °C; element concentration unit respectively was mg/kg

Table 4.2 Data of the experiment simulated natural conditions at the earth surface for trace metals desorption from the soil in the Hetao Area (pH = 5.6)

Element	Concentration	Desorption time (h)										
		2	4	6	10	14	22	30	46	62	94	126
V	In solution	0.0146	0.0113	0.01	0.0095	0.0089	0.0079	0.0082	0.0068	0.0047	0.0039	0.0042
	In blank	0	0	0	0	0	0	0	0	0	0	0
Cr	In solution	0.0393	0.0375	0.0352	0.0309	0.029	0.0249	0.0248	0.0234	0.0184	0.0143	0.0093
	In blank	0.0056	0.0056	0.0056	0.0056	0.0056	0.0056	0.0056	0.0056	0.0056	0.0056	0.0056
Mn	In solution	0.0238	0.0113	0.0111	0.011	0.0139	0.013	0.0167	0.0114	0.0105	0.0107	0.0107
	In blank	0	0	0	0	0	0	0	0	0	0	0
Co	In solution	0.0266	0.0255	0.0221	0.0228	0.0212	0.0208	0.0165	0.0152	0.0115	0.006	0.0045
	In blank	0	0	0	0	0	0	0	0	0	0	0
Ni	In solution	0.0385	0.0361	0.0338	0.0293	0.0244	0.0214	0.0235	0.0202	0.0134	0.0083	0.0032
	In blank	0.0035	0.0035	0.0035	0.0035	0.0035	0.0035	0.0035	0.0035	0.0035	0.0035	0.0035
Cu	In solution	0.0028	0.0025	0.0016	0.0003	0.0012	0.0004	0.0026	0.0009	0.0021	0.0008	0.0008
	In blank	0	0	0	0	0	0	0	0	0	0	0
Pb	In solution	0.1123	0.0982	0.0947	0.0933	0.0874	0.0678	0.0674	0.0501	0.057	0.0302	0.0184
	In blank	0	0	0	0	0	0	0	0	0	0	0
Fe	In solution	0.3143	0.0783	0.0677	0.0537	0.0455	0.0374	0.0572	0.0268	0.0374	0.0222	0.0245
	In blank	0	0	0	0	0	0	0	0	·0	0	0

Note Sample weight was 2.0 g; the concentration of anion in the solution of the simulated experiment was respectively (mg/kg): Cl^- = 0.1, SO_4^{2-} = 10.0, NO_3^- = 2.0, and HCO_3^- = 1.0; the pH of the solution = 5.6; the temperature of working circumstance was 13–16 °C; element concentration unit respectively was mg/kg

Table 4.3 Data of the experiment simulated natural conditions at the earth surface for trace metals desorption from the soil in the Hetao Area (pH = 7)

Element	Concentration	Desorption time (h)										
		2	4	6	10	14	22	30	46	62	94	126
V	In solution	0.0135	0.0122	0.0127	0.0122	0.0123	0.0079	0.0074	0.006	0.0065	0.0054	0.0045
	In blank	0	0	0	0	0	0	0	0	0	0	0
Cr	In solution	0.0412	0.0361	0.0335	0.0323	0.0285	0.0261	0.0251	0.021	0.0169	0.0145	0.0086
	In blank	0.0056	0.0056	0.0056	0.0056	0.0056	0.0056	0.0056	0.0056	0.0056	0.0056	0.0056
Mn	In solution	0.014	0.0095	0.0295	0.0092	0.01	0.0066	0.0069	0.006	0.0083	0.0078	0.009
	In blank	0	0	0	0	0	0	0	0	0	0	0
Co	In solution	0.0279	0.0279	0.032	0.0223	0.0211	0.0195	0.0169	0.0144	0.0096	0.0071	0.0068
	In blank	0	0	0	0	0	0	0	0	0	0	0
Ni	In solution	0.0372	0.0392	0.0325	0.0295	0.0238	0.0235	0.0233	0.022	0.0127	0.0087	0.0053
	In blank	0.0035	0.0035	0.0035	0.0035	0.0035	0.0035	0.0035	0.0035	0.0035	0.0035	0.0035
Cu	In solution	0.0048	0.0036	0.0052	0.0026	0.0023	0.0013	0.004	0.0029	0.0029	0.0011	0.0017
	In blank	0	0	0	0	0	0	0	0	0	0	0
Pb	In solution	0.111	0.1083	0.1093	0.0808	0.0764	0.0693	0.0626	0.0648	0.0462	0.0369	0.0384
	In blank	0	0	0	0	0	0	0	0	0	0	0
Fe	In solution	0.2127	0.1098	0.1589	0.0829	0.0923	0.035	0.0374	0.0339	0.0502	0.0327	0.0257
	In blank	0	0	0	0	0	0	0	0	0	0	0

Note Sample weight was 2.0 g; the concentration of anion in the solution of the simulated experiment was respectively (mg/kg): $Cl^- = 0.1$, $SO_4^{2-} = 10.0$, $NO_3^- = 2.0$, and $HCO_3^- = 1.0$; the pH of the solution = 7; the temperature of working circumstance was 13–16 °C; element concentration unit respectively was mg/kg

Table 4.4 Indexes of the desorption of trace metals from the soil in the Hetao Area under conditions on the earth surface (element concentration unit: mg/kg)

		CpH[a]	C_0^b	a[c]	Original concentration	After desorption	Desorption ratio (%)	Efficient speciation in total (%)	Exchangeable ratio in total (%)
pH 4	Cr	0.0357	0.0056	0.0133	61.56	6.06	90.156	29.00%	9.35%
	Co	0.0259	0	0.0162	10.65	7.85	26.291	1.28%	0.32%
	Ni	0.0426	0.0035	0.0196	34.37			20.36%	0.66%
pH 5.6	Cr	0.0353	0.0056	0.0104	61.56	55.72	9.587	29.00%	9.35%
	Co	0.0263	0	0.0144	10.65	9.55	10.329	1.28%	0.32%
	Ni	0.0365	0.0035	0.0179	34.37	29.79	13.262	20.36%	0.66%
	Pb	0.1059	0	0.0134				15.25%	1.13%
pH 7	Cr	0.0363	0.0056	0.0110	61.56	53.93	12.394	29.00%	9.35%
	Co	0.0272	0	0.0110	10.65	9.14	14.178	1.28%	0.32%
	Ni	0.0354	0.0035	0.0153	34.37	28.09	18.272	20.36%	0.66%
	Pb	0.097	0	0.0079				15.25%	1.13%

[a]The CpH is the initial desorption amount of the element under a certain pH condition

[b]C_0 means the background concentration of the desorption solution used in the experiment, and the real content of the metal element released from the soil is that the released content of the metal element calculated from the formula $\boldsymbol{Y} = \boldsymbol{C}_{\mathrm{pH}}\ \mathbf{e}^{-ax}$ subtracts the value of C_0

[c]a is desorption coefficient of the element under a certain condition (pH value and temperature)

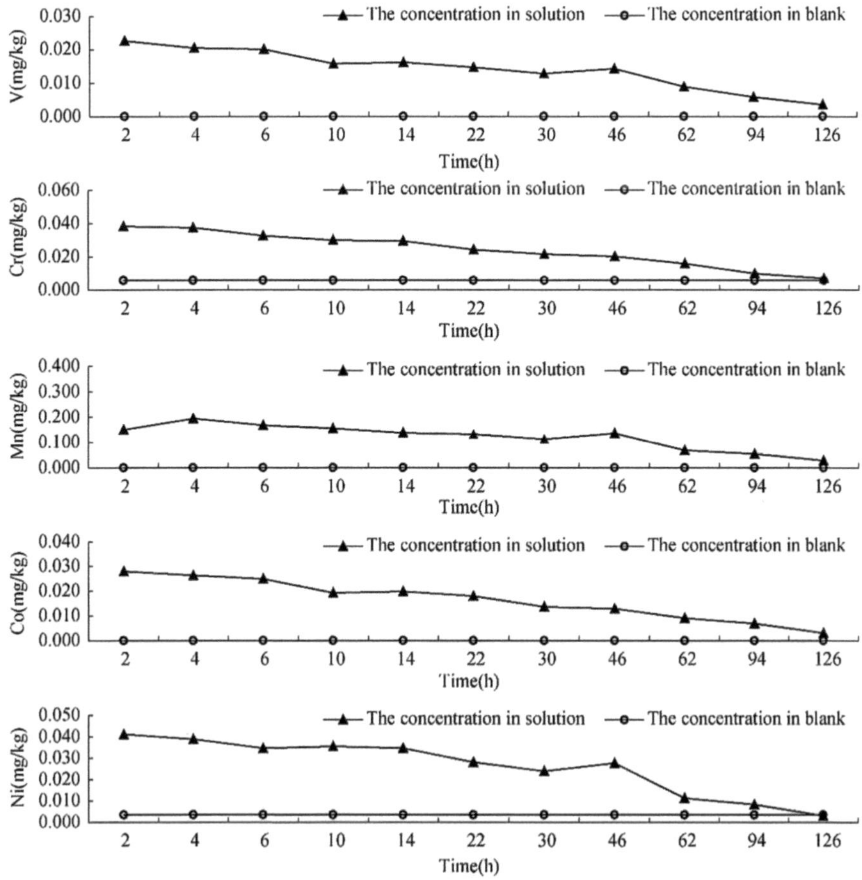

Fig. 4.1 Desorption of trace metals from soil versus time under pH 4 and 13–16 °C

centration in the experimental solution, and X is desorption time of the desorption reaction. The related parameters in the experiment of this study were obtained based on the experiment data shown in Table 4.4.

It should be pointed out that the desorption equation of trace metals from soil and related parameters on the desorption in this work are only an abstract result, and its general significance still needs to be confirmed by a large number of systematic work.

3. **Discussion on the desorption amount for the trace metals**

The experimental results show that under the natural conditions of the earth surface, the rate of element desorption from soil into solution has a good correlation with effective speciation as well as exchangeable concentration of element in the soil at low pH value. When pH value increases, the correlation gets weaker, until negative correlation arises. Therefore, the concentration of exchangeable speciation of ele-

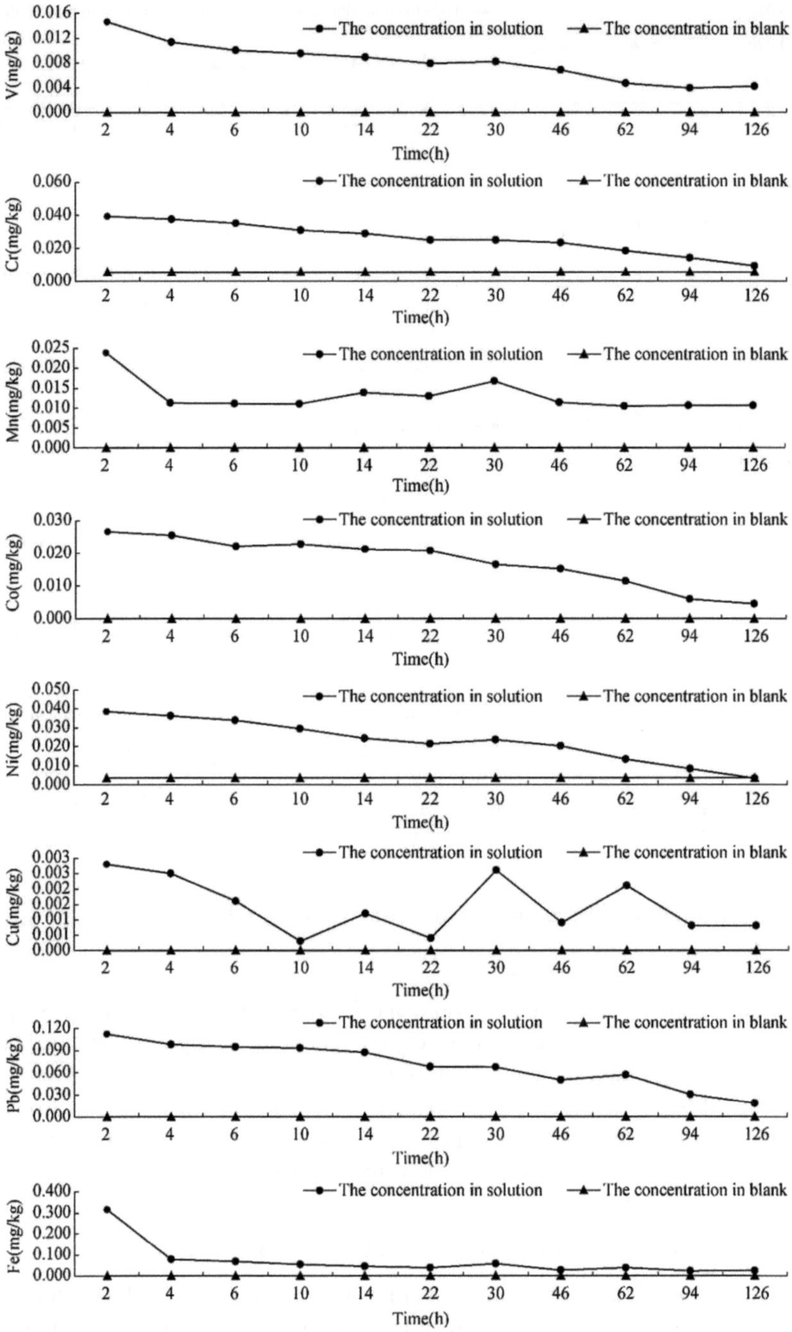

Fig. 4.2 Desorption of trace metals from soil versus time under pH 5.6 and 13–16 °C

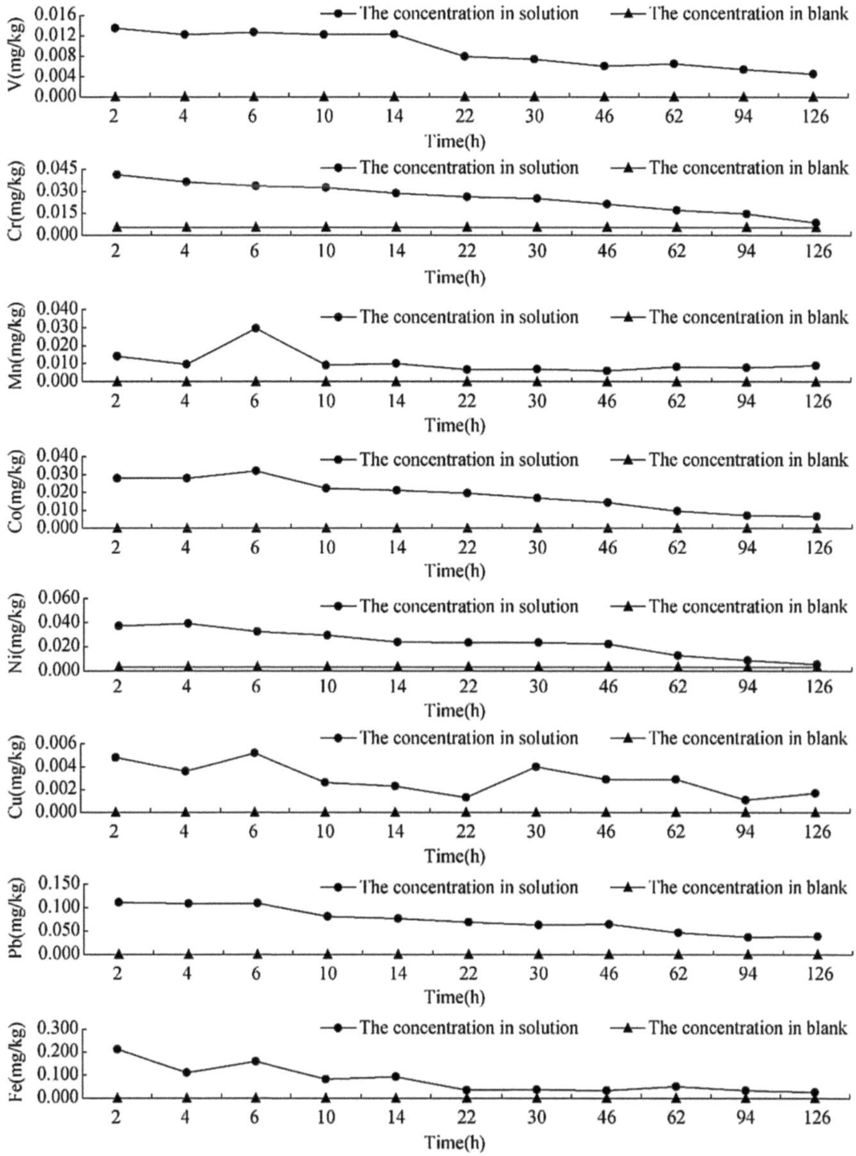

Fig. 4.3 Desorption of trace metals from soil versus time under pH 7 and 13–16 °C

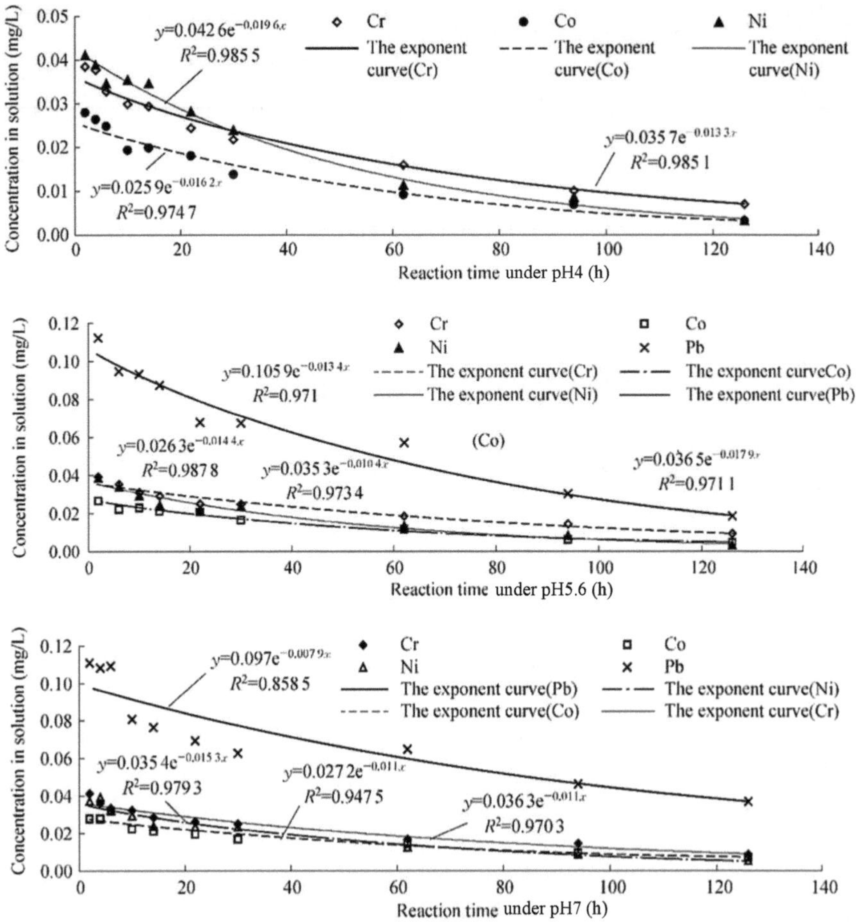

Fig. 4.4 Experimental curves of the desorption reaction for the trace metals from soil to the solution under 13–16 °C, 1 atm, and respectively at pH 4, pH 5.6 and pH 7

ment and the pH value of the solution might be the restricting factor for the element release from soil to environment. As shown in Figs. 4.5, 4.6 and 4.7, the desorption rate (the ratio of element desorption concentration and the total concentration in the sample) changes more as the pH value change, and this change within different elements is more too as the pH value changes. The desorption rate is higher at a low pH value. The desorption rate of pH-neutral solution is higher than that of weak acid solution. The element desorption rate at weak acid solution (pH = 5.6) is lowest among three pH value in these simulated experiments (pH = 4, pH = 5.6, and pH = 7). This is a very interesting phenomenon, which offers us an instructive direction to further study on this issue.

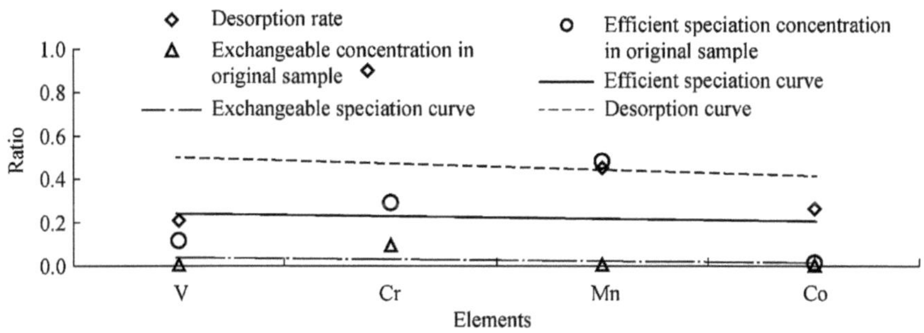

Fig. 4.5 Correlation between the desorption rate and the speciation concentration of the metal in soil (pH = 4)

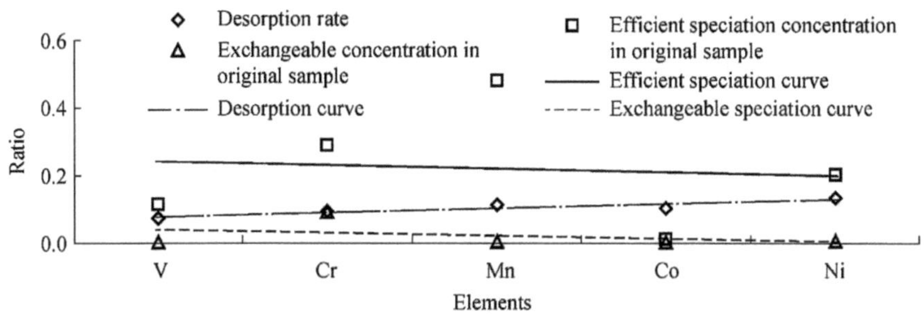

Fig. 4.6 Correlation between the desorption rate and the speciation concentration of the metal in soil (pH = 5.6)

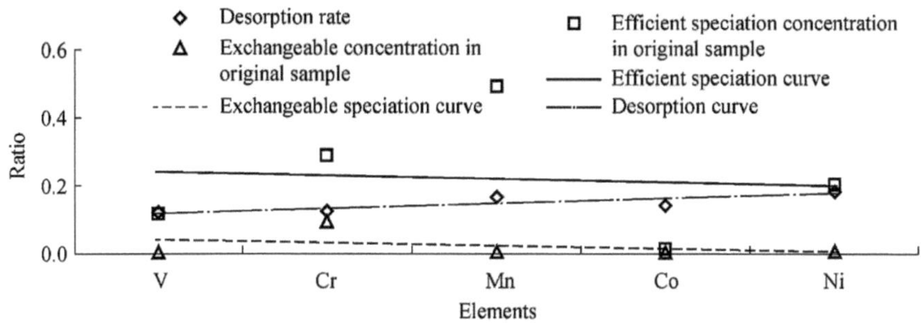

Fig. 4.7 Correlation between the desorption rate and the speciation concentration of the metal in soil (pH = 7)

The above facts show that the most outstanding change among elements of these experiments for the desorption rate (release rate) from soil arisen from Cr, which the biggest desorption rate was reached about 90% of the total amount (pH = 4) and the lowest desorption rate was less than 10% of the total amount (pH = 5.6), and this amount was equivalent to the exchangeable speciation concentration of it. The changes of the other elements are in the middle status of these extremes of Cr (Table 4.4).

4.2.1.2 The Simulated Experiments on the Adsorption for Trace Metals in the Soil System

The soil Sample No. 11 from sampling profile A of Hetao Area was a simulation sample, and elements adsorbed in solution for Cu^{2+}, etc., (V, Cr, Co, Ni, Zn, As, Pb) at pH = 3, pH = 4, and pH = 5.6, and for Pb^{2+} single element in solution at pH = 5.6 were simulated in the experiments, respectively. Each experiment sample was 2.0 g, and their pH value all was 7.0. The prepared solution with metal ions dynamically (adjustable switch control and filter paper) and slowly went through the sample and on time sampled and tested every metal ion concentration in the solution. When the metal ion concentration in the experimental solution was equal to the original solution concentration (before through the sample solution), namely think the sorption reaction reached balance and then measured and calculated the related parameters, respectively.

1. **Simulation condition**

Referring to metal concentration data of high background concentration groundwater in the study area in dive water and surface water, analysis data of water quality of the area (Tang Jie et al. in 1996) [2], and the element concentration of groundwater in arsenic poisoning area of Bangladesh [3, 4], the experimental solutions for the ion concentrations were respectively prepared as follows (unit: μg/g):

For the solution with elements Cu^{2+}, etc., $V^{5+} = 0.043$, $Cr^{6+} = 0.084$, $Fe^{2+} = 0.115$, $Co^{2+} = 0.069$, $Ni^{2+} = 0.074$, $Cu^{2+} = 0.038$, $Zn^{2+} = 0.229$, $As^{5+} = 0.340$, $Pb^{2+} = 0.140$.

For the solution with Pb^{2+} single element: $Pb^{2+} = 0.462$ μg/g.

The metal solution used to prepare the experimental solution was the national standard solution developed from National Iron and Steel Materials Testing Center. Experimental temperature was 13–16 °C (room temperature), and the pressure was atmospheric pressure (1 atm). The pH values of solutions were 3, 4, and 5.6, respectively, for the experiments.

The pH value was adjusted by acetic acid and ammonia, and measuring instrument was PHs-3C precision pH meter (PHs-3C, Shanghai Leichi). Arsenic concentration was analyzed by atomic fluorescence spectrometer AF-610, and detection limit was 8×10^{-5} mg/kg (AF-610, Beijing Rili). Other metal concentration was analyzed by ICP-AES JY380, and detection limit (mg/kg) was 8×10^{-4} for V, 5.7×10^{-4} for Cr, 4.2×10^{-4} for Co, 1×10^{-2} for Ni, 2×10^{-3} for Cu, 4×10^{-3} for Zn, 3.4×10^{-3}

for Pb, respectively (ICP-AES JY380, France). All experiments were completed at the Geochemical Laboratory of the Department of Earth Sciences and the State Key Laboratory for Mineral Deposits Research, Nanjing University.

2. **Results and discussion**

Experimental data are shown in Tables 4.5, 4.6, 4.7 and 4.8.

The results show that the sorption of the soil to most metal in the solution has not reached the sorption reaction equilibrium except As and Zn in the experiment period of this study (63 h). Among elements that did not reach the sorption equilibrium (V, Cr, Co, Ni, Cu, and Pb), the sorption capacity of V, Co, Ni, and Cu at high pH value is generally stronger than that at low pH value, as shown in Figs. 4.8, 4.9 and 4.10. Under the earth's surface conditions, the sorption parameters of soil to trace metals are shown in Table 4.9. Combined with above signals from sorption curves, it suggests that the time to reach the sorption equilibrium at low pH is shorter than that at the high pH value. These phenomena indicated that the ability of sorption and retention of solid particles to metal ions in acid condition may be limited with only weaker actions.

The sorption to single element in a certain time is greater than the sorption to multi-element, as shown in Fig. 4.10 (for Pb) and Fig. 4.11. Because Pb^{2+} concentration of mother liquor in two experiments is different and Pb^{2+} sorption failed to reach sorption equilibrium in two experiments, the sorption capacity (maximum sorption, sorption law with time) of soil in two experiments is difficult to estimate. However, these experimental data revealed that the sorption capacity of soil to metals also varies with different metals and their concentrations within a dozen of hours.

Table 4.5 Concentrations of Cu^{2+}, etc., absorbed by the soil from the Hetao Area under pH 3 and 13–16 °C

Element	Concentration	Absorbing time (h)												
		0	0.5	1	2	3	5	7	11	15	23	31	47	63
V	C after absorbing[a]	0.0436	0.0453	0.0384	0.0472	0.0447	0.0484	0.0320	0.0193	0.0240	0.0388	0.0295	0.0209	0.0164
	C in mother solution	0.0436	0.0436	0.0436	0.0436	0.0436	0.0436	0.0436	0.0436	0.0436	0.0436	0.0436	0.0436	0.0436
Cr	C after absorbing	0.0879	0.0492	0.0501	0.0447	0.0425	0.0365	0.0352	0.0455	0.0413	0.0364	0.0345	0.0319	0.0320
	C in mother solution	0.0879	0.0879	0.0879	0.0879	0.0879	0.0879	0.0879	0.0879	0.0879	0.0879	0.0879	0.0879	0.0879
Co	C after absorbing	0.0681	0.0589	0.0600	0.0589	0.0538	0.0414	0.0282	0.0575	0.0541	0.0452	0.0499	0.0458	0.0449
	C in mother solution	0.0681	0.0681	0.0681	0.0681	0.0681	0.0681	0.0681	0.0681	0.0681	0.0681	0.0681	0.0681	0.0681
Ni	C after absorbing	0.0742	0.0682	0.0695	0.0690	0.0572	0.0659	0.0440	0.0685	0.0608	0.0537	0.0557	0.0525	0.0482
	C in mother solution	0.0742	0.0742	0.0742	0.0742	0.0742	0.0742	0.0742	0.0742	0.0742	0.0742	0.0742	0.0742	0.0742
Cu	C after absorbing	0.0448	0.0174	0.0213	0.0214	0.0165	0.0138	0.0591	0.0310	0.0282	0.0204	0.0245	0.0268	0.0316
	C in mother solution	0.0448	0.0448	0.0448	0.0448	0.0448	0.0448	0.0448	0.0448	0.0448	0.0448	0.0448	0.0448	0.0448

(continued)

Table 4.5 (continued)

Element	Concentration	Absorbing time (h)												
		0	0.5	1	2	3	5	7	11	15	23	31	47	63
Zn	C after absorbing	0.1456	0.1357	0.1034	0.1429	0.3728	0.6229	1.0220	0.0882	0.1086	0.0606	0.0596	0.0592	0.1858
	C in mother solution	0.1456	0.1456	0.1456	0.1456	0.1456	0.1456	0.1456	0.1456	0.1456	0.1456	0.1456	0.1456	0.1456
As	C after absorbing	0.3562	0.3397	0.3275	0.3300	0.2810	0.2240	0.1076	0.3357	0.3480	0.2892	0.3505	0.3448	0.3500
	C in mother solution	0.3562	0.3562	0.3562	0.3562	0.3562	0.3562	0.3562	0.3562	0.3562	0.3562	0.3562	0.3562	0.3562
Pb	C after absorbing	0.1833	0.1083	0.1202	0.1488	0.1152	0.0970	0.1068	0.1027	0.0852	0.0785	0.0746	0.0736	0.0547
	C in mother solution	0.1833	0.1833	0.1833	0.1833	0.1833	0.1833	0.1833	0.1833	0.1833	0.1833	0.1833	0.1833	0.1833

[a]C means the concentration of the element in the reaction; sample weight was 2.0 g; the pH of the solution = 3; the temperature of working circumstance was 13–16 °C; element concentration unit was mg/kg

Table 4.6 Concentrations of Cu^{2+}, etc., absorbed by the soil from the Hetao Area under pH 4 and 13–16 °C

Element	Concentration	Absorbing time (h)												
		0	0.5	1	2	3	5	7	11	15	23	31	47	63
V	C after absorbing[a]	0.0422	0.0768	0.0428	0.0613	0.0500	0.0616	0.0519	0.0377	0.0210	0.0350	0.0149	0.0278	0.0184
	C in mother solution	0.0422	0.0422	0.0422	0.0422	0.0422	0.0422	0.0422	0.0422	0.0422	0.0422	0.0422	0.0422	0.0422
Cr	C after absorbing	0.0836	0.0433	0.0439	0.0388	0.0383	0.0403	0.0343	0.0308	0.0313	0.0275	0.0249	0.0248	0.0271
	C in mother solution	0.0836	0.0836	0.0836	0.0836	0.0836	0.0836	0.0836	0.0836	0.0836	0.0836	0.0836	0.0836	0.0836
Co	C after absorbing	0.1034	0.7653	0.5749	0.7859	0.6799	0.7124	0.0654	0.4290	0.3086	0.4722	0.1992	0.5166	0.3486
	C in mother solution	0.1034	0.1034	0.1034	0.1034	0.1034	0.1034	0.1034	0.1034	0.1034	0.1034	0.1034	0.1034	0.1034
Ni	C after absorbing	0.0733	0.0538	0.0533	0.0569	0.0559	0.0560	0.0518	0.0387	0.0542	0.0407	0.0478	0.0485	0.0414
	C in mother solution	0.0733	0.0733	0.0733	0.0733	0.0733	0.0733	0.0733	0.0733	0.0733	0.0733	0.0733	0.0733	0.0733
Cu	C after absorbing	0.0733	0.0663	0.0711	0.0642	0.0631	0.0654	0.0582	0.0483	0.0585	0.0453	0.0516	0.0546	0.0447

(continued)

Table 4.6 (continued)

Element	Concentration	Absorbing time (h)												
		0	0.5	1	2	3	5	7	11	15	23	31	47	63
	C in mother solution	0.0733	0.0733	0.0733	0.0733	0.0733	0.0733	0.0733	0.0733	0.0733	0.0733	0.0733	0.0733	0.0733
Zn	C after absorbing	0.0410	0.0089	0.0099	0.0102	0.0152	0.0126	0.0117	0.0221	0.0226	0.0148	0.0232	0.0168	0.0229
	C in mother solution	0.0410	0.0410	0.0410	0.0410	0.0410	0.0410	0.0410	0.0410	0.0410	0.0410	0.0410	0.0410	0.0410
As	C after absorbing	0.1236	0.1458	0.1171	0.1020	0.1886	0.0959	0.0800	0.1313	0.0737	0.1141	0.0604	0.1082	0.1148
	C in mother solution	0.1236	0.1236	0.1236	0.1236	0.1236	0.1236	0.1236	0.1236	0.1236	0.1236	0.1236	0.1236	0.1236
Pb	C after absorbing	0.1833	0.1083	0.1202	0.1488	0.1152	0.0970	0.1068	0.1027	0.0852	0.0785	0.0746	0.0736	0.0547
	C in mother solution	0.1833	0.1833	0.1833	0.1833	0.1833	0.1833	0.1833	0.1833	0.1833	0.1833	0.1833	0.1833	0.1833

[a]C means the concentration of the element in the reaction; sample weight was 2.0 g; the pH of the solution = 4; the temperature of working circumstance was 13–16 °C; element concentration unit was mg/kg

Table 4.7 Concentrations of Cu^{2+}, etc., absorbed by the soil from the Hetao Area under pH 5.6 and 13–16 °C

Element	Concentration	Absorbing time (h)												
		0	0.5	1	2	3	5	7	11	15	23	31	47	63
V	C after absorbing[a]	0.0427	0.0488	0.0258	0.0294	0.0240	0.0268	0.0240	0.0261	0.0158	0.0266	0.0173	0.0215	0.0130
	C in mother solution	0.0427	0.0427	0.0427	0.0427	0.0427	0.0427	0.0427	0.0427	0.0427	0.0427	0.0427	0.0427	0.0427
Cr	C after absorbing	0.0836	0.0370	0.0342	0.0335	0.0323	0.0312	0.0288	0.0249	0.0250	0.0194	0.0194	0.0144	0.0154
	C in mother solution	0.0836	0.0836	0.0836	0.0836	0.0836	0.0836	0.0836	0.0836	0.0836	0.0836	0.0836	0.0836	0.0836
Co	C after absorbing	0.0685	0.0362	0.0374	0.0487	0.0434	0.0375	0.0398	0.0324	0.0519	0.0305	0.0458	0.0354	0.0458
	C in mother solution	0.0685	0.0685	0.0685	0.0685	0.0685	0.0685	0.0685	0.0685	0.0685	0.0685	0.0685	0.0685	0.0685
Ni	C after absorbing	0.0736	0.0469	0.0437	0.0528	0.0474	0.0460	0.0459	0.0351	0.0562	0.0335	0.0463	0.0404	0.0499
	C in mother solution	0.0736	0.0736	0.0736	0.0736	0.0736	0.0736	0.0736	0.0736	0.0736	0.0736	0.0736	0.0736	0.0736
Cu	C after absorbing	0.0384	0.0066	0.0057	0.0061	0.0048	0.0048	0.0030	0.0037	0.0049	0.0051	0.0062	0.0056	0.0286

(continued)

Table 4.7 (continued)

Element	Concentration	Absorbing time (h)												
		0	0.5	1	2	3	5	7	11	15	23	31	47	63
	C in mother solution	0.0384	0.0384	0.0384	0.0384	0.0384	0.0384	0.0384	0.0384	0.0384	0.0384	0.0384	0.0384	0.0384
Zn	C after absorbing	0.2286	0.0499	0.0939	0.0955	0.2055	0.2086	0.0561	0.0930	0.0005	0.1134	0.0452	0.0521	0.0520
	C in mother solution	0.2286	0.2286	0.2286	0.2286	0.2286	0.2286	0.2286	0.2286	0.2286	0.2286	0.2286	0.2286	0.2286
As	C after absorbing	0.3400	0.2874	0.2475	0.3045	0.2455	0.2201	0.2421	0.2084	0.2996	0.2152	0.3017	0.2743	0.3248
	C in mother solution	0.3400	0.3400	0.3400	0.3400	0.3400	0.3400	0.3400	0.3400	0.3400	0.3400	0.3400	0.3400	0.3400
Pb	C after absorbing	0.1348	0.0975	0.1007	0.13	0.0891	0.0935	0.0743	0.0698	0.0681	0.0648	0.0478	0.0454	0.0237
	C in mother solution	0.1348	0.1348	0.1348	0.1348	0.1348	0.1348	0.1348	0.1348	0.1348	0.1348	0.1348	0.1348	0.1348

[a]C means the concentration of the element in the reaction; sample weight was 2.0 g; the pH of the solution = 5.6; the temperature of working circumstance was 13–16 °C; element concentration unit was mg/kg

Table 4.8 Concentrations of Pb^{2+} absorbed by the soil from the Hetao Area under pH 5.6 and 13–16 °C

Elements	Concentration	Absorbing time (h)												
		0	0.5	1	2	3	5	7	11	15	23	31	47	63
Pb	C after absorbing[a]	0.4619	0.2149	0.2066	0.2141	0.2502	0.0935	0.2321	0.0778	0.3046	0.3826	0.2473	0.2356	0.2740
	C in mother solution	0.4619	0.4619	0.4619	0.4619	0.4619	0.4619	0.4619	0.4619	0.4619	0.4619	0.4619	0.4619	0.4619

[a]C means the concentration of the element in the reaction; sample weight was 2.0 g; the pH of the solution = 5.6; the temperature of working circumstance was 13–16 °C; element concentration unit was mg/kg

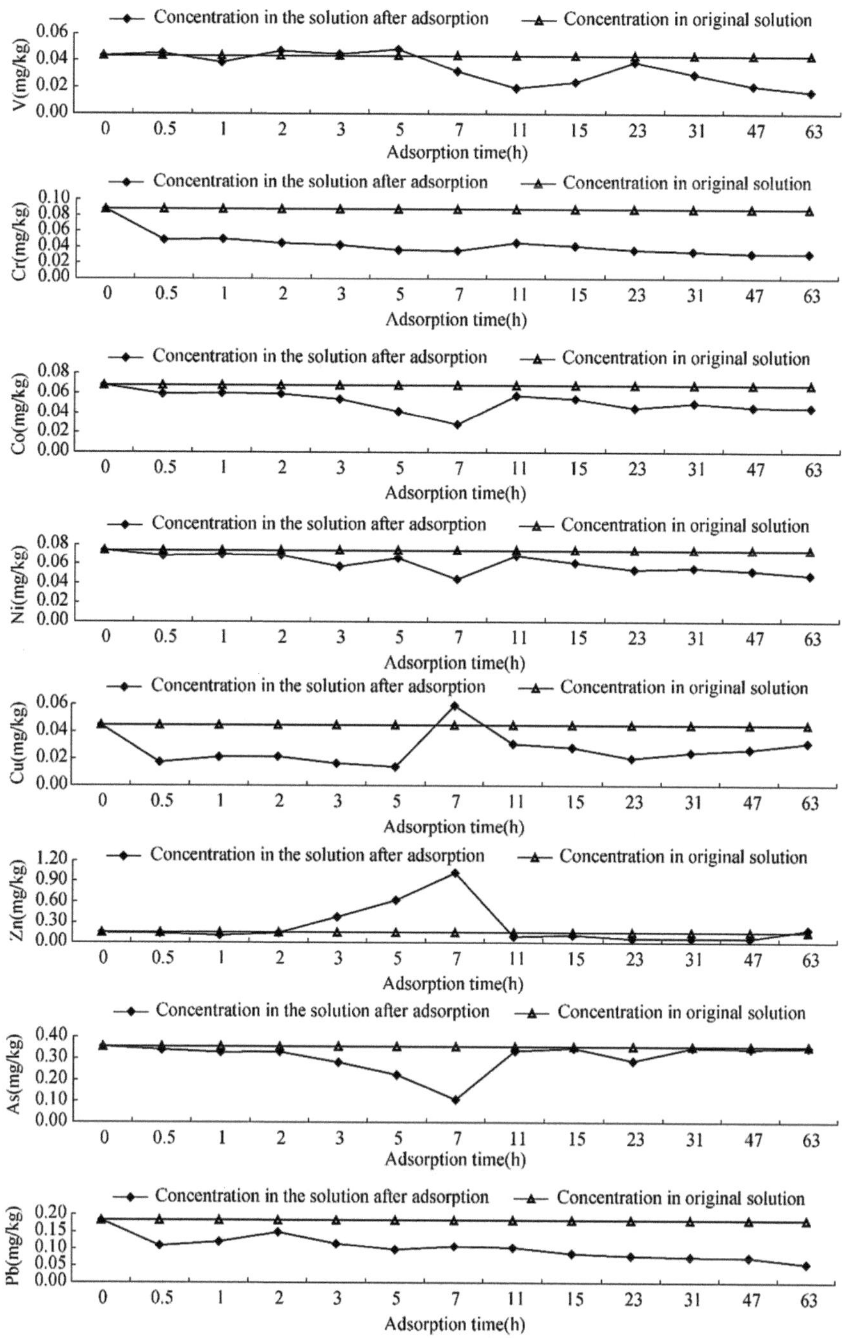

Fig. 4.8 Curve of absorption experiment for trace metal Cu^{2+} etc. absorbed by the soil from the Hetao Area under pH 3 and 13–16 °C

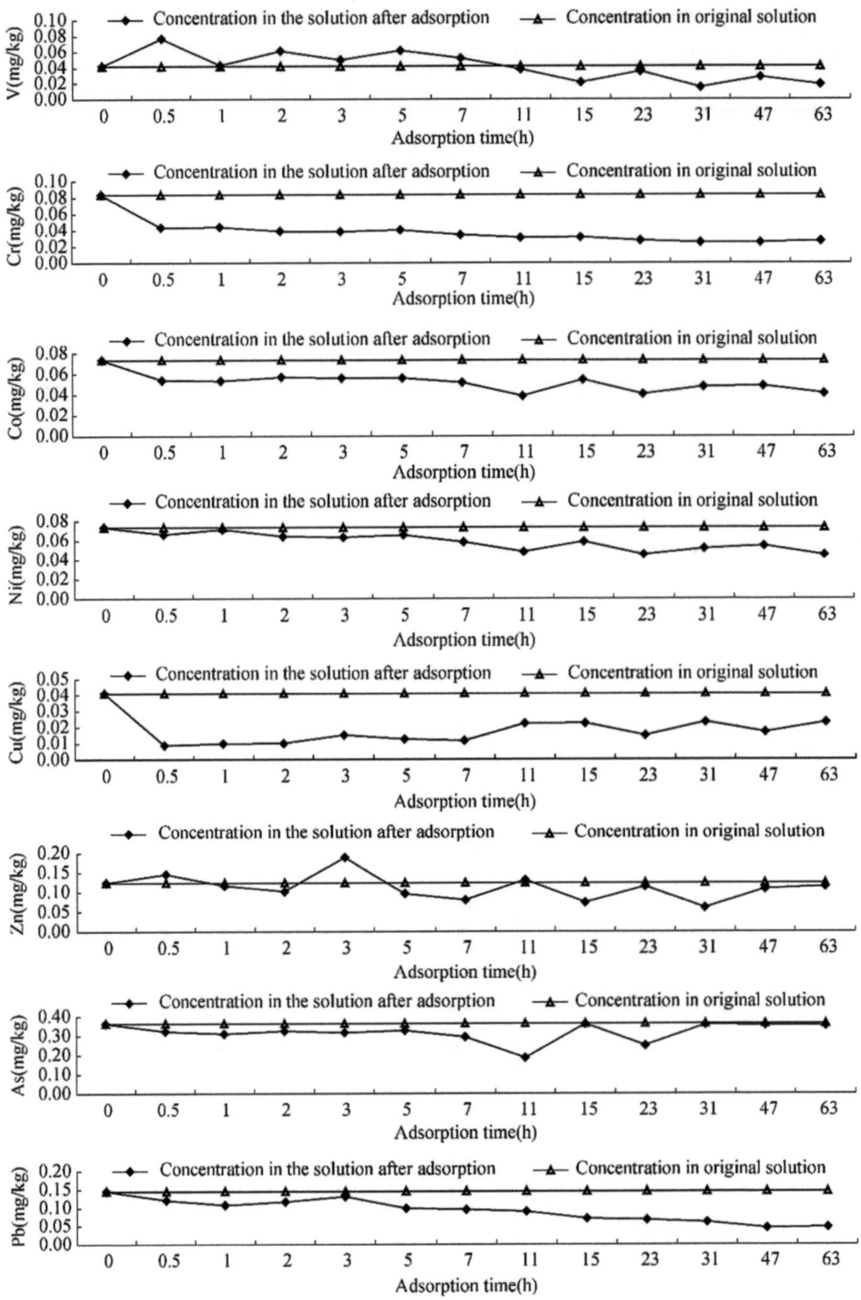

Fig. 4.9 Curve of absorption experiment for trace metal Cu^{2+} etc. absorbed by the soil from the Hetao Area under pH 4 and 13–16 °C

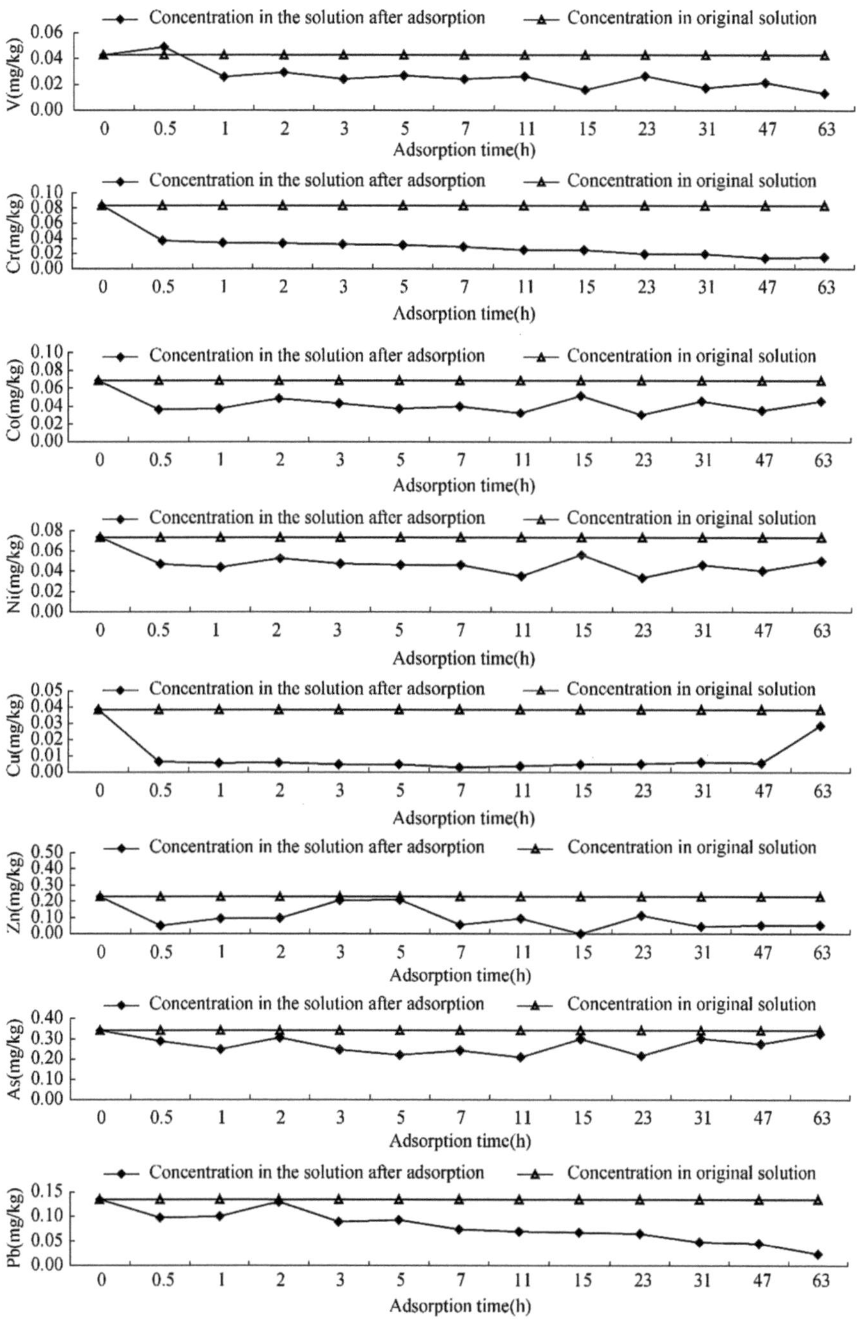

Fig. 4.10 Curve of absorption experiment for trace metal Cu^{2+} etc. absorbed by the soil from the Hetao Area under pH 5.6 and 13–16 °C

Table 4.9 Absorption indexes of the trace metals absorbed by the soil in the Hetao Area under the conditions on the earth surface (element concentration unit: mg/kg)

	Elements	Experiment time (h)	To the equilibrium time (h)	Original concentration	After absorption	Absorption ratio (63, mg/kg)
pH 3, with the solution of ions	V	63	>63	46.1999	52.0799	5.8801
	Cr	63	>63	61.5598	74.9331	13.3732
	Co	63	>63	10.6533	11.2933	0.6400
	Ni	63	>63	34.3732	35.5700	1.1968
	Cu	63	>63			
	Zn	63	≈63			
	As	63	31	7.2393	9.8000	2.5607
	Pb	63	>63			
pH 4, with the solution of ions	V	63	>63	46.1999	49.4665	3.2667
	Cr	63	>63	61.5598	108.0000	46.4402
	Co	63	>63	10.6533	11.6666	1.0133
	Ni	63	>63	34.3732	38.0132	3.6400
	Cu	63	>63			
	Zn	63	≈63			
	As	63	31	7.2393	10.1000	2.8607
	Pb					
pH 5.6, with the solution of ions	V	63	>63 h	46.1999	103.2000	57.0001
	Cr	63	>63 h	61.5598	98.6300	37.0702
	Ni	63	>63 h	34.3732	38.1900	3.8168
	Cu	63	≈63 h			
	Zn	63	>63 h			
	As	63	≈63 h	7.2393	7.7000	0.4607
	Pb	63	>63 h			
pH 5.6, with Pb^{2+} solution	Pb	63	>63	122.0000	129.5730	7.5730

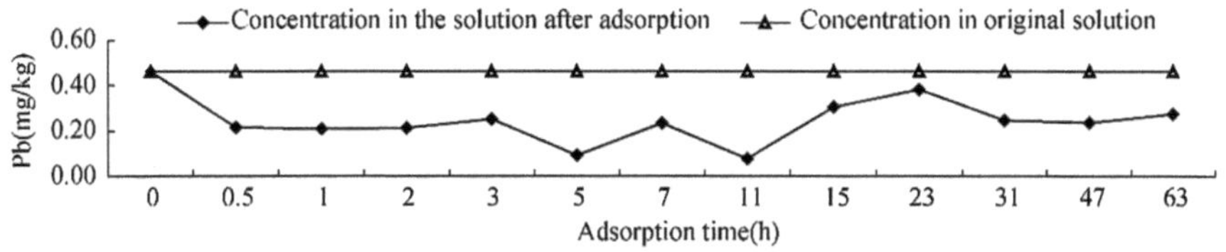

Fig. 4.11 Curve of absorption experiment for trace metal Pb^{2+} absorbed by the soil from the Hetao Area under pH 5.6 and 13–16 °C

When metals in solution are less and their concentrations are higher, the sorption of solid substances to metal increased accordingly. The sorption ability of solid substances to metal changes with the kinds of the metals and their concentrations, respectively. This means that the sorption ability of solid substances to metal is elastic and the sorption potential is considerable under certain conditions.

4.2.2 Discussion on the Facts Revealed by the Experiments

To sum up, the desorption and sorption of trace metals in the soil from regional and urban areas under the earth surface conditions by natural precipitation are represented by the simulated experiments, and from the experiment results, some facts on the trace metal sorption and desorption in soil can be deduced which might imply some rules as follows:

A. Under the influence of precipitation, the desorption rate of different elements from the soil is roughly similar; in other words, the change of element concentration in desorption solution versus time is roughly similar for different element, which might be an inherent nature of element desorption at common temperature and pressure reacting in natural precipitation.
B. At the beginning period of the desorption (first two hours of this simulation experiment), the desorption amount of trace metals is different with the solution pH value and the kind of element, and the concentration of the element in the desorption solution in initial period of the desorption reaction may be in a functional relationship with pH value and element kind.
C. The time of desorption reaction to reach the equilibrium is different for different elements and different pH values under the experimental conditions. The equilibrium time at lower pH value would be shorter. Elements are easy to desorp from the soil at acidic medium and high temperature.
D. The sorption capacity of soil for metals changes with the metal kind and concentration in the reaction solution (experimental solution) within a dozen of hours. When the kinds of metals are less and their concentrations are higher in the reaction solution, the sorption capacity of soil for the element would increase accordingly. The sorption ability of soil for metals in lower pH system is weaker

than that in higher pH system of soil, and under lower pH condition, the sorption reaction reaches the equilibrium earlier than that under higher pH condition. Exchangeable fraction of speciation concentration of element in soil and the pH value of the solution are restricting factors for soil to release element to the environment, and pH value is the most important, as shown in Figs. 4.5, 4.6 and 4.7. The desorption rate (the concentration of the element released from the sample accounts for percentage of the total concentration of the element in the sample) varies obviously under the different pH values and under the same pH value but different element conditions. The desorption rate in weak acid solution (pH = 5.6) is lowest for the elements released in these experiments simulated three pH conditions. Under earth surface conditions, trace metal desorption from soil in the natural aqueous solution would obey the following rule:

$$\boldsymbol{Y} = \boldsymbol{C}_{\mathrm{pH}}\mathbf{e}^{-ax}$$

Here, $\boldsymbol{Y}$ is element concentration in the aqueous solution, and X is desorption time. The $\boldsymbol{C}_{\mathrm{pH}}$ is the initial desorption amount of the element under a certain condition (pH value and temperature), which can be obtained by experiment. The $\boldsymbol{a}$ is desorption coefficient of the element under a certain condition (pH value and temperature), which means the characteristics and nature of the element to be desorbed from soil to the solution, and it can be obtained by fitting desorption curve from experiment data.

E. The desorption rate of Cr and Co in the soil of Hetao Area is larger than the efficient speciation concentration that might be a very interesting issue to further study and discuss.

4.3 The Simulated Experiments on the Release Mechanism of Trace Metals to Environment from Natural Substances (Rock or Mineral)

4.3.1 The Stimulated Experiment on the Trace Metal Desorption from Particles (Rock, Mineral) at the Earth Surface Conditions

The solid particles (rock, mineral, or compounds) on the earth's surface are the main original carrier of trace metals. In the view of environmental science, these are the initial sources of metal elements. The elements going through its original carrier rock or mineral particles (compounds) to release into the environment in the forms of solution, gas, and solids will lead to a series of environmental or biological impacts, and then, the environmental and biological effects would be resulted in or came up. In these processes, water is the most important as the medium and agent to take part in the reaction processes. Elements released from solid particles to the environment work going through the reaction between water and the medium, their migration, transformation, and biological absorption in the system are also inseparable from the water [5]. Therefore, simulation experiment for the release mechanism of trace metals from the solid particles under natural conditions on the earth surface that is under normal temperature atmospheric pressure, precipitation conditions, and participation of air and organisms is the key to understand the process and phenomenon of element release from its original carriers into the environment. This is the basic idea and theoretical foundation in these experimental design and simulations.

1. **Samples, conditions, and the step of the stimulation experiments**

The samples were taken from the lead, zinc, and sulfur mine of Dongshengmiao in the Hetao Area, Inner Mongolia. The samples which were well distributed for minerals and other components were selected, and they were cut into regular diamonds (1 × 1 × 0.2 cm, the total surface area was 2.80 cm^2), polished for every side, and calculated the surface area according to the actual size before using for the experiment.

Experimental solution was simulated by the composition of local precipitation, containing (μg/ml): $Cl^- = 0.1$, $SO_4{}^{2-} = 10.0$, $NO_3{}^- = 2.0$, $HCO_3{}^- = 1.0$. The pH value of the solution was simulated as follows: pH = 3, pH = 4, pH = 5.6, and pH = 7, respectively. Experimental temperature was 13–16 °C (room temperature) and 50 °C (only stimulation for pH = 5.6 case). The experimental time was in 1030 h.

The sample was put into a beaker with quantitative experimental solution after washing clean and drying. Let the less surface of the sample (0.2 cm) contact with the bottom of the beaker, and to keep sample upright, timing quantitatively took the solution and measured the concentrations of elements in the solution sample. The glass containers used in the experiments were immersed with 10% HCl for 24 h and then washed and dried before they were used. All reagents and water of the experiments were guaranteed. The condition details of the experiments were shown in Table 4.10.

2. **Analysis for the trace metals**

Anion concentration was measured by Diorex-300 Ion Chromatograph (Diorex-300, USA), and pH testing instrument was pHS-3 c precision pH meter ((PHs-3C, Shanghai Leichi)). Except As and Cd, other element concentration was measured all by Element II ICP-MS, and element detection limit was as follows (mg/kg): V – 1×10^{-7}, Cr – 1×10^{-6}, Mn – 1×10^{-6}, Co – 5×10^{-6}, Ni – 1×10^{-6}, Cu – 1×10^{-6}, Zn – 1×10^{-6}, Sb – 1×10^{-7}, Pb – 1×10^{-7}, Bi – 1×10^{-8}, Fe – 1×10^{-6}, and Al – 1×10^{-6} (Element II, Finnigan MAT). As and Cd measured by AF-610A atomic fluorescence spectrometer with detection limit of was 8×10^{-5} mg/kg for As, 0.02×10^{-5} mg/kg for Cd (AF-610, Beijing Rili). All experiments were completed in the Element Geochemistry Laboratory of the Department of Earth Sciences, Nanjing University. As and Cd were analyzed at the Center of Modern Analysis, Nanjing University; other elements and parameter measure were completed at the State Key Laboratory for Mineral Deposits Research, Nanjing University.

Table 4.10 Experimental conditions simulated the conditions on the earth surface for metal release from solid grain (minerals, rock)

Experiment number	Sample feature (Lead and zinc ore)		Anion concentration in the solution of experiment (mg/kg)				Feature of the solution	
	Weight (g)	Area (cm^2)	Cl^-	SO_4^{2-}	NO_3^-	HCO_3^-	pH	Temperature (°C)
1	0.72	2.80^2	0.10	10.80	2.00	1.00	3	13–16
2	0.87	2.87	0.10	10.80	2.00	1.00	4	13–16
3	0.74	2.80	0.10	10.80	2.00	1.00	5.6	13–16
4	0.73	2.80	0.10	10.80	2.00	1.00	7	13–16
5	0.74	2.61	0.10	10.80	2.00	1.00	5.6	50

Experiment number	Element concentration in the sample (mg/kg)										
	V	Cr	Mn	Co	Ni	Cu	Zn	As	Sb	Pb	Bi
1	124.4	42.42	12,140	24.66	48.42	1729	5713	104.3	40.36	244.2	53.54
2	124.4	42.42	12,140	24.66	48.42	1729	5713	104.3	40.36	244.2	53.54
3	124.4	42.42	12,140	24.66	48.42	1729	5713	104.3	40.36	244.2	53.54
4	124.4	42.42	12,140	24.66	48.42	1729	5713	104.3	40.36	244.2	53.54
5	124.4	42.42	12,140	24.66	48.42	1729	5713	104.3	40.36	244.2	53.54

4.3.2 Results and Discussion

The results of the simulated experiments above are shown in Tables 4.11, 4.12, 4.13, 4.14 and 4.15.

The data indicate that there are obvious effects from pH and temperature respectively on the element release amount (the element concentration amount resolved from the sample per unit area), as shown in Figs. 4.12, 4.13, 4.14, 4.15 and 4.16. In the same pH value, the release concentrations for most elements at high temperature are higher than that at low temperature. As shown in Fig. 4.14, the results at pH = 5.6 and T = 13 °C are obviously different with that at pH = 5.6 and T = 50 °C shown in Fig. 4.16.

The features above mainly reflect the differences between the amount of initial dissolution or release of the elements, which are the measured values of the reaction between sample and solution for 120 h of this work, and that the change rate of dissolution or release of most elements with time is generally consistent at different temperature conditions. These features are shown obviously in the experimental result for V, Cr, Mn, Ni, Cu, Sb, Fe, and other elements. However, Zn and Al are exceptive, and their release amounts reduce at high temperature instead. The change trend in Al dissolve curve is almost same at the different temperatures, and the release amount at higher temperature is lower in which the time to reach the dissolution equilibrium is ahead of that at low temperature. Zn, except for higher release amount at lower temperature its initial release, is the maximum in the measured values in 120 h in this experiment, and its dissolve equilibrium may be completed after about 240 h; at the higher temperature (50 °C), the release reaction will reach the maximum release after 840 h for Zn release. These are as shown in Figs. 4.14 and 4.16. The chemical and physicochemical mechanism in above phenomenon can be as the specific research topics to deepen in this direction in the future work.

At the same temperature and different pH values, the release amounts of V, Cr, Sb, and Fe are higher at low pH value than that at higher pH value, with an obvious change characteristics. The change rate of the release (the trend line slope of release curve) is similar at different pH values, and the time to reach the dissolve equilibrium is close too at the different pH values. The influence of pH value on the release of the element mainly reflects in the differences of the initial release process (the measured values of the reaction after 120 h in this experiment), as shown in Figs. 4.12, 4.13, 4.14 and 4.15.

The features of Mn, Cu, As, Pb, Bi, and Al are relatively special for their release from the samples in the experiment affected by the pH condition.

Mn release amount at low pH value is obviously higher than that at high pH value, and the release at the case of pH = 3 was 10 times of that at the case of pH = 5.6 in the same time in this experiment as shown in Tables 4.11 and 4.13. Its release amount reduces gradually with the increase of pH value; when the pH value is 5.6, the release reached the maximum amount after 840 h (Fig. 4.14). The release up to a maximum at a pH of 4 or lower (pH = 4 and pH = 3 in this experiment) needs more than 360 h, as shown in Figs. 4.12 and 4.13.

Table 4.11 Data of the experiment simulated the conditions on the earth surface for trace metal release from the solid grain (pH 3)

Element	120 h		360 h			840 h			1030 h		
	Dissolved[a]	Release[b]	Dissolved	Released	Accumulation for release[c]	Dissolved	Released	Accumulation for release	Dissolved	Released	Accumulation for release
V	0.2200	0.0786	0.4456	0.0805714	0.1591429	0.9408	0.1768571	0.336	0.8133	−0.045536	0.2904643
Cr	0.5700	0.2036	0.2918	−0.099357	0.1042143	1.191	0.3211429	0.4253571	0.9984	−0.068786	0.3565714
Mn	18.0300	6.4393	40.0392	7.8604286	14.299714	66.73215	9.5331964	23.832911	75.07915	2.9810714	26.813982
Co	0.3150	0.1125	0.1845	−0.046607	0.0658929						
Ni	1.2050	0.4304	0.8037	−0.143321	0.2870357	1.8927	0.3889286	0.6759643	1.89155	−0.000411	0.6755536
Cu	0.9100	0.3250	0.6618	−0.088643	0.2363571	0.51375	−0.052875	0.1834821	0.3722	−0.050554	0.1329286
Zn	47.5350	16.9768	37.4349	−3.607179	13.369607	23.4939	−4.978929	8.3906786	23.1994	−0.105179	8.2855
As	3.3350	1.1911	2.9417	−0.140464	1.0506071	0.8985	−0.729714	0.3208929	0.1193	−0.278286	0.0426071
Sb	0.6650	0.2375	0.5367	−0.045821	0.1916786	1.94835	0.5041607	0.6958393	1.1037	0.302	0.3942
Pb	2.1600	0.7714	3.6256	0.5234286	1.2948571	9.06105	1.9412321	3.2360893	7.2242	−0.656018	2.5800714
Bi	0.0600	0.0214	0.2934	0.0833571	0.1047857	1.2093	0.3271071	0.4318929	1.4608	0.0898	0.5217
Fe	623.0000	222.5000	1632.20	360.42857	582.92857	59,945.385	20,826.138	21,409.066	70,762.385	3863.2143	25,272.28
Al	17.5600	6.2714	22.9680	1.9314286	8.2028571	1.7682	−7.571357	0.6315	1.7682	0	0.6315

[a]Dissolved meant the accumulation metal amount (μg)
[b]Released meant the amount of the metal released from the sample and the unit was $\mu g/cm^2$
[c]Accumulation for release meant the total amount of the metal released from the sample and the unit was $\mu g/cm^2$; experiment temperature was 13–16 °C with common pressure; area of the sample used in the experiment was 2.8 cm^2

Table 4.12 Data of the experiment simulated the conditions on the earth surface for trace metal release from the solid grain (pH 4)

Element	120 h		360 h			840 h			1030 h		
	Dissolved[a]	Release[b]	Dissolved	Released	Accumulation for release[c]	Dissolved	Released	Accumulation for release	Dissolved	Released	Accumulation for release
V	0.1350	0.0470	0.4911	0.1241	0.1711	0.4468	−0.0154	0.1557	0.4338	−0.0045	0.1511
Cr	0.5550	0.1934	0.4611	−0.0327	0.1607	1.3137	0.2971	0.4577	1.5028	0.0659	0.5236
Mn	0.1750	0.0610	6.6713	2.2635	2.3245	10.3509	1.2821	3.6066	10.7955	0.1549	3.7615
Co	0.4050	0.1411	0.2681	−0.0477	0.0934	0.1233	−0.0505	0.0429	0.1824	0.0206	0.0635
Ni	0.5550	0.1934	0.4887	−0.0231	0.1703	1.9521	0.5099	0.6802	1.6781	−0.0955	0.5847
Cu	0.3250	0.1132	1.2891	0.3359	0.4492	0.8454	−0.1546	0.2946	0.2835	−0.1958	0.0988
Zn	28.9650	10.0923	63.6079	12.0707	22.1630	20.4368	−15.0422	7.1208	19.4697	−0.3370	6.7839
As	4.5850	1.5976	4.1529	−0.1506	1.4470	−0.3988	−1.5860	−0.1390	−1.2794	−0.3068	−0.4458
Sb	0.4550	0.1585	0.2021	−0.0881	0.0704	1.4851	0.4470	0.5174	1.3845	−0.0351	0.4824
Pb	2.2650	0.7892	0.7569	−0.5255	0.2637	3.2803	0.8792	1.1430	6.2021	1.0180	2.1610
Bi	0.0700	0.0244	0.3676	0.1037	0.1281	1.6980	0.4635	0.5916	1.0040	−0.2418	0.3498
Fe	492.60	171.64	1071.92	201.8542	373.43	36,566.24	12,367.36	12,740.85	33,197.54	−1173.764	11,567.09
Al	13.2800	4.6272	16.1470	0.9990	5.6261	1.1305	−5.2322	0.3939	1.1305	0.0000	0.3939

[a]Dissolved meant the accumulation metal amount (μg)
[b]Released meant the amount of the metal released from the sample and the unit was μg/cm^2
[c]Accumulation for release meant the total amount of the metal released from the sample and the unit was μg/cm^2; experiment temperature was 13–16 °C with common pressure; area of the sample used in the experiment was 2.8 cm^2

Table 4.13 Data of the experiment simulated the conditions on the earth surface for trace metal release from the solid grain (pH 5.6)

Element	120 h		360 h			840 h			1030 h		
	Dissolved[a]	Release[b]	Dissolved	Released	Accumulation for release[c]	Dissolved	Released	Accumulation for release	Dissolved	Released	Accumulation for release
V	0.1250	0.0446	0.3387	0.0763	0.1210	0.4250	0.0308	0.1518	0.5136	0.0316	0.1834
Cr	0.4550	0.1625	0.3493	−0.0378	0.1248	1.5106	0.4147	0.5395	1.4940	−0.0059	0.5336
Mn	1.0650	0.3804	3.3529	0.8171	1.1975	5.9958	0.9439	2.1414	5.2457	−0.2679	1.8735
Co	0.1400	0.0500	0.1694	0.0105	0.0605	0.3955	0.0807	0.1412	0.3264	−0.0247	0.1166
Ni	0.9400	0.3357	0.2680	−0.2400	0.0957	1.6621	0.4979	0.5936	1.3850	−0.0990	0.4946
Cu	0.8000	0.2857	1.1290	0.1175	0.4032	0.5130	−0.2200	0.1832	0.2084	−0.1088	0.0744
Zn	100.4500	35.8750	43.3422	−20.3956	15.4794	18.7732	−8.7747	6.7047	8.6116	−3.6291	3.0756
As	3.8350	1.3696	3.9469	0.0400	1.4096						
Sb	0.2450	0.0875	0.5345	0.1034	0.1909	1.7476	0.4332	0.6241	1.5460	−0.0720	0.5521
Pb	1.8050	0.6446									
Bi	0.1600	0.0571									
Fe	5.8750	2.0982	20.4591	5.2086	7.3068	247.43	81.0602	88.3670	117.01	−46.5789	41.7881
Al	14.3800	5.1357	18.6372	1.5204	6.6561	1.2492	−6.2100	0.4461	1.2492	0.0000	0.4461

[a]Dissolved meant the accumulation metal amount (μg)
[b]Released meant the amount of the metal released from the sample and the unit was $\mu g/cm^2$
[c]Accumulation for release meant the total amount of the metal released from the sample and the unit was $\mu g/cm^2$; experiment temperature was 13–16 °C with common pressure; area of the sample used in the experiment was 2.8 cm^2

Table 4.14 Data of the experiment simulated the conditions on the earth surface for trace metal release from the solid grain (pH 7)

Element	120 h		360 h			840 h			1030 h		
	Dissolved[a]	Release[b]	Dissolved	Released	Accumulation for release[c]	Dissolved	Released	Accumulation for release	Dissolved	Released	Accumulation for release
V	0.1450	0.0518	0.2479	0.0368	0.0885	0.4609	0.0761	0.1646	0.4319	−0.0104	0.1542
Cr	0.3300	0.1179	0.2728	−0.0204	0.0974	0.9653	0.2473	0.3447	1.8826	0.3276	0.6724
Mn	1.8550	0.6625	1.4867	−0.1315	0.5310	2.4882	0.3577	0.8886	1.4777	−0.3609	0.5277
Co	0.1850	0.0661	0.2273	0.0151	0.0812	0.2162	−0.0040	0.0772	0.1699	−0.0165	0.0607
Ni	49.3350	17.6196	3.2959	−16.4425	1.1771	4.3835	0.3884	1.5655	4.2954	−0.0314	1.5341
Cu	0.7450	0.2661	1.0199	0.0982	0.3643	0.8201	−0.0714	0.2929	0.2557	−0.2016	0.0913
Zn	49.3350	17.6196	30.8315	−6.6084	11.0113	10.0554	−7.4201	3.5912	4.5851	−1.9537	1.6375
As	3.2200	1.1500	2.5346	−0.2448							
Sb	0.4650	0.1661	0.1843	−0.1003	0.0658	1.3574	0.4190	0.4848	1.8784	0.1861	0.6708
Pb	0.0750	0.0268									
Bi	0.0650	0.0232									
Fe	7.7050	2.7518	12.6431	1.7636	4.5154	150.3715	49.1887	53.7041	25.7685	−44.5011	9.2030
Al	14.1600	5.0571	14.6864	0.1880	5.2451	1.3008	−4.7806	0.4646	1.3008	0.0000	0.4646

[a] Dissolved meant the accumulation metal amount (μg)
[b] Released meant the amount of the metal released from the sample and the unit was μg/cm^2
[c] Accumulation for release meant the total amount of the metal released from the sample and the unit was μg/cm^2; experiment temperature was 13–16 °C with common pressure; area of the sample used in the experiment was 2.8 cm^2

Table 4.15 Data of the experiment simulated the conditions on the earth surface for trace metal release from the solid grain (pH 5.6)

Element	120 h		360 h			840 h			1030 h		
	Dissolved[a]	Release[b]	Dissolved	Released	Accumulation for release[c]	Dissolved	Released	Accumulation for release	Dissolved	Released	Accumulation for release
V	0.2923	0.1120	0.8261	0.2045	0.3165	0.9733	0.0564	0.3729	0.8943	−0.0303	0.3426
Cr	0.1443	0.0553	0.0429	−0.0389	0.0164	0.3149	0.1042	0.1207	1.1199	0.3084	0.4291
Mn	5.2170	1.9989	9.7242	1.7269	3.7257	10.7614	0.3974	4.1231	6.0421	−1.8082	2.3150
Co	0.1702	0.0652									
Ni	0.2997	0.1148	2.0883	0.6853	0.8001	2.9570	0.3328	1.1330	1.6983	−0.4823	0.6507
Cu	0.8362	0.3204	1.1810	0.1321	0.4525	2.1309	0.3639	0.8164	0.7392	−0.5332	0.2832
Zn	7.7367	2.9643	9.0201	0.4917	3.4560	24.3856	5.8872	9.3431	5.6966	−7.1605	2.1826
As	3.2079	1.2291	−1.7259	−1.8903	−0.6613	1.0470	1.0624	0.4011	2.2969	0.4789	0.8800
Sb	0.0740	0.0284	2.9872	1.1162	1.1445	2.4209	−0.2170	0.9275	2.0706	−0.1342	0.7933
Bi	0.3034	0.1162	2.0306	0.6618	0.7780	1.6775	−0.1353	0.6427	2.0768	0.1530	0.7957
Fe	7.5591	2.8962	276.5913	103.0775	105.9737	386.5653	42.1356	148.1093	177.0693	−80.2667	67.8426
Al	6.4158	2.4582	1.9074	−1.7274	0.7308	1.9074	0.0000	0.7308	1.9074	0.0000	0.7308

[a]Dissolved meant the accumulation metal amount (μg)
[b]Released meant the amount of the metal released from the sample and the unit was μg/cm^2
[c]Accumulation for release meant the total amount of the metal released from the sample and the unit was μg/cm^2; experiment temperature was 50 °C with common pressure; area of the sample used in the experiment was 2.8 cm^2

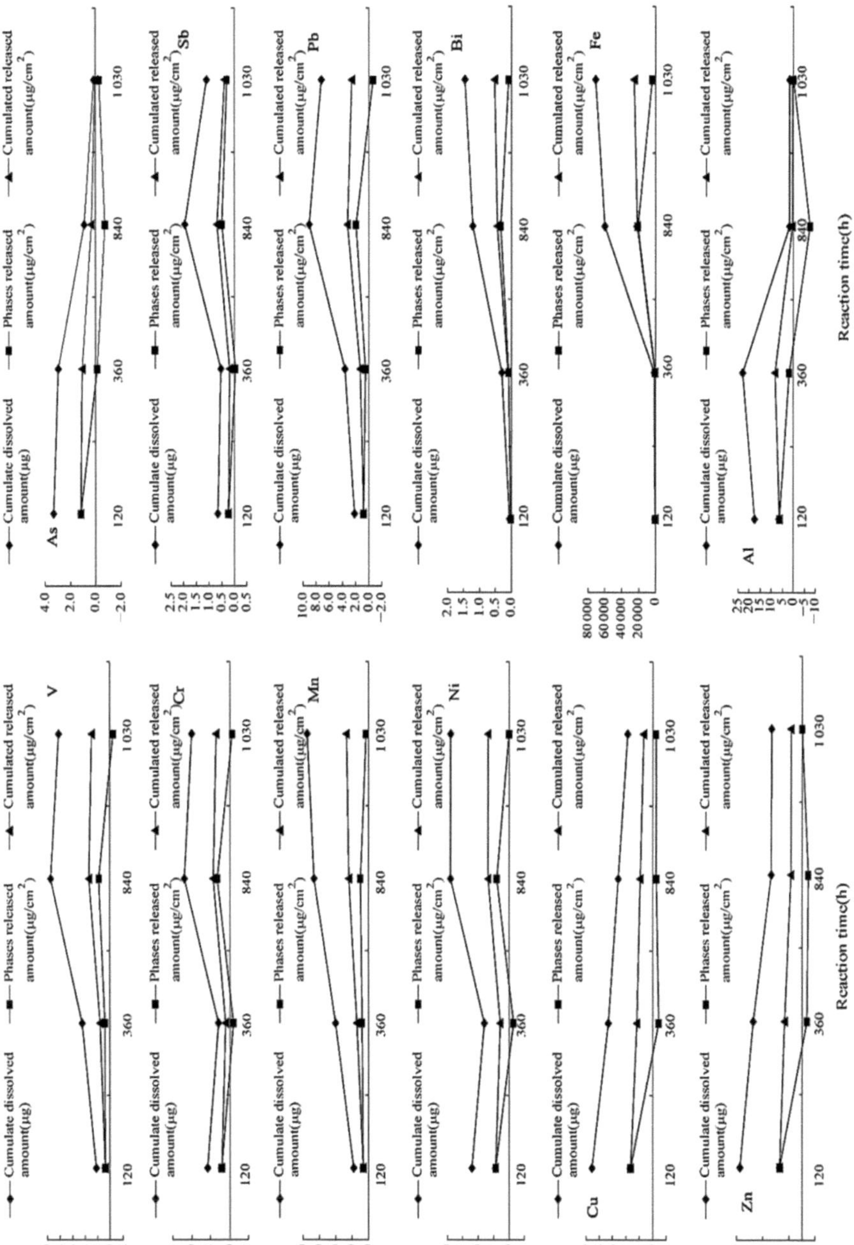

Fig. 4.12 Curve of the trace metal release from the solid grain to water simulated by the experiment under pH 3, 13 °C with 2.8 cm^2 surface area of the sample conditions

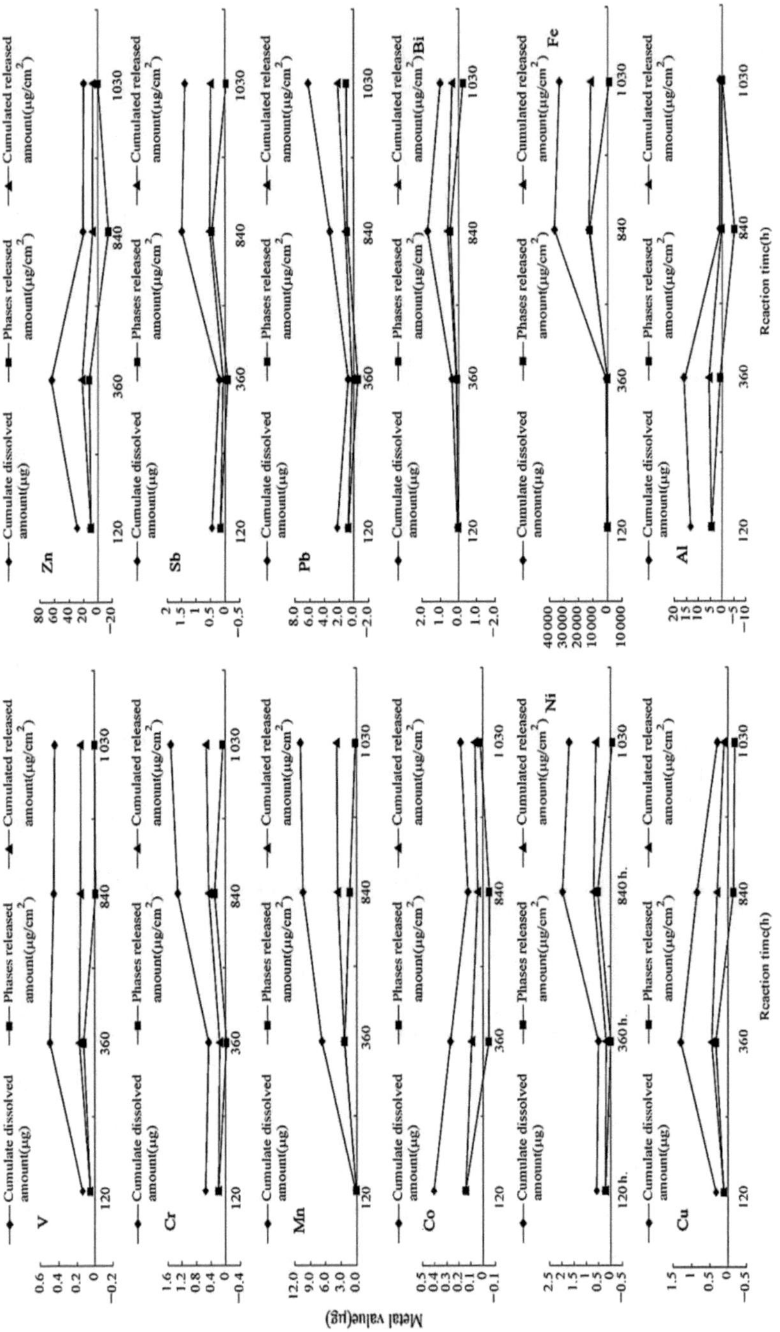

Fig. 4.13 Curve of the trace metal release from the solid grain to water simulated by the experiment under pH 4, 13 °C with 2.8 cm^2 surface area of the sample conditions

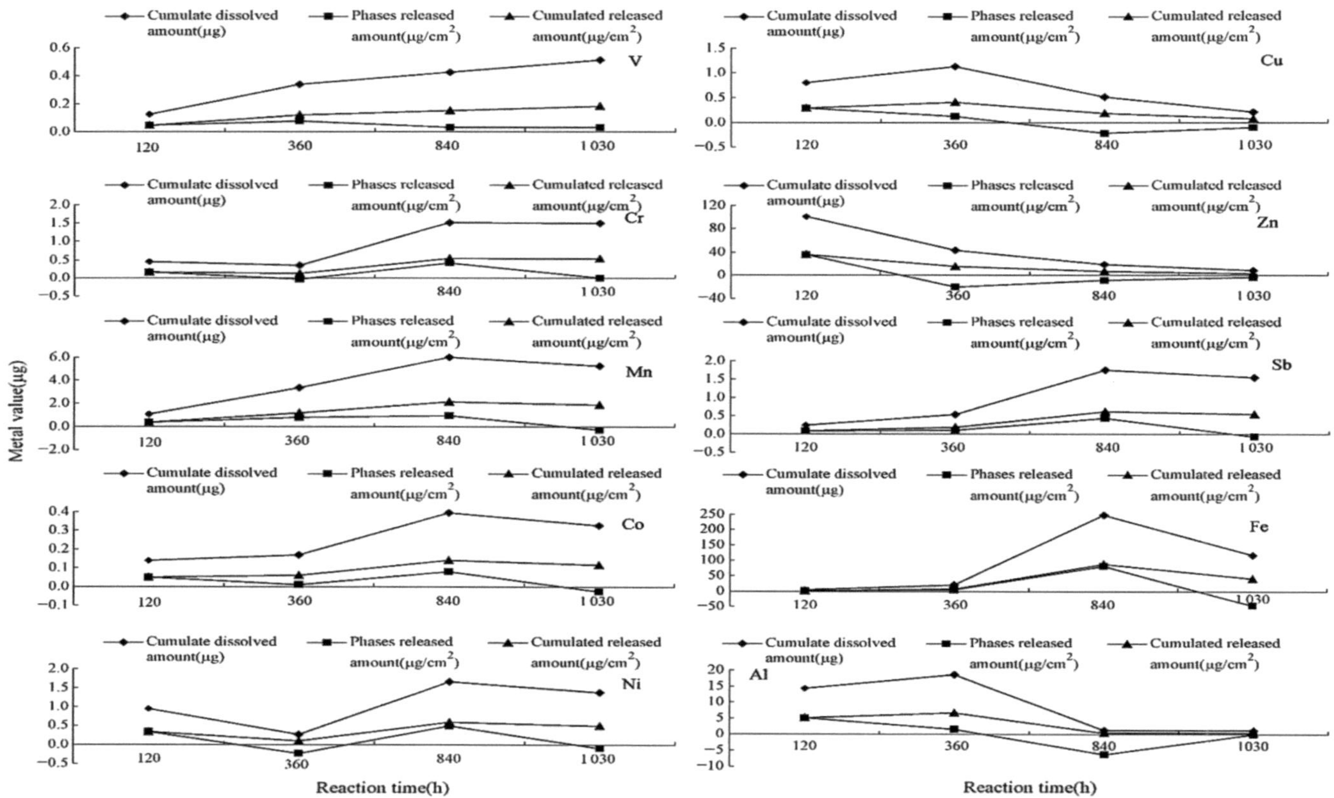

Fig. 4.14 Curve of the trace metal release from the solid grain to water simulated by the experiment under pH 5.6, 13 °C with 2.8 cm^2 surface area of the sample conditions

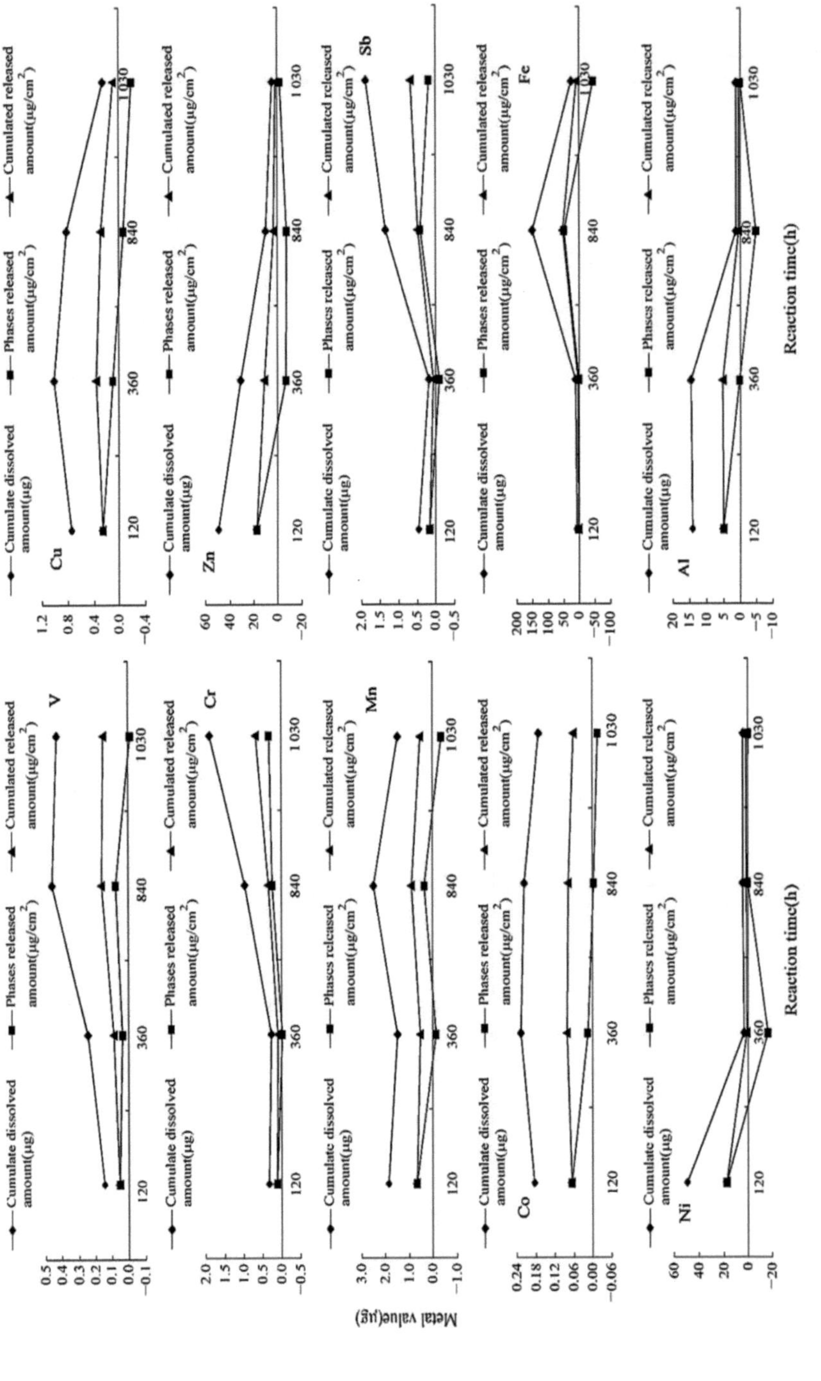

Fig. 4.15 Curve of the trace metal release from the solid grain to water simulated by the experiment under pH 7, 13 °C with 2.8 cm^2 surface area of the sample conditions

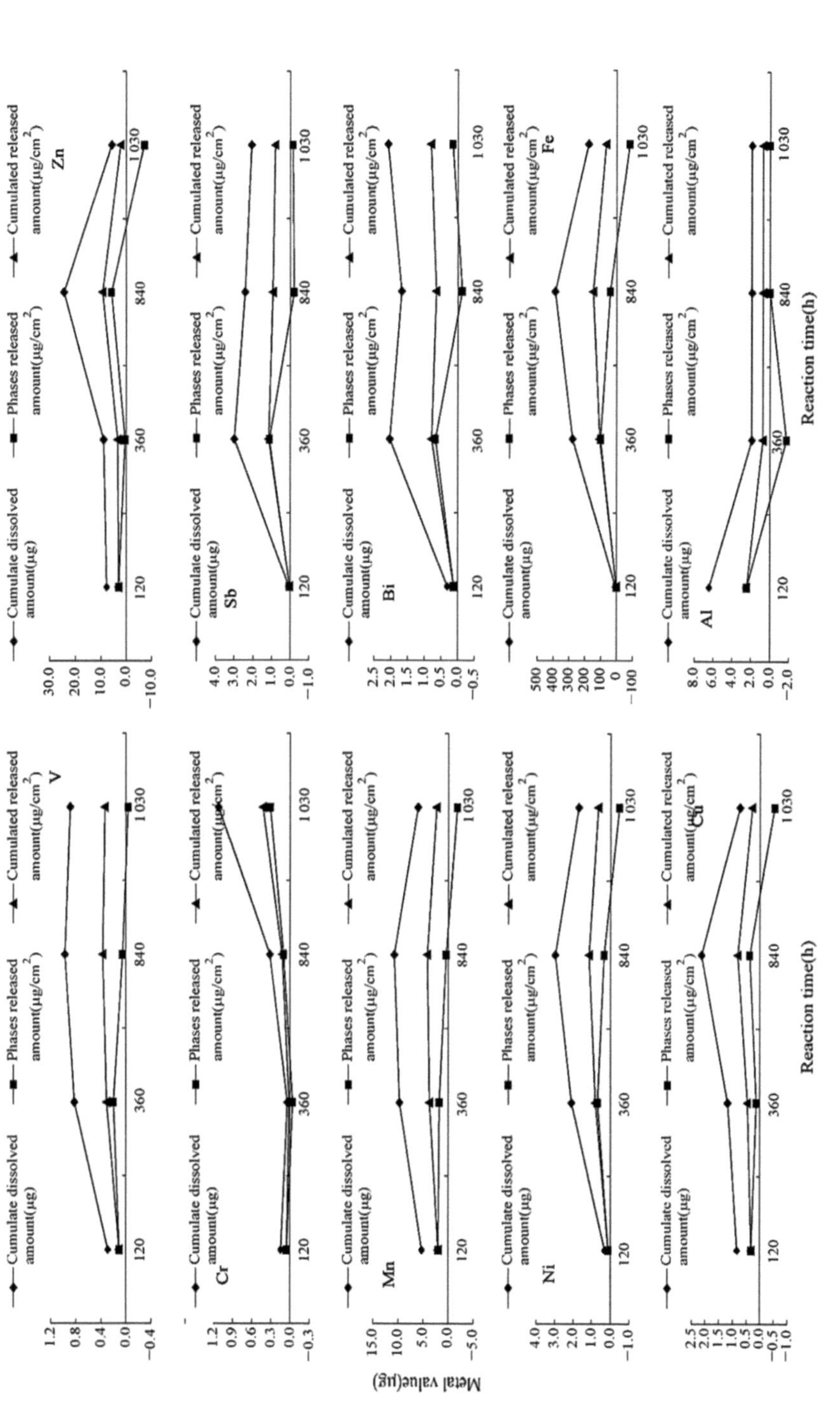

Fig. 4.16 Curve of the trace metal release from the solid grain to water simulated by the experiment under pH 5.6, 50 °C with 2.8 cm² surface area of the sample conditions

The Cu release amount at the initial stage (measured value after reaction for 120 h) at low pH value (pH = 3) reaches the maximum and at pH ≥ 4 conditions reaches the maximum after 360 h. After reaching the maximum of the release amount, the change of Cu concentration in the reaction solution with time is almost consistent under different pH conditions. The effect of pH value on the Cu release from the solid particles is mainly on the differences of the time for the release reaction to reach the equilibrium, and it is more rapid for the release reaction to reach the equilibrium at low pH value than that at higher pH values.

The As release can be measured at only low pH value (pH = 3) condition, and the initial release (for 120 h) is the maximum amount released of it. The As release is not to be measured (its release amount was less than the detection limit of the instrument used in the analysis as presented above) when the reaction system is at pH ≥ 4 conditions in this experiment.

The Pb release amount is earlier to reach the maximum at low pH value (pH = 3) than that at high pH (pH = 4). The time to reach the maximum release of Pb at pH = 3 was 840 h, and it is more than 1030 h to reach the release maximum amount at pH = 4. The release amount and release curve slope (the change rate of the release amount with time) of Pb does not change significantly at the different pH values, as shown in Fig. 4.16.

The release amounts of Bi and Al are not sensitive to pH value, and their releases at the different pH values in this experiment are almost consistent, as shown in Figs. 4.12 and 4.13. This indicated that pH value is of little influence on the release of Bi and Al from the solid particles at earth's surface conditions.

The experimental data show that under the simulated conditions, there are a positive correlation between the metal dissolution from unit surface area of solid samples and element concentration in the samples (particles), and this correlation is not affected by the pH values of the reaction system, as shown in Figs. 4.17, 4.18, 4.19 and 4.20. This phenomenon indicates that the metal concentration in the surface water is related with the direct contacted rock or minerals by solution and metal element concentration in these minerals or compounds. The element release amount from the particles which are with high metal concentration is more than that with low metal concentration. Among the studied 13 metal elements, the initial release amount (the release amount in 120 h) at different pH values (several possible cases under the common earth surface conditions) in these experiments of this work is between 0.05 and 222.50 $\mu g/cm^2$, as shown in Table 4.16. In these release amounts, Co release amount under the pH = 5.6 condition reaches the lowest (0.05 $\mu g/cm^2$) and Fe release amount under pH = 3 reaches the highest (222.50 $\mu g/cm^2$) of the researched elements, respectively.

To sum up, the metal release from the solid particles (rock, minerals, and compounds) into the environment under common natural conditions is different under different influences for different elements, physical and chemical conditions for restriction on metal release.

According to the experimental data above, there are the following summaries to metal release from the solid particles into the environment under common natural conditions:

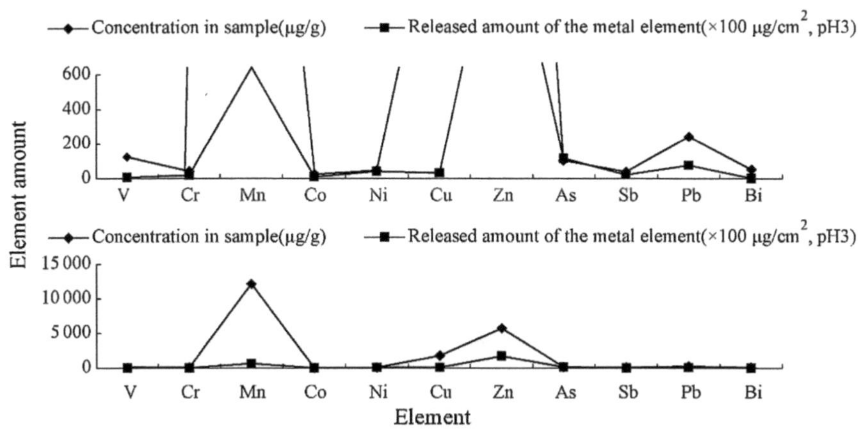

Fig. 4.17 Correlation of the metal concentration in the sample with the release concentration from the sample (pH = 3, t = 120 h)

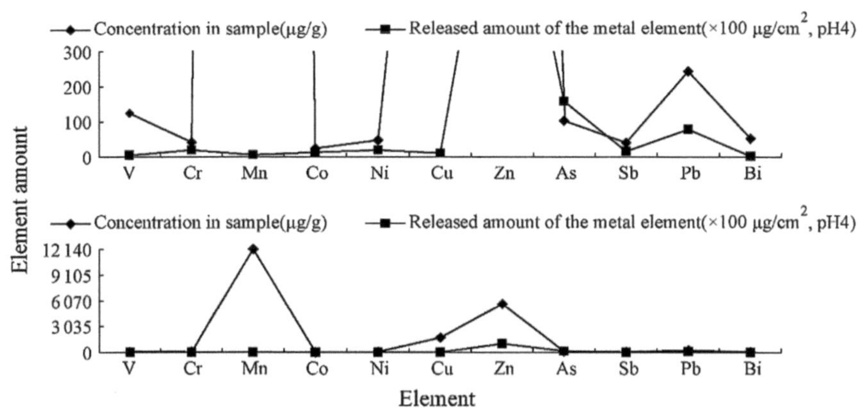

Fig. 4.18 Correlation of the metal concentration in the sample with the release concentration from the sample (pH = 4, t = 120 h)

A. In addition to Al and Zn, most of metal dissolution from solid particles in aqueous solution under common natural conditions increases with the increase of environmental acidity and temperature.
B. The time to reach the dissolution equilibrium is different for elements kinds, environmental pH values, and temperature in environment.
C. The influence from the environmental pH values and temperature on metal release from solid particles is mainly on the influence of the dissolution amount change and the dissolution reaction time to the equilibrium only at the initial reaction process (in 120 h of this work).
D. Under the natural earth surface conditions, the metal release from rock or mineral particles to the environment is with a positive correlation to the element con-

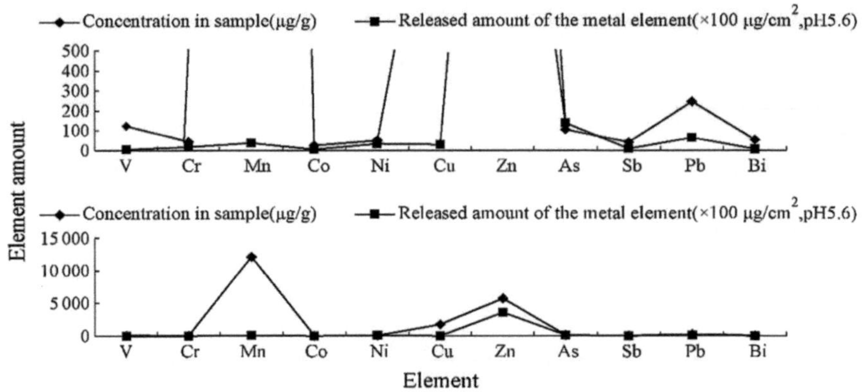

Fig. 4.19 Correlation of the metal concentration in the sample with the release concentration from the sample (pH = 5.6, t = 120 h)

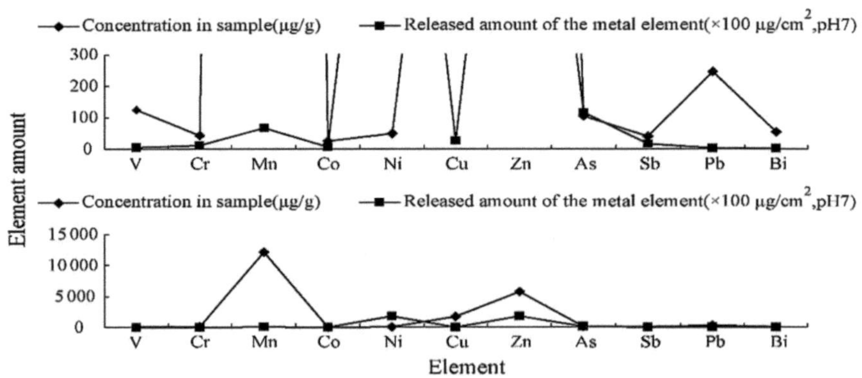

Fig. 4.20 Correlation of the metal concentration in the sample with the release concentration from the sample (pH = 7, t = 120 h)

centration in particles, and this correlation always arises whatever the change of environmental conditions (pH value and temperature) come up. This indicates that the area with metal element high geochemical background will be generally with higher concentration than that with low geochemical background in the surface media (water, soil, and sediment).

The above summaries are based on the assumption that sample components are in the uniform distribution on the solid sample surface; thus, the conclusions are only roughly in reason. In addition, in the experimental process there are many imperfect aspects; for example, some elements have not been achieved the dissolve equilibrium in the experiment, some elements detection limits of instruments cannot meet the requirements (such as As), and all of these may affect the results to be understood objectively and summarized for the law of metal element release under natural conditions. Therefore, the above summaries are only preliminary, and it is

Table 4.16 Comparison between the concentration of metal released from the sample and the concentration contained originally in the sample (unit: μg/cm^2)

Elements	Concentration in sample (μg/g)	Released amount[a] × 100 (pH 3)	Released amount × 100 (pH 4)	Released amount × 100 (pH 5.6)	Released amount × 100 (pH 7)
V	124.40	7.8571	4.7038	4.4643	5.1786
Cr	42.4200	20.3570	19.3380	16.2500	11.7857
Mn	12,140.00	643.93	6.0976	38.0357	66.2500
Co	24.6600	11.2500	14.1115	5.0000	6.6071
Ni	48.4200	43.0360	19.3380	33.5714	1761.96
Cu	1729.00	32.5000	11.3240	28.5714	26.6071
Zn	5713.00	1697.68	1009.23	3587.50	1761.96
As	104.30	119.11	159.76	136.96	115.00
Sb	40.3600	23.7500	15.8537	8.7500	16.6071
Pb	244.20	77.1430	78.9199	64.4643	2.6786
Bi	53.5400	2.1430	2.4390	5.7143	2.3214
Fe	310,200.00	22,250.00	17,163.76	209.82	275.18
Al	43,050.00	627.14	462.72	513.57	505.71

[a]The released amount of the metal was calculated from the value after release reaction 120 h

necessary to illustrate that the element release mechanism from the solid particles to environment in nature is an important direction in environmental science research. It is the hope that this work could help for enlightenment future work or in data accumulation to let human understand on the trace metal original release from the natural body early and deeply, and promote environment protection and human health to a better and deeper development.

References

1. Li J, Zheng CJ, et al. The handbook of environmental background data (Chemical composition data of the precipitation) (in Chinese). Beijing: China Environmental Science Press; 1989.
2. Yang J, Lin NF, Bian JM, Liu WZ, Zhang ZL. The study on environment geochemistry for arsenic in arsenic poisoning area of Hetao plain in Inner Mongolia (in Chinese with English summary). Hydrogeol Eng Geol. 1996;1:49–53.
3. Nickson R, Mcarthur JM, Burgess WG, Ahmedc KM, Ravenscroft P, Rahman M. Arsenic poisoning of Bangladesh groundwater. Nature. 1998;395:338.
4. Nickson R, McArthur JM, Ravenscroft P, Burgess WG, Ahmedc KM. Mechanism of arsenic release to groundwater, Bangladesh and West Bengal. Appl Geochem. 2000;15:403–13.
5. Zhang H. A comparison between heavy metals released from soil and Its efficient speciation extracted by sequential extraction procedure. Chin J Geochem. 2008;27:36–40.

Chapter 5
Probing into the Speciation of Trace Metals and Research Methods

5.1 The Significance and Developing Trends of the Research on Trace Metal Speciation

Under the background of the social industrialization fast development, environment problems have become a bottleneck for the social sustainable development worldwide. The widespread trace metal pollution becomes the emphasis and difficulty issue of environmental problems today, which is pollution in vast areas worldwide and serious consequences in each link of the natural environment and human activities.

With the development of environmental science, many scholars have realized that the total amount of trace metals in environment cannot well indicate the bioavailability, toxicity, and chemical activity of the trace metals in the environment. In fact, trace metals existed in the environment react with solid, liquid, and gaseous substances in the environment through physical and chemical processes to form various forms of the substances with the specific metal species. Therefore, the real species, chemical form, of trace metals in environment controls the environmental behavior and biological effect of trace metals in a greater extent. The systematic understanding and researches on the speciation of trace metals started in the 1970s were carried out by many scholars in past decades in series work with important results, and it also is a very interested topic today to attract a lot of scientists to study on these issues.

5.1.1 The Concept of the Trace Metal Speciation

Trace metal elements exist in the environment in different chemical forms, and the elements with different forms always embody to different chemical and geochemical behaviors in environment as the condition change of the environment. Biochemistry and toxicology research shows that the trace metal elements in different speciation are with different impacts on the environment and human health. Therefore, qualitative

H. Zhang, *Behaviors of Trace Metals in Environment*,
https://doi.org/10.1007/978-981-13-3612-6_5

and quantitative measurement for the different speciation of these elements is an important route to understand and evaluate on their biological toxicity, health risks for organisms, and the rules of their migration and transformation in the systems of the environment.

The concept of chemical speciation of the trace metal elements appears and develops with the development of the science and technology and the requirement arisen in the social application problems. Today, there are different understandings from scholars on the speciation and chemical form of the trace metal elements. The representative opinion on trace metal speciation had been from Swiss geochemist, Werner Stumm, Professor of Harvard University, in which it was believed that the chemical speciation of the trace metals was the actual form of the element existence of ions or molecules in the environment [1]. Chinese scholar Tang Hongxiao believed that the speciation of the trace metal elements and chemical forms included their valences, chemical combination forms, integrated forms, and structural states, which might be with different biological toxicities and environmental behaviors [2]. After the International Union of Pure Applied Chemistry (IUPAC) unifying in the norm for the specification on its relevant terms in speciation analysis in 2000, the main current understanding on the speciation of trace metal elements seems to appear in the field of environmental science and engineering. In this understanding, the speciation, chemical forms or species of the trace metal elements, is a unique form of the element such as isotopic composition, electron or redox state, compound or molecular structure, etc., and the speciation of an element refers to the distribution of specific chemical forms of this element in a certain system. The speciation analysis is the work of quantitative discerning and measuring for one or more chemical forms of the substance in a sample, and the sequential extraction refers to the process to classify and extract for a substance or a group of substances in the sample according to the physical or chemical properties (such as particle size, solubility, combining feature, and activations) [3].

5.1.2 The Trace Metal Speciation Analysis Methods and Mechanisms

At present, the analytical method of chemical speciation is still in exploring and developing stage, the main current methods on this issue can be divided into several approaches such as direct detecting, simulation calculation, simulation experiment (sequential extraction) method, and specific technical means separation method etc. reviewed in summary as follows:

5.1.2.1 Direct Detecting Method

According to the analyzing and measurement method, direct detecting method includes electrochemical method, chromatography, spectrometry, etc. Because of the limitation from the every analysis instrument, during speciation analysis, the combined analytical methods and instrumental techniques are usually used for mutual complementation, which combine to the separation and determination in one. For example, GC-QAAS, HPLC-AAS, HPLC-ICP—MS(AES), and MIP-AES are widely used in the chemical speciation analysis for the trace metal elements.

5.1.2.2 Simulation Calculation Method

Based on the chemical equilibrium in the reaction of a system, the establishment for the corresponding mathematical model to calculate the speciation of the substances in the reaction is a very important one of the methods in the speciation analysis. But the simulation calculation always needs to consider relationship of equilibriums and the possible interaction between different components in the system, and these complex computations require to establish the relevant thermodynamic and kinetic model to answer the questions. So, this method is mainly used in the speciation analysis for trace metals in water system. Now there are some related programs which can used for calculation for speciation analysis of trace metals, but it is still difficult to calculate well the metal element speciation in complex systems.

5.1.2.3 Simulation Experiment (Sequential Extraction) Methods

Sequential extraction is also called "partial extraction," "step extraction," "sequential extraction," or "continuous extraction." In the sequential extraction method, it is performed for simulating the change of variety of possible natural and man-made environmental conditions, and rational use of a series of selective reagents in accordance with the principle from weaker to stronger to dissolve the objective substances for the continuous dissolution of different absorption or containing for the trace metals in the mineral phases [4]. Because the advantages of this method are easy operation and wide application with more and useful information, this method is the most commonly used one in the trace metal element speciation analysis today.

In the sequential extraction method, the extraction or separation for trace metals is mainly dependent on the ability of chemical reagents to dissolve the metal elements combined in different statuses, and these chemical reagents are also known as the extraction agents. Commonly used extraction agents are neutral electrolytes (such as $MgCl_2$, $CaCl_2$), weak acid buffer solution (such as acetic acid or oxalic acid), chelating reagent (such as EDTA or DTPA), reducing reagent (such as $NH_2OH \cdot HCl$), oxidizing reagent (such as H_2O_2), and strong acid (such as HCl, HNO_3, $HClO_4$, HF), etc. Electrolytes, weak acids, and chelating agents can induce the metal elements to be into solution mainly from ion exchange, and strong acid and oxidizing agent can let the metal ions release into the solution mainly by means of destroying the matrix of the sample [5].

Since 1970s in view of extraction and separation of trace metal speciation in sediments and soils, many scholars have established a large number of methods, and according to their operation processes, these methods can be divided into two classes, separate species extraction by single reagent and sequential extraction of multiple species for the trace metals expressed as follows:

1. **The single reagent separate extraction for species of metals**

Single reagent extraction method is a kind of sequential extractions, and its feature is to use a certain extracting reagent to dissolve a specified species of the metals such as water-soluble fraction, transferable fraction, and bioavailable fraction. The separate extraction for a single species of trace metals is usually applied to the pollution survey when the trace metals greatly exceed the geochemical background values. Single reagent extraction method is generally used for the evaluation for the control mechanism of the release of specific pollutants such as desorption generated from adding salts or increasing ligands by adding competitive organic reagents. It is often used to extract the metal effective speciation for plants. Main single reagent speciation extraction methods are shown in Table 5.1.

From Table 5.1, it shows that the reagent scope used in the single reagent extraction method is very wide, including strong acid (aqua, nitric acid, and hydrochloric acid) and neutral salts with no buffer effects ($CaCl_2$ and $NaNO_3$). While other extracting agents, such as buffer salt solution or chelating agent, are generally in common use because they are able to form very stable water-soluble compounds from combining with most cations. In addition, some researchers use hot water to extract boron, and use NaOH to evaluate the effects on the release of trace metals from dissolved organic carbon in soil.

In some cases, the cation coordination can do also play a role in the ion combination, but the substances resolved by the neutral salt are mainly the part of the exchangeable cations of the substances. The trace metals dissolved from dilute acid are mainly exchangeable, carbonate species. The items dissolved from chelating agent include not only trace metals in the exchangeable, but also trace metals chelated by organics and trace metals immobilized on the hydroxide in the sample.

With the development of analytical techniques, more and more evidences show that the relationship between the part of exchangeable and the part of plant utilization metals is very corresponding, and the extraction method is developing in the direction of using less destructive agents [13].This kind of extraction agent is sometimes called "soft extraction agent," including no buffer salt solution, dilute acid, and chelating agent.

In definite soft extraction agents, DTPA, EDTA, CaCl, and $NaNO_3$ are widely used.

EDTA (0.005–1 mol/L) is a kind of strong chelating agent and can form stable chelate with many kinds of metal ions. But it was reported that EDTA could remove not only the metals combined with organic matter and that bound to the oxides,

Table 5.1 Extraction procedure for the certain single speciation of the metal element

Methods	Extractants	References
Acid extraction	$HNO_3 \cdot 0.43$–2.0 mol/L	[6]
	Aqua regia	[7]
	HCl $0 \cdot 1$–1.0 mol/L	[6]
	CH_3COOH 0.1 mol/L	[8]
	HCl 0.05 mol/L + H_2SO_4 0.0125 mol/L	[6]
Chelating agent extraction	EDTA 0.01–0.05 mol/L under certain pH conditions	[9]
	DTPA 0.005 mol/L + TEA 0.1 mol/L + $CaCl_2$ 0.01 mol/L	[10]
	CH_3COOH 0.02 mol/L + NH_4F 0.015 mol/L + HNO_3 0.013 mol/L + EDTA 0.001 mol/L	[11]
Buffer solution extraction	NH_4-$COOCH_3$ + CH_3COOH PH7 buffer solution (1 mol/L)	[12]
	NH_4-$COOCH_3$ + CH_3COOH PH 4.8 buffer solution (1 mol/L)	[6]
No buffer salt solution extraction	$CaCl_2$ 0.1 mol/L	[6]
	$CaCl_2$ 0.05 mol/L	[6]
	$CaCl_2$ 0.01 mol/L	[6]
	$NaNO_3$ 0.1 mol/L	[13]
	NH_4NO_3 1.0 mol/L	[6]
	$AlCl_3$ 0.3 mol/L	[13]
	$BaCl_2$ 0.1 mol/L	[14]

but also the metals bound in some secondary clay minerals. Therefore, EDTA is a non-selective extraction agent, and it can extract the unstable metals and the stable metals.

DTPA (0.005 mol/L) extraction method is also with good correlation for the metals extracted by it and the metals absorbed by plants, and it can be used to show the bioavailability of the metals for plant. The DTPA's complex ability for the metals is weaker than that of EDTA, but its extraction selectivity is higher than that of EDTA although it will sometimes lead to incomplete extraction for the metals. At present, DTPA is widely used in the extraction for trace metals possibly used by plants in soils. The DTPA buffer capacity at pH 7.3 can avoid the release of trace metals caused by the carbonate decomposition, which is considered to be the most suitable extracting agent for samples with limestone matrix. Moreover, there are also studies suggested that the DTPA is able to partially dissolve iron and aluminum compounds [6].

The $CaCl_2$ is recommended by many researchers because it is with about same concentration of Ca^{2+} with soil solution and the Ca^{2+} has the ability to promote the desorption of cations in soil solution [6]. But, studies also showed that $CaCl_2$ and $NaNO_3$ can only extract the metals exchanged out from the sample, and the amount of extraction always is very small.

The effective use of the single reagent extraction method for trace metals in agriculture has been proved by a lot of experiments and practices, and at present many countries have officially accepted or are considering to accept the extraction process of soil filtration for trace metals.

2. **The sequential extraction method for multiple species of metals**

The gradual sequential extraction method is commonly called continuous extraction. In general, the sequential extraction method is to use a series of reagents with continuously enhancing chemical their activities in the system to extract trace metals bound to particular chemical binding forms. Through imitating different conditions of the systems such as that of acidic or alkaline, oxidizing or reducing, and the chelating agent presence environment, the migration and release of the metals are studied and measured systematically, and it can provide more comprehensive information about the metal elements in the system. There are following advantages for this extraction method in the practical use:

A. The extraction process is similar to the leaching process of electrolyte solution for the polluted substances caused by natural and human factors in general nature conditions;
B. In theory, the sum of various fractions of the speciation gained from the continuous extraction shall be equal to the total amount of the metal element. So, the analysis results can be well self-checking calculation;
C. Through sequential extraction method, the environmental activity of trace metals in contaminated system under different environmental conditions can be obtained for judging their harmfulness and potential crises, and providing a scientific basis for the rational use and disposal of the substances [5].

In the process of continuous extraction, the use of extraction agent is generally in the following order: unbuffered saline solution, weak acid, reducing agent, oxidant, and strong acid. The trace metal species usually are exchangeable, are bound to carbonates, can be released under reduced conditions (such as those are bound to Fe and Mn oxides or hydrated oxides), are bound to the compounds capable of being oxidized (such as organic matter, sulfide), and the remaining part after the above extractions (residual). Some of the most commonly used sequential extraction agents and the corresponding trace metal speciation are listed as in Table 5.2.

Water-soluble trace metals can be confirmed in two methods. One is the trace metal concentration in interstitial water of sediment gained by centrifugation or dialysis, and this part of metals in the interstitial water is namely the water-soluble trace metals; the another is with serial operations with specialty reagents from the step-by-step extraction in the laboratory. There are two disadvantages in this method in use of H_2O as the extracting agent, one is that there is no the ability to buffer pH

Table 5.2 Sequential extraction procedure and the speciation for the element

Speciation extracted	Extractants
Water dissolved speciation	H_2O
Exchangeable and weak adsorption speciation	$NaNO_3$ 0.1 mol/L
	KNO_3 0.1 mol/L
	$MgCl_2$ 1 mol/L
	$CaCl_2$ 0.05 mol/L
	$Ca(NO_3)_2$ 0.1 mol/L
	NH_4OAc 1 mol/L pH = 7
Carbonate speciation	HOAc 0.5 mol/L
	HOAc/NAOAc 1 mol/L pH = 5
Fe–Mn hydrous oxide speciation	$NH_2OH \cdot HCl$ 0.04 mol/L + $HOAc/HNO_3$
	Sodium hydrosulfite + Na-Citrate + Natrium sulfurosum (DCB)
Organic speciation	H_2O_2
	NaOCl

value from the H_2O itself, and therefore, the pH value of the system cannot be timely controlled during the extraction process; second disadvantage is that the dissolved metal ions in the process of the extractions may produce serious re-adsorption within the substances in the system, which is possible to cause the extraction amount in false appearance in the process.

Exchangeable metal is extracted by using electrolyte of salts such as strong acid and strong alkali salt, or weak acid and weak base salt, with pH value around 7 to prevent the generation of hydroxide precipitation. The electrolyte solution can release metal cations adsorbed by electrostatic attraction. $CaCl_2$ will not change the pH value of the solution, and the divalent cations have better aggregation in suspension solution; the Ca is the main metal element in soil or sediment, which is an ideal extraction agent.

Carbonate metal is generally extracted at around pH 5 by using acetic acid or acetic acid—sodium acetate buffer solution. This extracting agent cannot destroy all carbonate (such as the containing dolomite carbonate) or cannot selectively play a role only in carbonate extraction. Sometimes, while acting with the carbonate they may partially remove the trace metals combined in organic matters.

The reductive extraction agent mainly affects on Fe–Mn oxide metals. The hydroxylamine in acid solution is widely used to dissolve such forms of the metals, and in fact it could not completely dissolve iron oxides. Ammonium oxalate seems to be the most effective agent (reaction in the dark), but even in very low pH conditions, it will still be precipitated with trace metals. Sodium hydrosulfite, sodium citrate, and sodium carbonate can dissolve oxides and hydroxides, but they can also destroy the silicate structure for the rich iron substances. Therefore, reductive extracting agent neither has ideal selectivity, nor completely extracts the metals bound in Fe–Mn oxides.

Oxidizing extraction agent in sequential extraction can destroy organic matter structures, and at the same time it can also oxide sulfide into sulfate. The most common oxide extraction reagents are H_2O_2 and NaOCl, and after the oxidation, species fraction is extracted, and the effect of using H_2O_2 to extract the metals bound in organic matter will be better.

Sequential extraction method has been an important method in the study on speciation extraction of trace metals, and scholars have carried out many researches on sequential extraction, which contain mainly Tessier extraction method, modified BCR extraction method, and other improvement methods based on the two methods as follows:

(1) **Tessier extraction method**

In 1979, Canadian geochemist Tessier proposed five-step sequential extraction method based on trace metal speciation analysis in sediments which is a very important research technique on speciation analysis, the toxicity, and biological availability for trace metals [15]. This method divided the speciation of metal elements into exchangeable, carbonate, ferro-manganese hydrated oxide, organic matter and sulfide, and residual fractions. Among these fractions defined by Professor Tessier, exchangeable species referred to the metal fraction which was available to be exchanged out from the metals adsorbed by clay minerals in sediments and other substances such as ferric hydroxide, manganese oxide, trace metals absorbed on humus; carbonate species referred to the trace metals combined with carbonate to precipitate into the system; ferro-manganese hydrated oxide referred to the part of trace metals which were combined by hydrous iron oxide and manganese oxide to form concretions in the system; organic matter and sulfide species referred to the trace metals combined or wrapped in organic matter with their special forms, or the part combined in sulfide in the system; the residual species (silicate) referred to the trace metals combined with silicium and aluminum silicate clay mineral and the metals in the lattices of other crystallographic minerals.

The environmental meaning of the speciation of trace metals in the environment divided into five types by Tessier et al. is as follows:

Exchangeable species (EXC) referring to trace metals mean that are absorbed on clay minerals in sediments and other substance grains such as ferric hydroxide, manganese oxide, humus. The trace metals of this fraction can be extracted by neutral salt reagent, and the performance of the extraction agent is mainly evaluated by the properties of cations; $MgCl_2$ is generally used as the neutral reagent. The proportion of exchangeable trace metals is usually small in the total concentration of the metals, and because of its direct bioavailable utilization, it is very important to evaluate the biological effects of trace metal contamination.

Carbonate species (CAR) referring to the trace metals mean that are in the form of precipitation of carbonate with water, which is usually extracted by using NaAc. The trace metals in this fraction will be released in the case of environmental pH reduction, so it is also an important part of the forms of trace metals in environment.

Iron manganese oxide or hydrated oxide species (RED) referring to the part of the trace metals mean that are of coprecipitation with iron manganese oxide, or absorbed by iron manganese oxide, or the hydroxide precipitation itself, which are usually extracted by $NaH_2OH \cdot HCl$–HOAc. Iron manganese oxide or hydrated oxide trace metals will be released in the case of changing the Eh value of the system, it is generally considered to be a component of the active trace metals (effective trace metals to biota), but in the vast majority of cases it is hard to meet its release conditions in nature.

Organic and sulfide species (OXI) referring to trace metals mean that are in the particulate matter entering or wrapped in organic matter in their special forms to be in sequestrate with organic matter or forming sulfide, which are extracted by using HNO_3, H_2O_2, and NH_4OAc as certain schedules (to prevent the extracted trace metals being re-adsorbed in the process of the extraction by adding NH_4OAc). The bioavailability of the trace metals in this fraction is extreme complex, because there are not only some parts that are readily available to be used by biota, but also some parts that are combined in some macromolecular organic complexation with low solubility and low bioavailability.

Residual silicate species (RES) referring to the fraction that is in the crystal lattice of quartz and clay minerals, belonging to the trace metals that are not normally used by biota which generally does not have biological effectiveness.

As of now, the Tessier's method for metal speciation analysis is still widely used in the speciation analysis of the trace metals in sediment and soil, and there is a lot of work carried out by researchers constantly to modify the extraction procedure for it. The extraction schedules and some of the major improvements in the analysis procedure of Tessier's method are listed as in Table 5.3, such as the improved methods from Forstner et al. and Meguelatti et al. [16, 17].

The modifications of Tessier's method carried by researchers are mainly focused on the effect and extent of the extraction agent actions. The ability of the function and adaptive mechanism in the selective conditions of system for the extracting agents used in Tessier method are in the following cases:

A. For exchangeable species: The strong exchange capacity of Mg^{2+} and weaker complexing forces of Cl^- combined each other can avoid dissolving of organic matter, silicate, and metal sulfides [18], although some minor dissolution of carbonate may be arisen (2–3%). But this dissolution can be prevented by shortening the extraction time when this circumstance is in occurrence. In the improving Tessier method, acetic acid/acetate is generally adopted to form the compound of metal and acetate ion, which is slightly more stable than that of metal chloride, and, because of the possible occurrence of buffer of this reagent in the system, it can also reduce the pH value change.
B. For carbonate species: The selection of NaOAc/HOAc buffer solution in Tessier method cannot totally dissolve dolomite, which is suitable for the samples with low concentration of carbonate only, for it will lead to dissolve carbonate in inadequacy. But by reducing the ratio of sample mass to the solution volume (m/v) or reducing acetate concentration, the results of the extraction can be

Table 5.3 Procedures of sequential extraction for the metal speciation suggested by Tessier, Förstner, and Meguelatti

Methods	1	2	3	4	5
Tessier method	Exchangeable	Carbonate	Fe–Mn hydrous oxide	Organic and sulfate	Silicate
	$MgCl_2$ 1 mol/L pH 7	NaOAc 1 mol/L pH 5	$NH_2OH{\cdot}HCl$ 0.04 mol/L25% HOAC	H_2O_2 8.8 mol/L NH_3OAc/HNO_3	$HF/HClO_4$
Forstner method	Exchangeable carbonate	Easily reducible	Stable and reducible	Organic and sulfate	Silicate
	NaOAc1 mol/L pH 5	$NH_2OH \cdot HCl$ 0.1 mol/L	NH_4O_X/HO_X 0.1 mol/L pH 3, reaction in darkness circumstance	H_2O_2 8.8 mol/L pH 7 NH_3OAc	HNO_3
Meguelatti method	Exchangeable	Organic and sulfate	Carbonate	Fe–Mn hydrous oxide	Silicate
	$BaCl_2$ 1 mol/L pH 7	H_2O_2 8.8 mol/L + HNO_3	NaOAc 1 mol/L pH 5	$NH_2OH \cdot HCl$ 0.1 mol/L 25% HOAc	Ashing + HF/HCl

improved. Some people use EDTA to more completely extract the trace metals bound to carbonate, even the metal elements of the organic matter in the system can be extracted too. Because of its super complexing ability, the extraction results for metals by EDTA are not very ideal.

C. For iron manganese oxide or hydrated oxide species: The extraction agent for this speciation is required to have appropriate reduction capability; meanwhile, it is also required to be able to generate soluble compounds with released metal elements. Commonly used alternative reagents include oxalic acid, sodium hydrosulfite, etc. Many researchers often make extraction for the metals bound to manganese oxide and that bound to iron oxide based on the dissolution of manganese oxides which are not affected by stirring time and the concentration of the reagent, while iron oxides require sufficient reaction time and extraction reagent concentration as well as lower acidity of the system. In the extraction, the amorphous oxide and the crystalline oxide of iron can evenly be discerned [19, 20]. The sodium hydrosulfite is strong reducing agent, which can totally dissolve the iron oxide at pH 7–8, which is commonly used in the determination for all iron oxides in the samples. Adding citric acid salt can avoid the precipitation of FeS. But sodium sulfate often contains more impurities, and in the analysis by flame atomic absorption spectrophotometer, the burner will be always blocked because of high salt concentration; thus, its practical application is to be limited [4].

D. For organic and sulfide species: Under the oxidation conditions, the organic matter is degraded to release the metal ion which is adsorbed or bound by it.

Due to the oxidation that may cause partial sulfide to be oxidized, this fraction of the species except containing organic matter also contains sulfide. The H_2O_2 and other reagents commonly used by Tessier method such as NaOH, or chloros all are not very ideal extraction agents. The by-product oxalic acid may be arisen in the oxidizing organic matter by H_2O_2 in system to cause the formation of insoluble salts, and the second time oxidation is required by H_2O_2. At the same time, there is a significant re-adsorption from this agent, so the metals in the system should be extracted again by NH_4OAc/HNO_3 solution after oxidation [15]. The NaOH solution also can destroy the organic matter, and at the same time it will also destroy the aluminum silicate to produce hydroxide precipitation [5]. The chloros can destroy organic matter under alkaline conditions without destroying the amorphous components and clay minerals, and it will produce MnO_4^- with manganese oxides to cause the metals unstable in water. In addition, the extraction time is about 15–30 min for this extraction, and the process of the extraction requires 2–3 times to achieve the complete dissolution for the organics [4]. Thus, this operation time is relatively longer.

In summary, Tessier's extraction method is one of the sequential extraction methods, but there are some insurmountable defects. Firstly, it lacks unified strict standard analysis method, the comparability of analysis results is poor, and the results from its continuous extraction procedure are defined only based on the operations; second, this method has no standard reference substances for analytical quality control in which results are unable to verify and compare within data form the procedures.

(2) **BCR extraction method**

In order to solve such problems that different analysis from operation procedures used by different scholars are lack of consistent procedure and standard reference, and also are lack of satisfactory comparability between the data from laboratories around the world, European Community Bureau of Reference (EC Standards, Measurements and Testing Programme (SM&T)/SM&T, currently named) proposed the BCR three-step extraction method based on Tessier method (Table 5.4) [21]. In the BCR extraction method, the shock speed in the extraction, optimum solid–liquid separation method after extraction, and how let the solid substances in the extraction solution always remain in suspended state in the extraction have been provisioned and described in detail [21]. The BCR three-step extraction method defines the extracted trace metal speciation into weak acid-extractable species, reducible species, and oxidizable species according to the extraction schedules.

At the same time, in order to strengthen the control for the quality of analysis, this method was also used to develop sediment reference materials BCR601, and more than 20 laboratories in 8 countries of the European Union were organized for the comparison experiment in 2 rounds. The correctness of this three-step extraction method had been well proved by the results compared between the results from the laboratories. Jeffrey et al. also verified the good reproducibility of BCR method through long-term studies [22]. In all the experiments above, for various detection techniques used for the determination for trace metals in extraction solution,

Table 5.4 Procedure of the sequential extraction for metal speciation suggested by BCR

Procedure	Extracted speciation	Extraction conditions (for 1 g sample)
1	Weak acid extraction (exchangeable, water and acid soluble)	The sample in tube is extracted with 40 mL of 0.11 mol/L HOAc at 20 °C pass a night with continuous agitation, then centrifuging under 1500 rpm for 20 min.
2	Reducible (iron and manganese hydroxide)	The residue from Step 1 is extracted with 40 mL of 0.5 mol/L $NH_2 \cdot OH \cdot HCl$, adjusting pH to 1.5 with HNO_3. At 20 °C pass a night with continuous agitation, then centrifuging under 1500 rpm for 20 min.
3	Oxidizable (organic matter and sulfides)	10 mL of 8.8 mol/L H_2O_2 (adjusted to pH 2.0–3.0 with HNO_3) is added to the residue from the Fraction 2 step, reacting for 1 h. Then heating it to 85 °C and adding second 10 ml 8.8 mol/L H_2O_2 (adjusted to pH 2.0 with HNO_3) to react for 1 h with agitating occasionally to volatilize it to several mls. Adding 50 ml 1 mol/L NH_3OAc into it to react (adjusted to pH 2.0 with HNO_3) pass a night at 20 °C with continually agitating. Then centrifuging under 1500 rpm for 20 min.

flame atomic absorption spectrometry (FAAS), electrothermal atomic absorption spectrometry (ETAAS), inductively coupled plasma atomic emission spectrometry (ICP-AES), and inductively coupled plasma mass spectrometry (ICP-MS), the data obtained all were used for comparison. The data from these work showed that the acceptable data ratio of ICP-MS method was the highest. BCR extraction method and reagents used in the extraction procedure are shown in Table 5.4.

Ure et al. proposed a four-step extraction BCR method in 1993, which increased the extraction of residual metal species [8], to examine the extraction effect of each step [23, 24].

Rauret et al. proposed the modified BCR sequential extraction procedure in 1999 based on the method proposed by Ure et al., and it increased the concentration of $NH_2OH \cdot HCl$ used in the reduced species extraction in Step 2 from 0.1 to 0.5 mol/L, changed the pH value from 2–5, modified the centrifugal force in each step of centrifugal treatment from 1500 g to 3000 g, confirming the reaction temperature at 22 °C $\pm$ 5 °C [25]. Moreover, the standard references BCR701 (Cd, Ni, Zn, Cr, Cu, and Pb were appointed) were developed according to the extraction procedure. The reproducibility of Cr, Cu, and Pb in the modified schedule was obviously improved, and the matrix effect was better than that of the original one, which was suitable for the analysis of soil and sediments in a wider range [26, 27]. Based on these, the standard reference material BCR601 had been stopped in use. Currently, the modified BCR extraction procedure has been adopted by more and more researchers, as shown in Table 5.5.

Table 5.5 BCR procedure of the sequential extraction for metal speciation modified by Rauret et al. [25]

Procedure	Extracted speciation	Extraction conditions (for 1 g sample)
1	Weak acid extraction (exchangeable, water and acid soluble)	The sample in tube is extracted with 40 ml of 0.11 mol/L HOAc at 22 ± 5 °C for 16 h with continuous agitation, then centrifuging under 3000 rpm for 20 min.
2	Reducible (iron and manganese hydroxide)	The residue from Step 1 is extracted with 40 ml of 0.5 mol/L $NH_2 \cdot OH \cdot HCl$, adjusting pH to 1.5 with HNO_3. At 22 ± 5 °C for 16 h with continuous agitation, then centrifuging at 3000 rpm for 20 min.
3	Oxidizable (organic matter and sulfides)	10 mL of 8.8 mol/L H_2O_2 is added to the residue from the Fraction 2 step (adjusted to pH 2.0–3.0 with HNO_3), reacting for 1 h. Then heating it to 85 ± 2 °C and adding second 10 ml 8.8 mol/L H_2O_2 (adjusted to pH 2.0 with HNO_3) to react for 1 h with agitating occasionally to volatilize it to several mls. Adding 50 ml 1 mol/L NH_3OAc into it to react (adjusted to pH 2.0 with HNO_3) for 16 h at 22 ± 5 °C with continually agitating. Then centrifuging under 3000 rpm for 20 min.
4	Residual	The residue from the Fraction 3 step is digested with 20 ml concentrated HNO_3 (70%), 8 ml $HClO_4$ (60%) and 32 ml HF (40%) until dry, using the following heating regime: 90 ± 2 °C for 6 h; 120 ± 2 °C for 10 h; 190 ± 2 °C for 6 h. 5 mL of 4 mol/L HCl is added into the remaining material at 70 ± 2 °C for 1 h, then dilute it to 80 mL with 0.3 mol/L HCl.

5.1.2.4 Specific Technical Methods in Separation for the Species of Trace Metals

Along with the progress of the analytical method and the deepening of understanding of researchers for the environmental behavior of trace metals, many new points on the speciation of trace metals have been put forward in recent decades, which have been confirmed by more and more research work. The diffusive gradients in thin-film technique (DGT) developed in recent years are the effective speciation determination method with good applicable meaning. This method is a new route adapted well in site to detect quantitatively the accumulation and measure for the effective species of trace metals, and as a new technique, it can be for the bioavailability or effective speciation of trace metals and nutrients (S, P) in soil, water sediment, and the interface between sediment and water samples to be measured in site quantitatively, based on specific technical material properties and technical means [28–30].

In the DGT technique, the Fick's first law is as a theoretical bases, describing steady-state diffusion as a theoretical principle, that is, in proportion to concentration gradients at the section of system and the diffusion material flow vertically through to the unit area of the section in diffusion direction (diffusion flux) in unit time. That is, the greater the concentration gradient, the greater the diffusion flux. According to this principle, the coupling phase (high molecular material or ion-exchange resin) can be separated with system solution by diffusion phase with certain thickness of a permeable ion, and the diffusion phase controlled ion diffusion process, quantitative accumulation, and measurement for the substance being monitored can be achieved [30]. In the elemental speciation analysis of soil system by DGT technique, the following formula is usually used to calculate the concentration of the bulk solution:

$$C_b = M \cdot \Delta g / D \cdot t \cdot A$$

In the formula, C_b is the concentration of trace metals in the solution after the diffusion reactions in the system, M is the diffusion capacity of trace metal diffusion to the DGT device, Δg is the thickness of the diffusion phase material (separated from the solution), D is the diffusion coefficient of trace metals in diffusion phase material, A is the area of ion to be tolerated to diffuse in DGT device, and t is diffusion time. M value can be obtained by the measurement for the diffusion device, Δg, A, and t are all measurable variables, and D is the diffusion coefficient of metal ions in the diffusion phase material at a certain temperature. In conventional applications, under a certain temperature, Δg, A, and D are all constants. So, from the measurement for M value, the C_b can be obtained because there is the function relationship between M and C_b as shown above [31, 32].

DGT technique is only applicable to the determination of the effective speciation concentration of the experimental samples. The key piece of the device is mainly composed of two parts, namely the diffusion phase material device and the combination phase material device. The external of the diffusion phase material device is directly connected with the sample, and the internal diffusion phase material device is closely connected with the combination phase material. Its role is to combine the monitored target objects of the solution, which is transferred by the diffusion phase material to the combination phase material, so that to reduce the concentration of the target objects to be concentration minimum in the solution at here. Its molecular structure of the combined phase material used by DGT technique contains some functional groups that can provide coordination electron pair such as azyl, sulfonic acid group, and carboxyl. These functional groups can rise coordination reaction with trace metal ions. The combination phase material of DGT device is divided into solid and liquid materials. The liquid phase material is easy to handle with the simple operation, and no elution is required during analysis, in which the analytical accuracy and precision of the DGT device are further improved. And so the further research on the DGT technique applied in metal speciation analysis has been developed well in recent years [33–35].

The features and advantages of DGT technique for metal speciation analysis are as follows:

A. DGT technique method can only measure the soluble metal species that can be passed through the diffusion phase material and accumulated in the combination phase material.
B. The mechanism of DGT technique measurement for metal speciation is different from that of other methods, which is a kind of dynamic sampling technique and not a balanced sampling technique. During sampling, a certain concentration gradient in the aquagel is maintained, which is only related to the dynamics of the substance being monitored and the characteristics of the diffusion phase material.
C. The average concentration of the effective speciation of the monitored substances can be obtained during the measurement period.
D. The effective speciation of the ultratrace amount of monitored substances can be measured.

In view of the usability of practical work to environmental issues, the DGT technique may be applied to wider systems. It can be used for the determination for the effective speciation of trace metals in water, sediment, soil, and other systems. The speciation of trace metals measured by DGT technique is only the part of metals which can be passed from the diffusion phase materials and combined in the combination material in soluble form of metals. And this part of the speciation is able to produce toxicity and harm to the biota, which is also the most concerned problem in the field of environmental science and engineering. This method is with the advantages of simplicity and convenience, no need for driven power, and long-time monitoring for pollutant concentration to understand its speciation changes to predict the bioavailability of the metals, so as to be a important complementary method for metal speciation analysis in soil and sediment system together with other analytical methods such as Tessier sequential extraction method and modified BCR sequential extraction method.

In addition, in recent years, the relevant work in this field has found that trace metal speciation is changing in the environmental system as the condition variation [36–38]. More and more researches have showed that trace metal speciation is constantly changing under certain conditions in the natural medium. The bioavailability of trace metals in media in somewhere and sometime may be in variation with environmental conditions; that is, the trace metal speciation parameter is constantly changing with continuous development of environmental conditions. There may be some kinds of transformation between different species under certain conditions. Medium composition, environmental physical and chemical conditions, microbial conditions, time, and so on all are factors that may affect the change of the speciation of trace metals [37, 38]. Therefore, speciation analysis of trace metals is an important and complex field in environmental science and engineering research, and there is still a lot of work to be studied.

5.1.3 The Questions in Speciation Analysis of Trace Metals

The speciation analysis methods presented all above provide the effective route for the study of trace metals in the speciation, pollution, and the secondary pollution risks, but due to the complexity of the problem itself of the speciation of trace metals and the limitation of the current human analytical technology, it should be said that these speciation analysis methods are still with some defects which cannot be overcome in the following:

1. **The selectivity of the extracting agent**

There is a big problem in both separate extraction and sequential extraction, that is, the limitation of the selectivity of the extraction agent, which may cause some overlap at a certain degree for species fractions in different extracting agents during dissolving out. This point has been confirmed by many researchers in theory analysis and experimental work. For example, after each step extraction, to extract again by use $BaClO_4$, operated by Bermond, the re-absorbed species of the metals were again also dissolved out [39]. Amnachalam et al. suggested that it seemed more reasonable to name the metal speciation based on the reagent used in the extraction experiment, such as ammonium acetate-extractable speciation or H_2O_2-extractable speciation [40].

Professor Tessier also considered that, because of great difference in carbonate solubility product for different metal elements, incomplete dissolution phenomena in the extraction process might arise [15]. Because of the limitation from the ability of the oxidant for the extraction for organic speciation, some hard to be oxidized organic matter could not be dissolved to release metal ions, while some sulfides with good solubility might have partial dissolution in the steps from the first step to the third step in the extraction procedure proposed by Tessier et al.

2. **Re-adsorption of trace metals in the extraction process**

Another drawback of the sequential extraction method is that it is impossible to avoid the metal ions to be secondary adsorbed; that is, metal ions are adsorbed by particles in the solution after being dissolved by extraction solution. The extent of its adsorption is related to the properties of metal elements, the nature of the sample, and the organic matter concentration in the sample.

Jeffrey et al. researched the method of avoiding the secondary adsorption for metals by means of adding chelating agent in the extracting solution. When the concentration of nitrilotriacetic acid sodium salt in the extracting agent was 200 mg/L, it could get a higher recovery rate of the extraction, and at the same time there was no effects on the non-target components in the sample extracted [41]. Shiowatana et al. and Chomchoei et al. designed a set of flowing extraction method, and the advantage of flowing extraction to quickly take away the dissolved metal ions for avoiding re-adsorption, so there was a good recovery rate in the added standard reference recovery [42]. In the study, it was also found that the existence of organic matter would promote the phenomenon of re-adsorption in the extraction stage of first two steps, and reducing the flowing rate and the concentration of the extracting solution might reduce the re-adsorption [43]. Bermond had made deep research on the role of H^+ in the sequential extraction method [39], and the results can be used to explain the problem of selectivity and re-adsorption. Bermond found that H^+ concentration in each extract reagent was in a great influence on the extraction efficiency, and in the extraction process, there were obvious H^+ consumption and the variation of pH value in the reaction terminal period of the system. The consumption of H^+ depends on the acidity of the extraction agent, and the greater the acidity of the extract is, the more H^+ consumed is too until a maximum value limitation of the acidity of the extraction agent. The H^+ consumption can be attributed to the following two reasons: firstly, the consumption of H^+ is due to the exchangeable cations in the particles of the solution being replaced; secondly, the presence of $CaCO_3$ and FeOOH in the soil will consume H^+. The result of the H^+ consumption is the change of the pH in the reaction terminal system, which can affect the extraction rate of the metal ions.

3. **The errors from operation due to time-consuming and tedious operating procedure**

Another drawback of sequential extraction for the metals is that it is time consuming. In Tessier method, 5 days and nights are required, while in BCR method it is 50 h. And more serious deficiency than time consuming is tedious operation steps, which would increase the possible errors from operation.

By comparison, separate extraction method is much more time-saver for the operation process, and it can effectively prevent the operation error caused by the increase of the extraction steps. Filip et al. used five equal soil samples, using the conditions of Tessier method for respective separate extraction, and the results obtained was no obvious differences with the results by Tessier method except for the large difference in the oxidation species [44].

Maiz et al., in view of rapid sequential extraction method aiming at the study on the availability of metal elements, divided the speciation of metal elements in soils into three types, and respectively by using 0.01 mol/L $CaCl_2$, 0.005 mol/L DTPA + 0.01 mol/L, $CaCl_2$ 0.1 mol/L triethylamine (TEA) mixed liquor, aqua and HF mixed acid extract agent extract them as certain procedures, to respectively extract flowing state (exchangeable species), fluidized state (including complex species, adsorbed, and carbonate species) and residual metals. This method for the species obtained all got very good results [45, 46].

The method of using Teflon reactor with microwave digestion can significantly speed up the extraction rate under the high temperature and pressure conditions, thus saving the reaction time. For example, microwave sequential extraction method proposed by E. Campos, under the premise of keeping the result same as that from traditional method, was used to extract completely carbonate, Fe–Mn oxide, and organic matter species only in 2 h [42]. In addition, online sequential extraction-flame atomic absorption spectrometry proposed by Shiowatana was considered to be a new way of thinking for rapid, simple, and accurate trace metal speciation analysis [42].

4. **The influences from the method of sample pretreatment**

Sample pretreatment will change the chemical speciation of the element, which depends on the sample processing method and the processing time. Zhang et al. has carried out different pretreatment on reservoir sediments respectively under the condition of oven drying under 85 °C, air drying under 20 °C, and freeze drying and compared the results from the same elemental speciation analysis method between the above three samples treated with different conditions and an untreated fresh reservoir sediment sample. The experimental results show that these three methods cannot maintain the original chemical speciation of metal elements, especially, the change on exchangeable and carbonate is more obvious. For air-dried sample, its speciation distribution of Zn ion would have significant change [47]. While the researches of Brewad et al. showed that the solubility of manganese and iron would decrease dramatically after air drying, the oven drying process can make the solubility of manganese increase several times [48].

Compared to the sediment in reductive environment, the oxidizing environment of the soil would cause the soil sample to be relatively stable in the pretreatment process. But, there are few literature reports on whether the soil sample pretreatment method may affect the speciation distribution of metal elements in the sample.

5. **The feature and limitation of the information revealed by DGT technique**

As a new kind of trace metal speciation analysis technique, DGT is with its special technical principles and technical means, the elemental speciation concentration obtained by this method is the comprehensive concentration of metallic particles with certain geometry size and migration ability in the system. This concentration is approximately equal to the effective speciation concentration (the sum of speciation concentration except residual species) from the sequential extraction method. Of which, the concentration of water-soluble and exchangeable species should be the main concentration. But there is no definite evidence to show the amount relationship of water-soluble, exchangeable, and other effective species concentrations in the speciation concentration obtained by this method. The boundaries of the information in different speciation are still fuzzy. Meanwhile, due to the high cost and large consumption of diffusion film materials, the analysis cost of DGT method is generally higher than that of other analytical methods.

In the process of determining the trace metal species by using DGT technique, there is no need to add any reagent in the system, therefore avoiding the problem of species distortion in the system brought by the chemical reagent addition. And it is with the advantage of easy to operate in water, sediment, and soil, and such technique is a very valuable method to be used in evaluating the bioavailability of trace metals in soil and sediment. The concentration data measured by DGT can be understood as the most likely the part with biological function in the total concentration of the trace metals in system.

In a word, a variety of single reagent extraction methods, classic Tessier five-step sequential extraction, BCR three-step sequential extraction and its improved methods, and DGT analysis method, all provide important method support for the study on bio-geochemical cycle, degree of environmental pollution, and the effect of trace metals. It can be understood that how to obtain the real value of the total amount of trace metals in the sample, how to determine the real concentration of various species (chemical forms) of trace metals without changing the nature of the samples in the extraction process, how to study the establishment of a wider general order for extracting chemical speciation analysis are still the frontier problems of environmental geochemistry, and it must be one of the hot spots and difficult points in environmental geochemistry field. To this end, the study on sequential extraction of chemical speciation analysis on soil, sediment, and other environmental media shall be focused on simple, efficiency, and easy-to-be-standardized method research. So as to better meet the needs of the current pollution control, environmental protection works.

5.2 The Pertinence on the Analysis Procedures of the Speciation Analysis for Trace Metals

5.2.1 The Speciation Analysis for the Trace Metals Dissolved in Water and Contained in Particulates

Since 1970s, speciation analysis of trace metals had become a hot research topic in the field of environmental science. Over the years, the scholars have carried out a large number of analysis and research work on trace metal speciation. Switzerland's

Stumm and Germany's Forstner had made special approaches to the speciation of trace metals. At present, there are different opinions about the definition of "chemical speciation," and as mentioned before, some people think that the chemical speciation is the practical form of an element in the presence of an ion or molecule in environment. Some scholars believe that the elemental speciation actually includes valence state, compound form, combined and structure forms, such as ionic state, complexing form, dissolved state, and particle form of metals [14, 49]. The so-called speciation actually includes such four aspects as valence state, compound form, combined form, and structure form, which may show different biological toxicity and environmental behavior [6]. The speciation analysis usually refers to the analysis for the valence state and the complexing form of the metals and element-related life. The speciation analysis of pollutants is to detect the various valence state, complexing form, and the concentration distribution of the pollutant elements in the sample by certain physical and chemical methods associated with testing instruments, so as to make the chemical speciation on the environmental behavior of the pollutants be distinguished at molecular level. The main purpose of trace metal speciation analysis is to determine the concentration of trace metals which are with biological toxicity.

The speciation of trace metals in water environment can be divided into dissolved species (dissolved in water phase) and particle species (suspended in the suspended phase or deposited in sediments). The speciation analysis method is accordingly divided into the speciation analysis of dissolved trace metals and the speciation analysis of trace metals in particle state.

1. **The speciation analysis for trace metals dissolved in water**

After filtration through 0.45 filter membrane, the water sample re-acidated is directly tested to obtain the concentration as total dissolved state concentration, and if the filtered water samples without acidification, the part of metal with electrical activity to be detected directly is called unstable state concentration; the rest part after above two steps is the concentration of combined state of the elements. The unstable state of elements includes free and simple inorganic complexing ions. By Chelex-100 ion-exchange resin exchange column and batch equilibrium method, the combined state of the elements can be divided again into middle unstable state (including the part of complexing weakly with organic matter or colloids), slow unstable state (including complexing part combined strong), and inert state (no sensitive to the resin but strongly associated with the organic matter or colloid in the water).

2. **The speciation analysis for trace metals contained in particulates**

Trace metals in the form of particles (suspended and sedimentary state) are usually separated by a variety of chemical extraction methods to separate the trace metals into speciation with different bioavailability. For the extraction and separation of trace metal speciation in sediments and soils, as mentioned above, some extraction methods established by scholars can be generally divided into separated extraction method for separated species and the sequential extraction method for all species as follows:

(1) **The separated extraction method**

The separate extraction method is characterized by the use of a certain extract reagent directly to dissolve a specific speciation of the metals, such as water-soluble species/transferable species, and biologically available species. This method can be applied to the investigation for the pollution when the trace metals are much more than the geochemical background values. Ure pointed that this method was to evaluate the part of the trace metals in the granular medium that could be absorbed and utilized by organisms including animals, plants, and microorganisms, or the part that could affect the biological activity [12]. Usually, this part of trace metals is called the effective species. According to the composition, the nature, and the extraction of target to the metal elements of the samples as well as the different purposes of extraction, the reagents used in this method are different, which are mainly divided into acid, chelating agent, neutral salt, and buffer. At present, these extracting agents such as DTPA, EDTA, $CaCl_2$, and $NaNO_3$ are the widely used ones.

(2) **Sequential extraction method**

The sequential extraction method is to use a series of reagents for continuously enhancing chemical activity to extract trace metals bound to a particular chemical group. By simulating different environmental conditions such as acidic or alkaline environment, oxidizing or reducing environment, and the environment of chelating agent presence, this method can study systematically the mobility or the release of metal elements, which can provide more comprehensive information of the current existence or occurrence of the elements.

Sequential extraction method has been an important method for speciation extraction of trace metals, and the scholars have made many researches on sequential extraction, including seven species classification, five species classification, and four species classification, as mentioned above. Gambrell pointed out that geochemical speciation of trace metals in sediments included seven types, such as water-soluble, easy exchange, inorganic compound precipitation, macromolecular humus binding, absorption or adsorption of hydroxide precipitation, sulfide precipitation, and residual species [50]. Wan pointed out that the trace metals in sediment should be divided into soluble, exchangeable, carbonate, Fe–Mn oxide, organic, and residual species [51]. Tang believed that chemical speciation classification should not be too cumbersome, and combined with the environmental conditions, it could be divided into active, buffer, and stable species [6]. Tessier used the method of sequential extraction to divide presence speciation of metals in solid particles into exchangeable, carbonate, reducible, oxidizable, and residual, and this method had been frequently used in environmental research [15, 52]. In addition, the four-step extraction method proposed by BCR/SM&T has been also widely used presently, dividing into acid-extractable, reducible, oxidizable, and residual species.

The results of sequential extraction method depending on physical and chemical properties of the medium and the concentration of elements in samples are usually with some obvious deviations. The main causes are the selective restriction from the extracted speciation, redistribution of trace metals in the extraction phases, arising overloading of chemical systems from too high trace metal concentration in the sample, etc. However, the sequential extraction method can still provide complete information for predicting the distribution, activity, and bioavailability of trace metals in the medium. According to the actual situation in the field of speciation analysis, in order to reduce the deviation caused by different continuous extraction methods, at present there are still different scholars putting forward their own views. For example, Rauret et al. had suggested that the sequential extraction method proposed by Tessier was suitable for the speciation analysis of trace metals [21]. It should be said that in the practical work today, the BCR sequential extraction method seems to be used more frequently.

5.2.2 The Speciation Analysis for the Trace Metals in Soil

In the environmental problem research, it often needs to face the cases of the solid medium such as the water sediment or soil. Therefore, the speciation analysis of trace metals in sediments and soils is most commonly used in daily work. The two different types of media are with different causes, and the physical and chemical properties of them are also different. But from the perspective of analysis for the speciation and their chemical composition, there is no essential difference within these two kinds of media. In the earth surface environment, these two kinds of materials are loose substance system based on inorganic aluminum silicate minerals. In addition, they also often contain a certain amount of organic matter, oxide (or its hydrate), and small amount of other salts, such as carbonate, sulfate, and phosphate. Therefore, from the perspective of extraction and separation of components, speciation analysis of trace metals in sediments and soils has no essential difference with the above speciation analysis of trace metals in water bodies or speciation analysis for trace metals in atmospheric particulates. This is also the reason that the speciation analysis method based on speciation study of trace metals in water sediment proposed by Tessier et al. is frequently used and improved. At present, there is no significant difference between the application of soil and sediment to the extraction and analysis procedure in the speciation analysis of trace metals.

Currently, on the basis of the comparison of the various methods and the causes discussion, commonly used methods in the speciation analysis of trace metals in soils are the same with the speciation analysis methods of the particulate trace metals in water system (sedimentary phase or particle state of metals), including five-step sequential extraction speciation analysis method proposed by Tessier et al. and four-step sequential extraction analysis method proposed by BCR/SM&T.

5.2.3 The Method for Speciation Analysis of Trace Metals Used in This Work

In this study, environmental media for speciation analysis of trace metals include soil, water, and water sediment. When selecting the speciation extraction and analysis method, the commonly accepted method was focused on. It is because that on one hand the speciation analysis of trace metals is a controversial issue, and on the other hand some new methods still have some limitations in the application and validation studies. At the end, the commonly accepted method as a main method was used in the work, and in the earlier stage (1999–2005) the five-step sequential extraction and analysis method proposed by Tessier et al. was used for speciation analysis of trace metals; then, (after 2005) the four-step sequential extraction analysis method proposed by BCR had been used in this work.

In the discussion based on above work in this book, in order to facilitate the comparison between data, according to the above situation some roughly merge or integration process is made for the quotation of these speciation character data. For example, the first two speciations in the five-step sequential extraction speciation analysis method proposed by Tessier et al., exchangeable and carbonate species, are regarded as the same of the acid-soluble species in the four-step sequential extraction analysis method proposed by BCR/SM&T. And the Fe–Mn oxide, organic, and residual species in the five-step sequential extraction speciation analysis method proposed by Tessier et al. are corresponding to the reduced, oxidation, and residual species in the four-step sequential extraction analysis method proposed by BCR/SM&T. To this processing, we do point out and propose to pay attention on it for readers.

5.3 Suggestions and Improving on the Trace Metal Speciation Analysis Procedures, the Method of Three-Step Speciation Analysis

5.3.1 The Meaning of the Method of Three-Step Speciation Analysis (Three-Step Procedure)

As mentioned above, there are still many deficiencies and defects in the current speciation analysis, such as incomplete reaction between target and action agent, the influences on the effectiveness of the extraction caused by the experimental process, the accuracy difference in the results from the ratio of sample and reagent amount, the accuracy of concentration from the particle size distribution and mineral composition in the samples, and the speciation change during sample preparation process. And different extraction methods and different reagents used in the extraction process all will bring an impact on the results, which lead to the lack of comparability between data gained from different extraction methods. In view of this, based on the results of previous studies and laboratory work carried out by and in the Pollutant Behavior Chemistry Group, Shanghai Jiaotong University, recent years, this work intends to explore a method to be convenient, applicable, and efficient for the speciation analysis of trace metals.

5.3.2 The Proposal of the Three-Step Procedure

In view of five-state extraction for trace metals of Tessier method, which includes exchangeable, carbonate, reducible, oxidizable, and residual species. The exchangeable trace metals bound in clay minerals and other components and the carbonate trace metals bound in carbonate minerals may be released into the system solution as the pH decrease of the environment, which belong to the parts of the trace metals available to biota. The parts of metals bound in Fe, Mn or Al oxides and their hydroxides as well as the trace metal complexing in organic matter or bound in sulfates can only be released when environmental Eh value changes. These parts of the metals, in usually condition they are relatively difficult to be used by biota compared that bound to exchangeable and carbonate species. Residual species of trace metals is the part to be trapped in clay crystal lattice of silicate minerals, on the normal earth surface conditions it may be difficult to be used by organisms, and it generally does not have biological activity. Therefore, in this study for trace metal speciation in soil and water sediment, the exchangeable trace metals and carbonate trace metals defined in Tessier method are here called as "easy bioavailable species" trace metals, the Fe–Mn oxide species trace metals and the organic species trace metals here are called as "bioavailable species" trace metals, and the residual species trace metal is here called as "non-bioavailable species" trace metals. Such trace metal speciation analysis methods seem to be closer to the actual situation, and in the practical work, it is beneficial to the direct understanding and evaluation of the biological effects of trace metals. In this work, this metal species division and corresponding analysis schedule and procedure was called as "three-step procedure."

According to the similarity of the release conditions of trace metal elements in nature from the medium, this method divides the speciation of trace metals, and at the certain extent it is to be clear for the degree of difficulty of the trace metals to be used by biota as well as their secondary pollution risk. It is hope that this method will be useful to play a positive means on quick and efficient evaluation for the trace metal pollution in environment. The concluded extraction procedure of the three-step procedure is shown in Table 5.6.

Table 5.6 Procedure of the three steps of the sequential extraction for metal speciation in soil and sediment (for 1 g dry sample)

Procedure	Extracted speciation	Extraction conditions
1	Easy bioavailable species	The sample in tube is extracted with 40 ml of 0.11 mol/L HOAc at room temperature for 16 h with continuous agitation, then centrifuging under 4000 rpm for 15 min. Then the supernatant taken to be constant volume and analyzed.
2	Bioavailable species	The residue from Step 1 is extracted with 10 mL of 8.8 mol/L H_2O_2 (adjusted to pH 2.0–3.0 with HNO_3) reacting for 1.5 h at room temperature, then under 85 ± 2 °C in water bath reaction for 4.5 h to let the solution to be 3 ml; Then adding 10 ml 8.8 mol/L H_2O_2 extracting for 1 h under 85 ± 2 °C in water bath to let the solution to be 3 ml; Then adding 40 ml 0.5 mol/L $NH_2 \cdot OH \cdot HCl$ at room temperature for 16 h with continuous agitation. Then adding 5 ml 5.5 mol/L NH_3OAc into it to react and centrifuging at 4000 rpm for 15 min. Then the supernatant taken to be constant volume and analyzed.
3	Non-bioavailable species	The residual from last step is digested by HCl–HF–$HClO_4$ (aqua regia) heated to be true solution for analysis (for As and Hg the digestion temperature below 100 °C).

5.3.3 The Comparison Experiments Between the Results Obtained Respectively from the Mainstream and the Three-Step Procedure Analysis Methods

In order to verify the applicability and analysis effects of the "three-step procedure" in a large range of medium, in this work three samples, raw coal, soil, and water sediment, are selected to be used in the comparison experiments. The representative methods of the current mainstream trace metal speciation analysis are selected, which are modified BCR sequential extraction method and separate extraction method (DTPA, 5Na salt used in this study) to be comparison with the "three-step procedure," so as to make a judgment for the effectiveness and applicability of the method.

Experimental samples include blind coal sample (Shanxi Dayan Coal Mine), soil sample (As-polluted soil in the Hetao Area, Inner Mongolia), and water–sediment sample (Huangpu River modern sediment, Shanghai). These samples were experimented respectively by the modified BCR sequential extraction procedure, DTPA—5Na salt extraction procedure, and three-step extraction procedure to make speciation extraction and analysis of trace metals.

Table 5.7 Procedure of the sequential extraction for metal speciation suggested by BCR

Procedure	Extracted speciation	Extractants	Extraction condition
1	Water and acid dissolution	40 ml/g(sample) 0.11 M CH_3COOH	16 h, room temperature, 4000 rpm 20 min
2	Reducible	40 ml/g(sample) 0.5 M, $NH_2OH \cdot HCl$	pH 5, 16 h, room temperature, 4000 rpm 20 min
3	Oxidizable	2 × 10 ml/g 8.8 M H_2O_2 evaporation to dry, 45 ml/g 1 M NH_4AC	pH 2, 4000 rpm 20 min
4	Residual	HCl–HF–$HClO_4$	Digestion (for As and Hg digestion temperature below 100 °C)

Table 5.8 DTPA-5Na procedure of the sequential extraction for metal speciation

Procedure	Extracted speciation	Extractants	Extraction condition
1	Bioavailable speciation	10 ml/g(sample) 0.05 M DTPA-5Na + 0.01 M $CaCl_2$ + 0.1 M TEA[b]	pH 7.3, 2 h at room temperature under 4000 rpm for 20 min
2	Non-bioavailable speciation	HCl-HF-$HClO_4$	Digestion (for As and Hg digestion temperature below 100 °C)

[a]DTPA is the abbreviation of diethylenetriamine pentaacetic acids
[b]TEA is the abbreviation of triethanolamine

Table 5.9 Three-step procedure of the sequential extraction for metal speciation

Procedure	Extracted speciation	Extractants	Extraction condition
1	Easy bioavailable speciation	40 ml/g(sample) 0.11 M CH_3COOH	16 h, room temperature, 4000 rpm 20 min
2	Bioavailable speciation	10 ml/g 8.8 M · H_2O_2, evaporation to dry (using same dosage to act two times), 40 ml/g(sample) 0.5 M $NH_2OH \cdot HCl$,5 ml/g 5.5 M NH_4AC	pH 2, 16 h, room temperature, 4000 rpm 20 min
3	Non-bioavailable speciation	HCl–HF–$HClO_4$	Digestion (for As and Hg digestion temperature below 100 °C)

All kinds of samples are in 1 g dry weight, and the operation procedures and reagents used respectively are shown as in Tables 5.7, 5.8 and 5.9.

Table 5.10 Detecting limits and the precision of ICP-AES analysis for elements

	Pb	Zn	Cu	As
Detecting limits (μg/g)	0.0600	0.2112	0.0012	0.045
Sensitivity (μg/g)	0.0200	0.0704	0.0004	0.015
Standard deviation (%)	0.0369	0.0193	0.7760	0.6914
Recovery (%)	102.7	100.20	100.40	101.10

5.3.3.1 The Relative Experimental Instruments and Related Parameters

In this study, the discussion on Pb, Zn, Cu, and As is respectively carried out. According to the analysis for the chemical characteristics of the studied elements, different instruments were used in the work. Main experimental instruments included IRIS Advantage 1000 full spectrum of inductively coupled plasma atomic emission spectrometer (ICP-AES, Thermo Jarrell Ash USA) used for Cu analysis, PE-5100 flame atomic absorption spectrophotometer (PE USA) used for Cu and Zn analysis, AFS-810 double channel atomic fluorescence spectrometer (AFS-810 Jida, Beijing Little Swan Instruments Co. Ltd. China) used for As and Pb analysis, AMA 254 mercury vapourmeter (Sigma AMA 254, Germany) used for Hg analysis, and 18,000 r/min temperature control and speed control centrifuge.

All chemical reagents used were analytical above grade reagents, and the experimental water was quartz sub-boiling distilled water. All laboratory wares, before using, were soaked in 10% nitric acid solution for more than 12 h and washed by quartz sub-boiling distilled water before drying for use.

In the experiment, working conditions of AES-ICP were as follows: RF power 1100 W, working frequency 27.12 Hz, creep pump speed 100 r/min, atomization pressure 220 kPa, carrier gas flow 1.85 L/min, and CID temperature 42 °C. The working conditions of atomic absorption spectrophotometer were as follows: lamp current 10 mA, spectral passband width 0.7 nm, flame type air—acetylene, acetylene flow 2 L/min, air flow 10 L/min, and burner height 5 mm. The working conditions of AFS-810 double channel atomic fluorescence spectrometer: negative voltage of photoelectric multiplier tube 270 V, atomic device temperature 200 °C, atomization device height 8 mm, lamp current 80 mA, carrier gas flow 400 mL/min, shielding gas flow 1000 mL/min, detection limit of As and Pb was same as 0.0001 g/kg, standard curve correlation coefficient for each element during detection were 0.99996 for As and 0.99989 for Pb. Detection limit of AMA 254 mercury vapourmeter was 0.0001 mg/kg, and the relative standard deviation was 0.03. The detection limits, precision, and recovery rates of the other studied elements are shown in Tables 5.10 and 5.11.

Table 5.11 Detecting limits and the precision of PE-5100 atomic absorption spectrometer analysis for elements

	Wavelength (nm)	Detecting limits (μg/g)	Sensitivity (μg/g)	Recovery (%)
Cu	324.8	0.05632	0.0770	100.83
Zn	213.9	0.0326	0.0180	99.64

Table 5.12 Speciation of the metals in soil from the Hetao Area analyzed by modified BCR's procedure (unit: mg/kg)

	Pb		Zn		Cu		As	
	C[a]	Ratio in total (%)[b]	C	Ratio in total (%)	C	Ratio in total (%)	C	Ratio in total (%)
Acid dissolving	1.99	2.63	0.00	0.00	0.00	0.00	12.72	0.13
Reducible	13.49	17.83	0.00	0.00	0.54	1.70	213.77	2.14
Oxidizable	0.00	0.00	0.00	0.00	0.26	0.83	1.15	0.001
Residual	60.19	79.54	127.20	100.00	30.66	97.47	9767.96	97.72

[a]C means concentration of the element
[b]Ratio in total means the ratio of the speciation concentration of the element in its total concentration in the sample

5.3.3.2 The Experimental Results and Discussion

1. **The speciation distribution of Pb, Zn, Cu, and As in the As-polluted soils in the Hetao Area, Inner Mongolia**

(1) **The results from the modified BCR method**

The experimental data from modified BCR analysis procedure are shown as in Table 5.12. The As and Pb elements in this batch of experimental data were measured by atomic fluorescence spectrometer, and the Cu and Zn elements were measured by flame atomic absorption spectrophotometer. The speciation of Pb, Zn, Cu, and As in the As-polluted soils in the Hetao Area, Inner Mongolia, analyzed by modified BCR method is with differences between different elements.

The concentration of Pb is mainly in the residue species, and the residual fraction concentration (79.54%) > reduced species concentration (17.83%) > acid-soluble species concentration (2.63%); the concentration of oxidizable Pb is lower than that of the instrument detection limit.

Zn concentration exists all in the residual species, and the reduced or acid dissolved species, oxidizable species Zn concentrations are all lower than the detection limit of the instrument.

Table 5.13 Speciation of the metals in soil from the Hetao area analyzed by DTPA-5Na's procedure (*unit* mg/kg)

	Pb		Zn		Cu		As	
	C[a]	Ratio in total (%)[b]	C	Ratio in total (%)	C	Ratio in total (%)	C	Ratio in total (%)
Bioavailable speciation	0.00	0.00	0.04	1.07	8.64	25.33	0.00	0.00
Non-bioavailable	52.52	100.00	3.44	98.93	25.48	74.67	9995.60	100.00

[a]C means concentration of the element
[b]Ratio in total means the ratio of the speciation concentration of the element in its total concentration in the sample

The concentration of Cu is mainly in the residue species, and the residual fraction concentration (97.47%) > reduced species concentration (1.70%) > oxidizable species concentration (0.83%). The concentration of acid-soluble Cu is lower than that of the instrument detection limit.

The concentration of As is mainly in the residue species, and residual fraction concentration (97.72%) > reduced species concentration (2.14%) > acid-soluble species (0.13%) > oxidizable species concentration (0.01%).

(2) **The results from the DTPA-5Na salt method**

The experimental data of the As-contaminated soil in Hetao Area, Inner Mongolia, by using DTPA-5Na salt method are shown in Table 5.13. The As and Pb elements in this batch of experimental data were measured by atomic fluorescence spectrometer, and the Cu and Zn elements were measured by flame atomic absorption spectrophotometer. The distribution characteristics for different elements in the As-contaminated soil in the Hetao Area, Inner Mongolia, by DTPA-5Na salt method are different.

Pb is completely in the presence of the non-bioavailable species (100%), and the concentration of the bioavailable species is not detected.

The Zn is mainly in the species of the bioavailable species, and the bioavailable species Zn accounts for 1.07% in the total concentration of the Zn in the sample. The concentration of the proportion of the non-bioavailable Zn is 98.93%.

The Cu is mainly in the non-bioavailable species, and the bioavailable species concentration of Cu accounts for 25.33% in the total concentration of Cu in the sample. The non-bioavailable species accounts for 74.67% of the total amount.

The As is completely in the species of the non-bioavailable species (100%), and the concentration of the bioavailable species is not detected.

Table 5.14 Speciation of the metals in soil from the Hetao area analyzed by three-step procedure (*unit* mg/kg)

	Pb		Zn		Cu		As	
	C[a]	Ratio in total (%)[b]	C	Ratio in total (%)	C	Ratio in total (%)	C	Ratio in total (%)
Easy bioavailable speciation	0.95	1.49	0.00	0.00	1.00	2.63	8.40	0.08
Bioavailable speciation	3.52	5.53	4.67	1.33	6.07	15.95	85.56	0.86
Non-bioavailable speciation	59.25	92.99	347.39	98.67	31.00	81.43	9901.64	99.06

[a]C means concentration of the element
[b]Ratio in total means the ratio of the speciation concentration of the element in its total concentration in the sample

(3) **The results from the three-step procedure analysis method**

The result data from three-step procedure are as shown in Table 5.14. The As and Pb in this batch of experimental data were measured by atomic fluorescence spectrometer, and the Cu and Zn were measured by flame atomic absorption spectrophotometer. Analysis of distribution features of elements in As-contaminated soil in the Hetao Area, Inner Mongolia, by three-step procedure is as described following with obvious characteristics.

The Pb is mainly in the fraction of the non-bioavailable species, and the concentration of Pb non-bioavailable species (92.99%) > bioavailable species (5.53%) > easy bioavailable concentration (1.49%).

The Zn is mainly in the species of the non-bioavailable species, the non-bioavailable concentration Zn species (98.67%) > bioavailable species (1.33%), and the easy bioavailable concentration was lower than the detected limit of the instrument.

The Cu is mainly in the species of the non-bioavailable, the non-bioavailable concentration of Cu species (81.43%) > bioavailable species (15.95%) > easy bioavailable species concentration (15.95%).

The As is mainly in the species of the non-bioavailable species, the non-bioavailable concentration of As species (99.06%) > bioavailable species (0.86%) > easy bioavailable species concentration (0.08%).

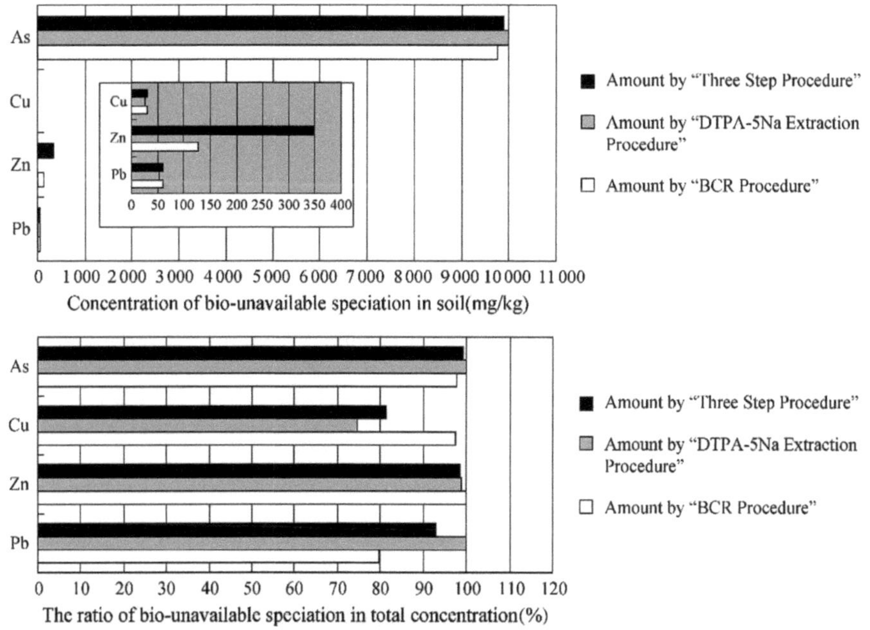

Fig. 5.1 Bio-unavailable speciation of trace metals in soil from the extraction methods

Table 5.15 Relative error from method BCR modified and DTPA-5Na for bio-unavailable speciation compared with the "three-step procedure" for soil (%)

Methods		Elements			
		Pb	Zn	Cu	As
Residual from modified BCR method	Error for concentration	1.59	63.38	1.10	1.35
	Error for the ratio in total	14.46	1.35	19.70	1.35
Bio-unavailable from DTPA-5Na salt method	Error for concentration	1.14	99.01	17.81	0.95
	Error for the ratio in total	7.54	0.26	8.30	0.95

(4) Comparison between the results from different analytical methods for the trace metal species in soil

The definition non-bioavailable species of three-step procedure is the same as that in BCR method and DTPA-5 Na method. As Fig. 5.1 shows, comparing with other methods, the result from three-step procedure shows, in addition to the large difference error of Zn concentration, Zn and other element proportions in their total amounts are all with the relative error less than 20% (<19.70%), as shown in Table 5.15.

Table 5.16 Speciation of the metals in the sediment of the Huangpu River analyzed by modified BCR's procedure (*unit* mg/kg)

	Pb		Zn		Cu		As	
	C[a]	Ratio in total (%)[b]	C	Ratio in total (%)	C	Ratio in total (%)	C	Ratio in total (%)
Acid dissolving	11.96	13.06	68.75	20.25	5.57	10.65	32.23	0.83
Reducible	18.02	19.68	48.93	14.41	4.29	8.19	365.05	9.37
Oxidizable	0.00	0.00	43.51	12.81	22.25	42.53	2.17	0.06
Residual	61.61	67.26	178.39	52.53	20.21	38.63	3496.40	89.75

[a]C means concentration of the element
[b]Ratio in total means the ratio of the speciation concentration of the element in its total concentration in the sample

2. **The distribution of Pb, Zn, Cu, and As in the sediments of the Huangpu River**

(1) **The results from the modified BCR method**

The experimental data are shown in Table 5.16. The As and Pb in this batch of experimental data were measured by atomic fluorescence spectrometer, and the Cu and Zn were measured by flame atomic absorption spectrophotometer. The speciation and distribution of the elements in the sediments of the Huangpu River analyzed by the modified BCR method is with the features described in the following.

The concentration of Pb is mainly in the residue species non-bioavailable species, the residual fraction of Pb (67.26%) > bioavailable species (concentration of reduced species, 19.68%) > easy bioavailable species (water- or acid-soluble concentration, 13.06%).

The concentration of Zn is mainly in the residue species, the residual fraction of Zn (52.53%) > bioavailable species concentration (reduced species/14.41%, oxidation species concentration/12.81%) > easy bioavailable species (acid-soluble concentration, 20.25%).

The concentration of Cu is mainly in the residue species, the oxidation species of Cu (42.53%) > residual fraction concentration (38.63%) > easy bioavailable species concentration (acid-soluble species concentration/10.65%) > bioavailable species concentration (reduced species/8.19%).

The concentration of As is mainly in the residue species, residual fraction of As (89.75%) > available species concentration (reduced species/9.37%) > easy bioavailable species concentration (acid-soluble species/0.83%), > bioavailable concentration (oxidation species/0.06%).

Table 5.17 Speciation of the metals in the sediment from the Huangpu River analyzed by DTPA-5Na's procedure (*unit* mg/kg)

	Pb		Zn		Cu		As	
	C[a]	Ratio in total (%)[b]	C	Ratio in total (%)	C	Ratio in total (%)	C	Ratio in total (%)
Bioavailable speciation	0.00	0.00	56.05	17.98	20.68	34.28	0.00	0.00
Non-bioavailable speciation	54.64	100.00	255.70	82.02	39.65	65.72	3895.84	100.00

[a]C means concentration of the element
[b]Ratio in total means the ratio of the speciation concentration of the element in its total concentration in the sample

(2) **The results from the DTPA-5Na salt method**

Experimental results by DTPA-5Na salt method are shown in Table 5.17. The As and Pb in this batch of experimental data were measured by atomic fluorescence spectrometer, and the Cu and Zn were measured by flame atomic absorption spectrophotometer. The 5Na-DTPA salt method used to analyze the distribution characteristics of Pb, Cu, As, and Zn in the sediment of the Huangpu River obtains the results with obvious features in the speciation distribution.

The Pb in the sediments of the Huangpu River is completely in the non-bioavailable species (100%), and the bioavailable species is lower than the instrument detection limit.

The Zn bioavailable species in the sediment of the Huangpu River accounts for 17.98% of the total Zn in the sample, and the non-bioavailable species concentration of Zn accounts for 82.02% of the total concentration of it in the sample.

The bioavailable species Cu in the sediments of the Huangpu River accounts for 34.28% of the total Cu in the sample, and the non-bioavailable species Cu concentration accounts for 65.72% in the total Cu in the sample.

The As in the sediments of the Huangpu River is completely in the non-bioavailable species (100%), and the bioavailable species As is lower than the detection limit of the instrument.

(3) **The results from the three-step procedure method**

The experimental results are shown in Table 5.18. The As and Pb in this batch of experimental data were measured by atomic fluorescence spectrometer, and the Cu and Zn were measured by flame atomic absorption spectrophotometer. The distribution characteristics of the elements in the sediments of the Huangpu River by three-step procedure are with the obvious itself features, and the concentrations of Pb, Zn, Cu, and As are mainly in the form of the non-bioavailable species.

The non-bioavailable species of Pb accounts for 9.62% of the total concentration of it in the sample, and the easy bioavailable species concentration of it is 11.61%

Table 5.18 Speciation of the metals in the sediment from the Huangpu River analyzed by the three-step procedure (*unit* mg/kg)

	Pb		Zn		Cu		As	
	C[a]	Ratio in total (%)[b]	C	Ratio in total (%)	C	Ratio in total (%)	C	Ratio in total (%)
Easy bioavailable speciation	9.55	11.61	63.58	11.03	5.11	7.99	23.94	0.61
Bioavailable speciation	7.22	8.77	83.95	14.57	19.93	31.18	110.06	2.83
Non-bioavailable speciation	65.48	79.62	428.78	74.40	38.89	60.83	3761.84	96.56

[a]C means concentration of the element
[b]Ratio in total means the ratio of the speciation concentration of the element in its total concentration in the sample

in the total concentration of Pb in the sample; bioavailable species of Pb is 8.77% accounting for the total concentration of Pb.

The non-bioavailable species of Zn accounts for 74.40% of the total concentration of it in the sample, and the bioavailable species of the Zn accounts for 14.57% of the total concentration of the Zn; the easy bioavailable species concentration of the Zn accounts for 11.03% of the total concentration of the Zn in the sample.

The non-bioavailable species of Cu accounts for 60.83% of the total concentration of it in the sample, and the bioavailable species of the Cu accounts for 31.18% of the total concentration of Cu in the sample; the easy bioavailable species concentration of the Cu accounts for 7.99% of the total concentration of the Cu in the sample.

The non-bioavailable species of As accounts for 96.56% of the total concentration of it in the sample, and the bioavailable species of the As accounts for 2.83% of the total concentration of it in the sample; the easy bioavailable As concentration accounts for 0.61% of the total concentration of it in the sample.

(4) **Comparison between the experimental results from different methods**

Comparison between experimental results of Cu, Zn, As, and Pb in the sediment of the Huangpu River by different analysis methods is as shown in Fig. 5.2.

The results of Fig. 5.2 show that the relative error of the non-bioavailable species concentrations by three-step procedure accounts for less than 16.56% of the total concentrations of them; in addition to the relative error of Pb by DTPA-5Na salt method, Zn by the BCR and DTPA-5Na salt methods and Cu by the BCR method are all little more than that (16.56%), as shown in Table 5.19.

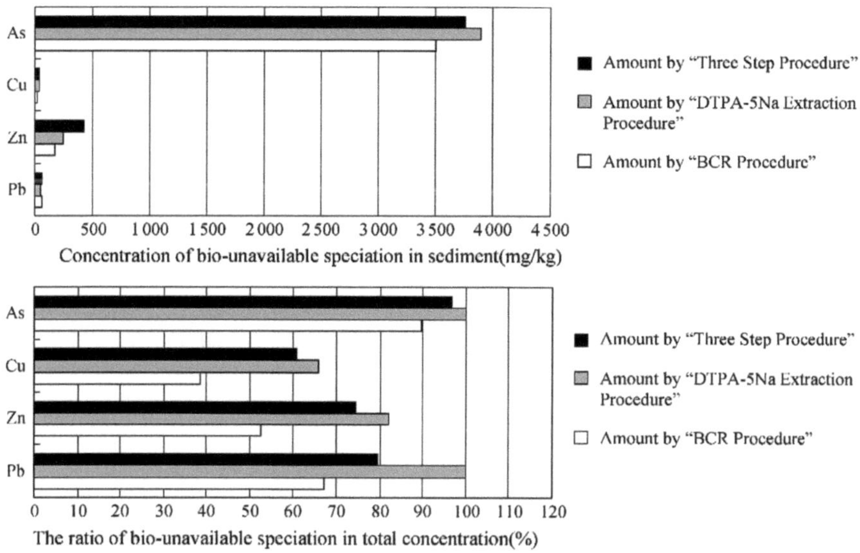

Fig. 5.2 Bio-unavailable speciation of trace metals in sediment from the extraction methods

Table 5.19 Relative error for methods BCR modified and DTPA-5Na compared with the "three-step procedure" from the sediment from the Huangpu River (%)

Methods		Elements			
		Pb	Zn	Cu	As
Residual from BCR modified	Error of concentration	5.91	58.40	48.03	7.06
	Error of ratio in the total	15.52	29.40	36.50	7.05
Non-bioavailable from DTPA-5Na	Error of concentration	16.56	40.37	1.95	3.56
	Error of ratio in the total	25.60	10.24	8.04	3.56

3. **The speciation distribution of Zn, Cu, Pb, and As in coal**

(1) **The results from the modified BCR method**

Modified BCR method analysis result data for coal are shown in Table 5.20. The As and Pb in this batch of experimental data were measured by atomic fluorescence spectrometer, and the Cu and Zn were measured by flame atomic absorption spectrophotometer. Analysis of distribution characteristics of elements in the anthracite from the Shanxi Dayan Coal Mine by modified BCR method is with obvious features for every element.

The Pb concentration is mainly in the residue species, and the proportion in the total concentration for every species of it is in the older as follows: the residual fraction of Pb (64.98%) > the reduced species concentration (24.40%) > the acid-soluble species concentration (10.62%). The oxidation species of Pb is not detected.

Table 5.20 Speciation of the metals in the coal from the Dayan Colliery, Shanxi, China, analyzed by modified BCR's procedure (*unit* mg/kg)

Speciation	Pb		Zn		Cu		As	
	C[a]	Ratio in total (%)[b]	C	Ratio in total (%)	C	Ratio in total (%)	C	Ratio in total (%)
Acid dissolving	8.85	10.62	6.58	43.35	0.00	0.00	20.67	2.25
Reducible	20.32	24.40	0.00	0.00	0.29	7.47	128.28	13.96
Oxidizable	0.00	0.00	4.17	27.46	0.82	21.46	0.68	0.07
Residual	54.11	64.98	4.43	29.20	2.72	71.07	769.10	83.71

[a]C means concentration of the element
[b]Ratio in total means the ratio of the speciation concentration of the element in its total concentration in the sample

The Zn is mainly in the acid-soluble species, and the proportion in the total concentration for every species of it is in the older as follows: the acid-soluble species concentration of Zn (43.35%) > the residual fraction concentration (9.20%) > the oxidation species concentration (27.46%). The reduced species of it is not detected.

The Cu is mainly in the residue species, and the proportion in the total concentration for every species of it is in the older as follows: the residual species fraction of Zn (71.07%) > the oxidation species concentration (21.46%) > the reduced species concentration (7.47%). The acid-soluble species of the Cu was not detected.

The As is mainly in the residue species, and the proportion in the total concentration for every species of it is in the older as follows: the residual species fraction of As (83.71%) > the reduced species concentration (3.96%) > the acid-soluble species (2.25%) > oxidation species concentration of it (0.07%).

(2) **The results from the DTPA-5Na salt method**

The analysis result data by the DTPA-5Na salt method are shown in Table 5.21. The As and Pb in this batch of experimental data were measured by atomic fluorescence spectrometer, and the Cu and Zn were measured by flame atomic absorption spectrophotometer. The speciation distribution characteristics of the elements are respectively with features for elements studied.

The Pb is completely in the non-bioavailable species (100%), and the bioavailable species concentration is not detected out.

The bioavailable species concentration of Zn accounts for 75.27% of the total Zn, and the non-bioavailable species concentration is 24.73%.

The bioavailable species concentration of Cu accounts for 75.27% of the total Cu, and the non-bioavailable species concentration is 59.32%.

The As is completely in the non-bioavailable species (100%), and the bioavailable species concentration of it is not detected.

Table 5.21 Speciation of the metals in the coal from the Dayan Colliery, Shanxi, China, analyzed by DTPA-5Na's procedure (*unit* mg/kg)

Speciation	Pb		Zn		Cu		As	
	C[a]	Ratio in total (%)[b]	C	Ratio in total (%)	C	Ratio in total (%)	C	Ratio in total (%)
Bioavailable	0.00	0.00	12.05	75.27	1.71	40.68	0.00	0.00
Non-bioavailable	51.77	100.00	3.96	24.73	2.50	59.32	918.72	100.00

[a]C means concentration of the element
[b]Ratio in total means the ratio of the speciation concentration of the element in its total concentration in the sample

Table 5.22 Speciation of the metals in the coal from the Dayan Colliery, Shanxi, China, analyzed by the three-step procedure (*unit* mg/kg)

Speciation	Pb		Zn		Cu		As	
	C[a]	Ratio in total (%)[b]	C	Ratio in total (%)	C	Ratio in total (%)	C	Ratio in total (%)
Easy bioavailable speciation	8.11	9.84	9.54	49.35	0.68	17.08	20.87	2.27
Bioavailable speciation	3.23	3.92	9.19	47.54	0.25	6.29	91.44	9.95
Non-bioavailable speciation	71.09	86.24	0.60	3.11	3.05	76.63	806.42	87.78

[a]C means concentration of the element
[b]Ratio in total means the ratio of the speciation concentration of the element in its total concentration in the sample

(3) The results from the three-step procedure method

The analytical results from the three-step procedure method for the coal are shown in Table 5.22. The As and Pb in this batch of experimental data were measured by atomic fluorescence spectrometer, and the Cu and Zn were measured by flame atomic absorption spectrophotometer.

Pb is mainly in the non-bioavailable species, and the proportion in the total concentration for every species of it is in the older as follows: non-bioavailable species concentration of Pb (86.24%) > the easy bioavailable concentration (9.84%) > bioavailable species concentration of it (3.92%).

The proportion in the total concentration for every species of Zn is in the older as follows: the easy available species concentration of Zn (49.35%) > the bioavailable species concentration (47.54%) > the non-bioavailable species concentration

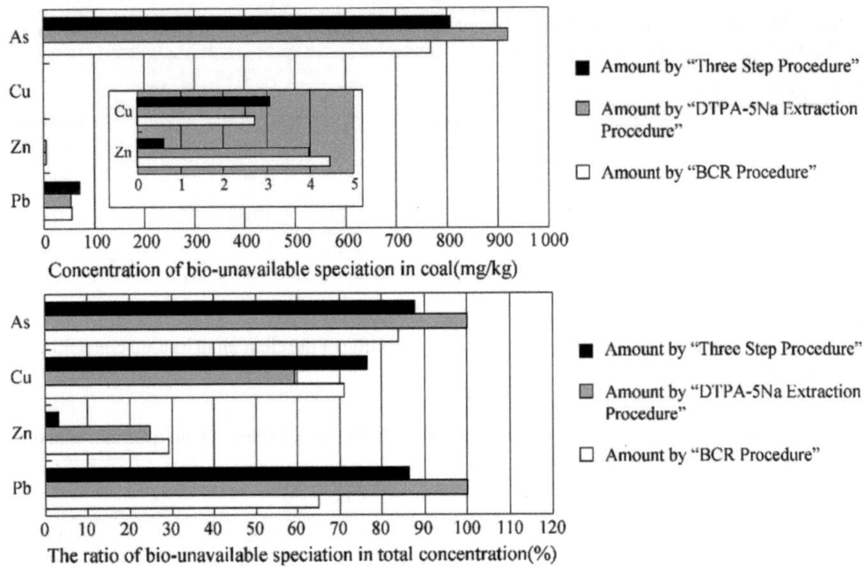

Fig. 5.3 Bio-unavailable speciation of trace metals in coal from the extraction methods

(3.11%), and this element might be with a greater risk of release from the medium studied.

The Cu is mainly in the non-bioavailable species concentration, and the proportion in the total concentration for every species of it is in the older as follows: the non-bioavailable species concentration of Cu (76.63%) > the easy bioavailable species concentration (17.08%) > bioavailable species concentration of it (6.29%).

The As is mainly in the non-bioavailable species, and the proportion in the total concentration for every species of it is in the older as follows: the non-bioavailable species concentration (87.78%) > the bioavailable species concentration (9.95%) > the easy bioavailable species concentration of it (2.27%).

(4) **Comparison between the results for Cu, Zn, As, and Pb in coal analyzed by different analysis methods**

The results from different methods for the speciation of the elements are as shown in Fig. 5.3.

The results of Fig. 5.3 show that the relative errors of the non-bioavailable species concentration and proportion accounting for in the total of the concentration of the elements are less than 16.56%, in addition to the relative errors of the Pb species and total concentration, the Zn species and total concentration in BCR and DTPA-5Na salt methods, the Cu species and total concentration in DTPA-5Na salt method, as shown in Table 5.23.

Table 5.23 Relative error of methods BCR modified and DTPA-5Na compared with the "three-step procedure" for analysis of the metal speciation in coal (%)

Methods		Elements			
		Pb	Zn	Cu	As
Residual from BCR modified	Error of concentration	23.89	638.33	10.82	4.63
	Error of ratio in the total	24.65	838.91	7.26	4.64
Non-bioavailable from DTPA-5Na	Error of concentration	27.18	560.00	18.03	13.93
	Error of ratio in the total	15.96	695.18	22.59	13.92

5.3.4 The Significance and Application of the Three-Step Procedure Method on Trace Metal Speciation Analysis

The three-step procedure method is an improved method for the sequential analysis of trace metals based on the sequential extraction of Tessier and the continuous extraction method of BCR. Among the main methods of speciation analysis of trace metals at present, the modified sequential extraction BCR method of the sequential extraction methods, DTPA-5Na salt method of the single speciation sequential extraction methods is selected compared with the three procedure methods for the results from the experiments on the same three kinds of samples. The three kinds of samples are representative soil, water sediment, and raw blind coal as presented above. So, the extraction effect of the three-step procedure method can be evaluated from the comparison between the results for the extracted bioavailable species concentration, the extracted amount of the metals respectively accounted for the total concentration from another two methods for the same sample.

From the comparison above, it shows that there are differences in the results of the analysis methods for the non-bioavailable species between different media and different elements, of which the data of soil agree better than that of the sediment and raw blind coal in the different analysis methods. The data of Zn in the media agreed with each other are poor, and the Cu in the sediment from the BCR method agreed with other methods is poor too. In addition to the above poor agreement of Zn and Cu, the differences of the data between different methods in different kinds of media are still within the range of significances with the relative error less than 25.6%.

Due to the fact that the 5Na DTPA salt method can only give the data of the easy bioavailable species concentration, there are differences in the definition for the species of the BCR method and three-step procedure method. In the actual application of the speciation analysis for trace metals, it is a special case or less used case for the 5Na DTPA salt method, and in the comparative study here the comparison only with exchangeable species, acid-soluble species, and easy bioavailable species from the methods are carried out. At present in the soil and water–sediment trace metal speciation studies, the most commonly used method is modified BCR method. The bioavailable species in the three-step procedure method is equivalent to the acid-

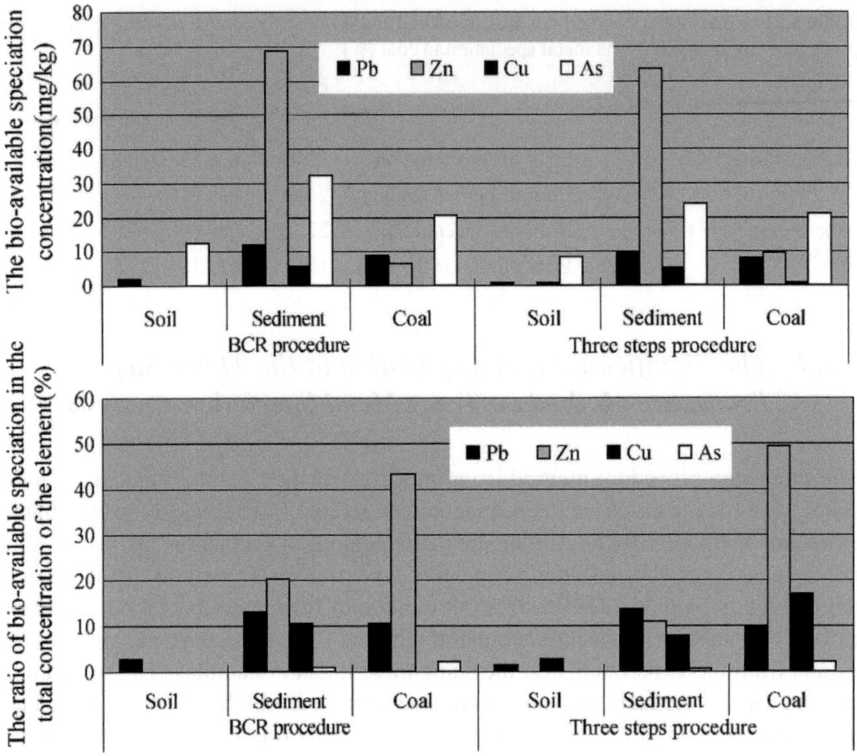

Fig. 5.4 Comparison between the analysis from the procedures of three steps and BCR modified for the bioavailable speciation

soluble species in the modified BCR method, and it is the most bioavailable part in the metal concentration in the environmental system, which is also the most concerned issue of the metals in present practical work. Analysis results of the easy bioavailable species of different medium from this work are shown in Fig. 5.4.

It is shown by Fig. 5.4 that in general the part of the easy bioavailable species of the metals in the media gained from the different two methods, three-step procedure method and the modified BCR method, is with a good accordance. The concentration data of the bioavailable species from three-step procedure method in the coal are higher than that from the modified BCR method, but lower than that in the water sediment by the modified BCR method, in which the concentration of each element by three-step procedure method is 0.96–31.03% higher than that from the modified BCR method for the coal and lower 7.52–25.72% than that by the modified BCR method for the sediment. The above feature shows that these errors of the two methods are within relatively acceptable range.

In the speciation analysis for trace metals, the most critical part of the problem is the concentration of the two parts, the easy bioavailable species concentration which is a direct indicator of the bioavailability of pollutants and also the most concerned part and the non-bioavailable species concentration which is the concentration of biological safety. Their amount size can be used as an indirect mark of the bioavailability of pollutants [53]. The comparison result shown by the above two part concentrations from different methods and different media suggests that no matter the data are obtained from which method the results of the analysis of the three-step procedure method all are relatively acceptable as a more straightforward procedure to be used in the vast request for the trace metal bioavailability evaluation in soil and sediment.

It should be explained that the current methods for speciation analysis of trace metals including the three-step procedure method here still remain in the stage of operating definition species. Based on the specific circumstances of the media and the complexity of the speciation problem itself as mentioned above, it is inevitable to be with errors between data gained from the methods. The operating procedures and analysis results of "three-step procedure" above show that this method is easy to operate and be suitable comparison with the current mainstream methods, and it is quite meaningful and worthy of further exploration and development. It can be mentioned that the significance of this study should be far more important than the data obtained in this study, due to the limited experimental data and the experimental conditions. It is hoped that the method of the three-step procedure for speciation analysis of trace metals can be as an inspiration for the route probing in the constantly improvement or a datum accumulation for the trace metal speciation analysis.

References

1. Stumm W, Brauner PA. A chemical speciation. In: Riley JP, Skirrow G, editors. Chemical oceanography. New York: Academic Press; 1975. p. 173–279 (Chapter 3).
2. Cai XY, Lin LH. A random simulation for the distribution of heavy metals in water environment. China Bull Geol Dis Treat. 1997;8(2):40–7 (in Chinese).
3. Zhou XY. A research on the method for the heavy metal speciation analysis. Master thesis. School of Environmental Science and Engineering, Shanghai Jiaotong University; 2006.
4. Gleyzes C, Tellier S, Astruc M. Fractionation studies of trace elements in contaminated soils and sediments: a review of sequential extraction procedures. Trends Anal Chem. 2002;21(6–7):451–67.
5. Das AK, Chakraborty R, Cervera ML, et al. Metal speciation in solid matrices. Talanta. 1995;42(8):1007–30.
6. Novozamsky I, Lexmond THM, Houba VJG. A single extraction procedure of soil for evaluation of uptake of some heavy metals by plants. Int J Environ Anal Chem. 1993;51:47–58.
7. Griepink B, Maier EA. The certification of the contents (mass fractions) of Ca, K, Mg, P, S, Zn, I, N and Kjeldahl-N in hay powder. Commission of the European communities community Bureau of reference. In: Applied metrology and chemical analyses, report; 1989.
8. Ure AM, Quevauviller P, Muntau H, et al. Speciation of heavy metals in soils and sediments—an account of the improvement and harmonization of extraction techniques undertaken under the auspices of the BCR of the commission-of-the-European-communities. Int J Environ Anal Chem. 1993;51(1–4):135–51.
9. Mulchi CL, Adamu CA, Bell PF, Chaney RL. Residual heavy metal concentrations in sludge-amended coastal plain soils, I. Comparison of extractans. Commun Soil Sci Plant Anal. 1991;22(9–10):919–41.

10. Lindsay WL, Norvell WA. Development of a DTPA soil test for Zn, Fe, Mn and Cu. Soil Sci Soc Am J. 1978;42:421–8.
11. Melich A. Mehlich 3 soil test extractant, a modification of Mehlich 2 extractant. Commun Soil Sci Plant Anal. 1984;15:1409–16.
12. Ure AM, Thomas P, Littlejohn D. Ammonium acetate extracts and their analysis for the speciation of metal ions in soils and sediments. Int J Environ Anal Chem. 1993;51:64–84.
13. Gupta SK, Aten C. Comparison and evaluation of extraction media and their suitability in a simple model to predict the biological relevance of heavy metal concentrations in contaminated soil. Int J Environ Anal Chem. 1993;51:25–46.
14. Juste C, Solda P. Changes in the cadmium, manganese, nickel and zinc bioavailability of a sewage sludge-treated sandy soil as a result of ammonium-sulfate, acid peat, lime or iron compound addition. Agronomie. 1988;8:897–904.
15. Tessier A, Campbell PGX. Bisson M sequential extraction procedure for the speciation of particulate trace metals. Anal Chem. 1979;51:844–51.
16. Forstner U. In: Lechsber R, Davis RA, Hermitte PL, editors. Chemical methods for assessing bioavailable metals in sludges. London: Elsevier; 1985.
17. Meguellati M, Robbe D, Marchandise P, Astruc M. Heavy metals in the environment. In: Proceedings of the international conference on heavy metals in the environment. Edinburgh: Heidelberg CEP Consultants, 1983. p. 1090.
18. Towner JV. Studies of chemical extraction techniques used for eluciding the partitioning of trace mentals in sediments. University of Liverpool; 1985.
19. Gao YZ, He JZ, Ling WT. The speciation of Cd and Cu in the soils from Hubai Province. Bull Central China Agricult Univ. 2001;20(2):143–7.
20. Sun JM, Wen QZ. The heavy metal forms existed in loses and ancient soil and influences from weathering and pedogenesis. Soil Bull. 1994;3(3):305–11.
21. Rauret G, Rubio R, Lopez-Sanchez JF. Optimization of Tessier procedure for metal solid speciation in river sediments. Trends Anal Chem. 1989;36:9–83.
22. Jeffrey RB, Hewitt IJ, Cooper P. Reproducibility of the BCR sequential extraction procedure in a long-term study of the association of heavy metals with soil components in an upland catchment in Scotland. Sci Total Environ. 2005;337:191–205.
23. Morera MT, Eeheverra JC, Mazkiarn C, et al. Isotherms and sequential extraction procedures for evaluating sorption and distribution of heavy metals in soils. Environ Pollut. 2001;113(2):135–44.
24. Huang Y, Chen YJ, Tao S. The influences from the rhizosphere circumstance of mycorrhizal plant to the speciation of Cu, Zn, Pb, Cd in polluted soil. Bull Appl Ecol. 2000;11(3):431–4.
25. Rauret G, Lopez-Sanchez JF, Sahuquillo A, et al. Improvement of the BCR three-step sequential extraction procedure prior to the certification of new sediment and soil reference materials. J Environ Monit. 1999;1:57–61.
26. Sahuquillo A, Rigol A, Rauret G. Overview of the use of leaching/extraction tests for risk assessment of trace metals in contaminated soils and sediments. Trends Anal Chem. 2003;22(3):152–9.
27. Kaherine FM, Davidson CM. Comparison of original and modified BCR sequential extraction procedures for the fractionation of copper, iron, lead, manganese and zinc in soils and sediments. Anal Chim Acta. 2003;478:111–8.
28. Davison W, Zhang H. In situ speciation measurements of trace components in natural waters using thin-film gels. Nature. 1994;367:546–8.
29. Davison W, Zhang H. In situ measurement of labile species in water and sediments using DGT. In: Varney M, editor. Chemical sensors in oceanography. Gordon and Breach; 2001. p. 283–300.
30. Fan HT, Sun T, Sui DP, et al. Two technics for passive sampling in suit of environmental monitoring, thin film diffusion balancing technique and thin film diffusion gradient technique. China Chem Bull. 2009;72:421–6.
31. Jean Philibert. One and a half century of diffusion: Fick, einstein, before and beyond. Diffus Fundam. 2005;2(1):1–10.

32. Philibert J. One and a half century of diffusion: Fick, einstein, before and beyond. Diffus Fundam. 2006;4(6):1–19.
33. Li W, Teasdale PR, Zhang S, et al. Application of a poly (4-styrenesulfonate) liquid binding layer for measurement of Cu^{2+} and Cd^{2+} with the diffusive gradients in thin films technique. Anal Chem. 2003;75(11):2578–83.
34. Fan H, Sun T, Li W, et al. Sodium polyacrylate as a binding agent in diffusive gradients in thin-films technique for the measurement of Cu^{2+} and Cd^{2+} in waters. Talanta. 2009;79:1228–32.
35. Li W, Zhao H, Teasdale PR, et al. Trace metal speciation measurements in waters by the liquid binding phase DGT device. Talanta. 2005;67:571–8.
36. Scheckel K, Luxton T, Elbadawy A, Impellitteri CA, Tolaymat T. Synchrotron speciation of silver and zinc oxide nanoparticles aged in a kaolin suspension. Environ Sci Technol. 2010;44:1307–12.
37. Catalano JG, Huhmann BL, Luo Y, Mitnick EH, Slavney A, Giammar DE. Metal release and speciation changes during wet aging of coal fly ashes. Environ Sci Technol. 2012;46:11804–12.
38. Zhang H. The factors for transformation between the fractions of speciation of trace metals in lake sediments. Acta Geol Sin. 2017;91:753–4.
39. Bermond A. Limits of sequential extraction procedures re-examined with emphasis on the role of H+ ion reactivity. Anal Chim Acta. 2001;445(1):79–88.
40. Arunachalam J, Ernons H, Krasnodebska B, et al. Sequential extraction studies on homogenized forest soil samples. Sci Total Environ. 1996;181(2):147–59.
41. Jeffrey L, Howard ShuJ. Sequential extraction analysis of heavy metals using a chelating agent (NTA) to counteract resorption. Environ Pollut. 1996;91(1):89–96.
42. Shiowatana J, Tantidanai N, Nookabkaew S, et al. A flow system for the determination of metal speciation in soil by sequential extraction. Environ Int. 2001;26(5–6):381–7.
43. Maiz I, Arambarri I, Garcia R, et al. Evaluation of heavy metal availability in polluted soils by two sequential extraction procedures using factor analysis. Environ Pollut. 2000;110(1):3–9.
44. Filip M, Verloo MG. Single extraction versus sequential extraction for the estimation of heavy metal fractions in reduced and oxidized dredged sediments. Chem Speciat Bioavailab. 1999;11(2):43–51.
45. Maiz I, Esnaola MV, Millin E. Evaluation of heavy metal availability in contaminated soils by a short sequential extraction procedure. Sci Total Environ. 1997;206(2–3):107–15.
46. Campos E, Barahona E, Lachica M, et al. A study of the analytical parameters important for the sequential extraction procedure using microwave heating for Pb, Zn and Cu in calcareous soils. Anal Chim Acta. 1998;369(3):235–43.
47. Zhang SZ, Wang S, Shan XQ. Effect of sample pretreatment upon the metal speciation sediments by a sequential extraction procedure. Chem Speciat Bioavailab. 2001;13(3):69–75.
48. Breward N, Peachey D. The development of a rapid scheme for the elucidation of the chemical speciation of elements in sediments. Sci Total Environ. 1983;29(1–2):155–62.
49. Noble AD, Hughes JC. Sequential fractionation of chromium and nickel from some serpentinite-derived soils from the eastern Transvaal. Commun Soil Sci Plant Anal. 2008;22(19–20):1963–73.
50. AFNOR (Association Française de Normalization). Annual report. Paris: AFNOR; 1994. p. 250.
51. UNICHIM (Ente Nazionale Italiano di Unificazione). Annual report UNICHIM. Milan: UNICHIM; 1991.
52. Houba VJG, Lexmond ThM, Novozamski I, van der Lee JJ. State of the art and future developments in soil analysis for bioavailability assessment. Sci Total Environ. 1996;178:21–8.
53. Zhang H, Zhou XY. Speciation variation of trace metals in coal gasification and combustion. Chem Speciat Bioavailab. 2009;21(2):93–7.

Chapter 6
Main Conclusions and Perspectives on the Work

Based on a typical case study of regional trace metal pollution and urban trace metal pollution, combined with relevant simulation experiments and methodological exploration and discussion, the environmental behavior, distribution pattern, pollution characteristics, and mechanism of trace metal pollution caused by natural and anthropogenic factors are studied in this work. Regular facts confirmed, disclosed, or deduced by specific research contents or work have been elaborated and summarized with analyses and discussion in specific chapters. In order to further reflect the internal relationship between the main ideas of this work and pollution problems, its main research and understanding are here systematically summarized and introduced as follows.

6.1 The Environmental Behaviors and Pollution of the Trace Metals in Area Pollution

6.1.1 The Case Studies of Trace Metals and Pollution in Soil and Groundwater

The migration of elements with water from high background area to low-concentration area at the lower reaches causes the regional As and other trace metal pollution and resident poisoning in Hetao Area. The migration process is significantly affected by the local human mining activities. It is a typical case of natural environmental pollution caused by human mining activities. The main signs are:

A. The concentrations of As, Zn, Pb, Cu, and Sb in soil, groundwater, and hair of local residents in Hetao Area decrease slowly from high to low in the background area of upstream to downstream, and there is a very obvious positive correlation among the concentrations of the soil, diving, and resident hair of the area.

H. Zhang, *Behaviors of Trace Metals in Environment*,
https://doi.org/10.1007/978-981-13-3612-6_6

The concentration of Hg, Mo, and Bi in the upstream undeveloped high-concentration abnormal zone has no obvious trend change law in the downstream soil.

B. The available concentrations of As, Zn, Pb, Cu, Sb, and other trace metal elements in soil decrease slowly from upstream to downstream. From the Yin Mountains in the upper reaches of the study area to the downstream 72 km, the variation of bioavailable speciation content (the percentage of the bioavailable speciation concentration in the total concentration of the element) of each element (AA′ section) is as follows:
As 27.32–6.67%, Zn 20.32–24.16%, Pb 31.82–16.67%, Cu 30.27–22.01%.

C. The regularity of the ratios of Sr and Pb isotopes shows that the element composition and concentration in the phreatic water are obviously affected by the weathering substances and their migration from the upstream. The phreatic water in the contaminated area or the diseased area of the population has been obviously contaminated by the upper ore field water or the surface water rich in the weathering substances in the upstream. Along the direction of flowing water (145°), the degree of the contamination is gradually weakened away from mining area of upstream at the Yin Mountains, which is with high background of the relative elements.

6.1.2 *Formation Mechanism of Regional Trace Metal Pollution and Its Behavior in Related Media*

In this study, the spatial distribution and concentration variation of trace metal elements show that the high-concentration areas of As, Zn, Pb, Cu, Sb, and other trace metal elements in phreatic water and soil are concentrated in the vicinity of large Cu, Zn, Pb, and S deposits with long mining history in the upper reaches. The incidence and severity of poisoning in the local population correspondingly are with that of the above elements.

Regional trace metal pollution is a natural phenomenon caused by the geochemical cycle of elements. It is an inevitable trend of element distribution toward homogenization. It has a strong natural inertia. The one of preconditions of this phenomenon is that there are significant differences in the concentration of trace metal elements in different zones. The high-concentration zone is located upstream of the local phreatic flow direction, which is the regional hydrogeological recharge area, and the low-concentration zone is located downstream of the phreatic flow, which is the regional hydrogeological discharge area. The natural flow of phreatic water and the evolution process of natural material distribution in the direction of homogeneous development are the basic mechanisms for the formation of the regional trace metal pollution. Aqueous solution is an indispensable medium for this mechanism. The composition and physicochemical conditions of aqueous solution (mainly pH value for surface precipitation) are the change factors, which are susceptible to artificial activities, and this has important influence on the pollution intensity and diffusion range of the regional trace metal pollution. Regional trace metal pollution in contaminated soil, groundwater, and organisms will appear in the above mechanism of action corresponding to the changes in the concentration of pollution elements, which is the objective record of the pollution process.

6.2 The Environmental Behaviors and Pollution of the Trace Metals in the Modern Metropolises

Urban trace metal pollution is an environmental phenomenon that trace metal elements superimpose on the natural environment under the influence of man-made factors, and this pollution is entirely caused by artificial activities. The main processes of metals from the initial smelting of their raw materials to the manufacture, use, even consumption, and wear of products in cities are the main origin of the pollution. Large-scale modern production and high-quality life in modern cities make this process happen at an unprecedented rate and in an unprecedented number. Therefore, the urban is the most important source of pollution in man-made trace metal pollution.

Urban water body is the record carrier of trace metal pollution in the urban. Because of the complexity of the source body, its pollution characteristics also have their own characteristics. Generally, it only reflects the local situation. The characteristics and changes of trace metals in urban rivers, lakes, and sediments are indirect and comprehensive records of the process and effect of urban trace metal pollution, which can reflect the overall status and trend of urban trace metal pollution. These records can be used to retrieve or trace the process and characteristics of pollutants discharged from the source to the environment. The change is a comprehensive reflection of the pollution trend of trace metals in cities. Trace metal pollution in sediments, as a record of specific urban pollution sources, is generally characterized by high pollution intensity and distinct correspondence with the source. It is a key indicator to understand the mechanism and environmental effects of urban trace metal pollution and to evaluate urban trace metal pollution in depth.

6.2.1 The Behavior and Pollution of Trace Metals in Urban Rivers

1. The behavior and pollution of PCBs and trace metals in Suzhou Creek, Shanghai

In the study of trace metal pollution in Suzhou Creek, the concentration and distribution of PCBs in sediments are studied. The PCBs and trace metals and pollution of the Suzhou Creek in Shanghai are as follows:

(1) **Distribution and evolution of PCBs in surface sediments of the Suzhou Creek**

The concentration of PCBs in the surface sediments of the urban section of the Suzhou Creek ranges from 4.4×10^{-9} to 14.8×10^{-9}, which has not reached the pollution level yet, but has a tendency to increase with time. The concentration of PCBs in sediments showed a clear change law, that is, with the increase of distance away from coastal industrial pollution sources, the concentration of PCBs in sediments showed a downward trend.

(2) **Distribution and evolution of trace metals in the Suzhou Creek sediment system**

In the period of this study, the trace metal pollution in the Suzhou Creek is more prominent in Zn, Hg, and Cd, especially in suspended solid particles and surface sediments, which is one of the important hidden dangers of trace metal pollution in the Suzhou Creek.

The pollution of As, Hg, and Cd in river water phase is serious, which has exceeded the national surface water quality standard of Class III of China. The rate of exceeding the standard is 100%. Among them, Cd pollution is the most serious. Compared with other cities in China, the concentrations of As, Hg, and Cd in the Suzhou Creek are higher, but the concentrations of Pb are lower. The total distribution of trace metal elements As, Hg, Cd, and Pb is not very uniform, but the trace metal (except Hg) concentration was relatively high along the road from the site of Shanghai Grease Plant to the site of Wuzhenlu Bridge. This area is once an industrial concentration area, and the population is dense; a large number of industrial and domestic wastewater discharges are the main reason for the current phenomenon of the Suzhou Creek.

As, Hg, Cd, and Pb in the Suzhou Creek are mainly existed in water-soluble form, and most of them flows into the Huangpu River with the current. Among them, Cd is with the widest influence range and has a greater cumulative contribution to the lower reaches. It is roughly estimated that the contribution of the Suzhou Creek to the trace metals in the Huangpu River is Pb 0.457 t/a, Cd 0.115 t/a, As 1.749 t/a, Hg 0.048 t/a, and Zn 24 t/a. Trace metals in the water-borne of the Suzhou Creek are important sources of trace metals in the Huangpu River, which seriously affect the water quality of the Huangpu River. The accumulation degree of Hg in suspended solid particles and sediments is the highest, followed by As, and Cd accumulation is the lowest. Hg accumulated in the sediment may be released when the concentration of it in river water is low or other environmental condition changes. It may be one of the important hidden dangers of the trace metal pollution in the Suzhou Creek.

Except for Pb, the concentrations of other trace metals in the suspension substances are higher, and the concentrations ranged from Cd 3.19 to 16.39 mg/kg, As 4.55 to 6.34 mg/kg, Hg 0.465 to 1.627 mg/kg, and Zn 753.49 to 2889.34 mg/kg. The distribution of Zn, Cd, and Hg corresponded to the distribution of coastal industries, and the tributary of Pengyuepu Creek contributed more to the As pollution of the Suzhou Creek.

The concentrations of trace metals in suspended solids and surface sediments are well correlated, and the concentrations and the speciation distribution of trace metals are also significantly correlated, which indicates that the migration and transformation of trace metals in suspended solids and sediments are intrinsically dependent on the hydrodynamic conditions and environmental conditions of the Suzhou Creek. Except for Pb and As, the concentrations of other trace metal elements in suspended solids are higher than those in sediments, indicating that there are still important trace metal sources affecting the water quality evolution of the Suzhou Creek. Some of these metals will pollute the downstream water with the river water flowing into the Huangpu River, and some will sink under certain conditions. Falling into the sediment system causes pollution superposition.

The Fe–Mn oxide speciation of Pb, Zn, and As is dominant in the sediments of the Suzhou Creek. They are easily released from the sediments when the redox potential decreases or the water is hypoxic, resulting in secondary pollution to the water. Cd mainly exists in exchangeable and carbonate speciation, indicating that Cd in sediments has high mobility and biological activity, and has high potential ecological hazards. Various metals in suspension particles mainly exist in residual speciation; organic speciation is the speciation with higher concentration except residual speciation, in which Cd and Cu concentrations are the highest; carbonate metal concentration is lower; Fe–Mn oxide speciation is higher with Zn concentration; exchangeable speciation and carbonate speciation are the strongest with Cd accumulation ability, migration, and biological effect. Compared with the sediments, the proportion of exchangeable, carbonate, and Fe–Mn oxide speciation of most metals in suspended solids is lower, especially for the Fe–Mn oxide, and the proportion of residual speciation is obviously higher. This may be due to the release of trace metal elements from suspended solids during their interaction with water or the complicated geochemical processes during long-term solid–liquid interaction and sedimentation.

(3) **Understanding of trace metal pollution and its evolution in sediments of the Suzhou Creek**

The concentration of trace metals in the sediments of the Suzhou Creek is slightly lower in the surface and bottom, but higher in the middle. This shows that the concentration of trace metals in the early Suzhou Creek is low. With the development of coastal industry and urbanization, wastewater containing a large number of trace metals is discharged into the Suzhou Creek, resulting in the accumulation of trace metal pollutants and gradually forming a high concentration of sediment pollution at the middle layer. The concentrations of trace metals in the sediments of the Suzhou Creek decrease gradually after the control step of external pollution sponsored by Shanghai Government. The concentrations of Hg, Cu, and As in the sediments near the surface sediments are still in a downward trend, indicating that the external pollution has been effectively controlled. However, the concentrations of Zn and Pb in the section of Shanghai grease plant site increase gradually during the process of approaching the surface layer, indicating that there are still pollution factors of Zn and Pb in the Suzhou Creek in the near future.

According to the preliminary evaluation of trace metal pollution in the sediments of the Suzhou Creek by the accumulation index and potential ecological pollution index, the pollution degree of each metal element is different. At present, the main pollution elements of the Suzhou Creek are Cd and Hg, which should be paid more attention. Trace metal pollution is the most serious in the area of Zhonghuaxinlu Bridge. The pollution in Shanghai grease plant site section is moderate. The pollutions in the area of Jiangnan Paper Mill and Wuzhenlu Bridge sites are slight. The pollution has a certain relationship with the distribution characteristics of coastal industry. At the same time, the pollution of trace metals in the tributary of Suzhou Creek (Pengyuepu Creek) is more serious, which has an important contribution to the pollution of trace metals in Suzhou Creek.

2. The behavior and pollution of trace metals in Huangpu River, Shanghai

The Huangpu River in Shanghai has gone through a period of serious pollution and long harnessing. With the increasing harnessing efforts, the results are becoming increasingly obvious. But there are also new problems that need great attention. The concentrations and changes of trace metals, PAHs, and related indicators in the Huangpu River water and sediments show that the current trace metal pollution has been effectively reduced and restrained, while organic pollutants show an increasing trend. The following are the circumstances (Tables 6.1, 6.2 and 6.3).

The data of this study show that the organic pollution of the Huangpu River is aggravating with the increase of domestic wastewater. Although the concentration of organic pollutants is lower than that of rivers in the USA, Canada, and some rivers of China, the increasing trend of organic pollutants has pointed out problems for the prevention and control of water environmental pollution in Shanghai, and this is also an important aspect for the environmental protection of urban rivers in the world.

The variation trend of organic matter concentration in the sediments of the Huangpu River in Shanghai confirms the fact that the organic pollution of the Huangpu River tends to aggravate. The PAHs mean value of sediment samples is 1.26 mg/kg, and TOC average is 1.34%. At present, the concentrations of PAHs and

Table 6.1 Concentrations of metals in the Huangpu River (μg/L)

Elements	Criterion value[a]	Criterion value[b]	Concentration of the Huangpu River water (8 samples)
Cu	1000	13	60.45
Pb	50	65	9.56
Cd	5	2.0	0.12
Cr^{6+}	50	16	8.75 (total Cr)
Hg	0.1	1.4	0.31
COD_{Cr}	15,000		20,000–30,000

[a]Surface water criterion grade III established by China Environmental protection Ministry at 2002 [4]
[b]Freshwater criterion established by US EPA at 2006 [5]

Table 6.2 Metals in the sediments of urban rivers or harbors from countries (mg/kg)

	Cu	Pb	Cr	Cd	Hg
Sydney Harbor pass urban and industry region, Canada	19.0–110.0	25.5–408.0	47.0–86.0	0.16–0.94	0.09–0.48
Seine River pass Paris, France	98.0	107.0	12.0	0.42	
Shahe River pass Chendu City, China	95.71	65.29		0.543	0.283
Estuary of the Yellow River pass industry region, China	6.7-56.9	13.2–38.1	33.7–88.8		
The Huangpu River pass Shanghai, China	33.43	24.75	77.23	0.36	0.22
Criterion[a]	390	450	260	5.1	0.41

[a]Sediment Quality Chemical Criterion (EB/OL) established by US Geological Survey at 2006 [6]

Table 6.3 Concentration of PAHs and TOC in river sediment of countries

		PAHs (ng/g)	TOC (%)	References
US	The Mill River pass Connecticut state	590–39,000		[7]
Canada	Sydney Harbor pass urban and industry region	4770–246,400	2.91–12.13	[8]
China	Estuary of the Pearl River pass industry region of Pearl River Delta, China	93.8–4307.0		[9]
	Pearl River pass industry region	255.9–16,670.3		[10]
	The Huangpu River pass industry region, Shanghai	1256.13	1.341	Study area of this work

TOC in the Huangpu River are still lower than those in the USA, Canada, and other countries, but the main indicative parameters of these organic pollution change [1, 2], indicating that the organic pollution of the Huangpu River has appeared signs of increasing. The P monitoring data of the Huangpu River also show that the change of total phosphorus in river water is consistent with that of domestic wastewater discharge, indicating that the increase of organic pollutants in the Huangpu River is mainly related to domestic sewage discharge. These phenomena need to be emphasized.

3. The behavior and pollution of trace metals in Nanjing section of the Yangtze River

The trace metal pollution process not only reflects the concentration of sediments in urban water bodies, but also records the evolution characteristics of the concentration with time. The total concentration and speciation of trace metal elements in water system are the key factors in the formation and evolution of this kind of pollution; that is, the participation of water and the dissolution of elements by water are the basic premises. The release of trace metals from pollutants into the environment (water) and the reentry into the sedimentary facie are two basic links in the formation of pollution in sediments.

The distribution of trace metal elements in the modern tidal flat of Nanjing section of the Yangtze River has no obvious change with offshore distance on the same sedimentary plane. In the vertical section, the concentration of trace metal elements in the sediments increases from low to high according to the sedimentary sequence. The change of trace metal concentration in the Yangtze River is the dominant factor for the change of sediment concentration. At present, the concentration of trace metal elements in the Yangtze River in Nanjing is increasing or the concentration of trace metal elements in the Yangtze River sediments is increasing day by day, which is the reflection of the increasing amount of trace metal discharged into the Yangtze River by the source body in the upper reaches. The obvious change of trace metals in the modern sediments (Bagua Island tidal flat) in the lower reaches of Nanjing, which is close to Nanjing city, shows that the pollution of trace metals in Nanjing has become more and more serious in recent years. Therefore, the concentration of trace metal elements in the Yangtze River in Nanjing section is increasing in recent years. The acceleration rates of elements in the modern sediments of the Yangtze River in Bagua Island, Nanjing, are as follows (mg/kg/cm): Cu 0.083, Pb 0.067, Cd 0.004, Sb 0.025, Hg 0.0006, Co 0.05, Ni 0.067, Cr 0.217.

6.2.2 The Behavior and Pollution of Trace Metals in Urban Lakes

Pollutants in lake sediments are the deposits of pollutants in water with the deposition of particulate matter and the adsorption of pollutants by sediments. Compared with rivers, lakes are relatively stable sedimentary environments, which can better preserve the material records of human disturbance to the natural environment. The main source of trace metals in the Dianshan Lake sediments of Shanghai is human production and life activities. The obvious coupling relationship between trace metals such as nitrogen, phosphorus, organic carbon, and organochlorine pesticides and their changes in the sediments of the Dianshan Lake is the objective record and evidence of the impact of human activities on the natural environment. The abnormal importation of these substances has already made a interference to the normal material interchange between natural water, sediments, and even organisms in the sediments. In order to develop industry and improve people's life, metals, organic compounds, and synthetic chemicals are used extensively, and pesticides and fertilizers are applied extensively and continuously to increase crop yields. These conditions eventually lead to the accumulation of related substances in the natural system and even lead to the formation of pollution, seriously affecting the quality of the ecosystem. The fact that the changes of nitrogen, phosphorus, organic carbon, pesticides, and other substances in sediments of the Dianshan Lake imported from human production and life into the natural environment are obviously related to the trend of trace metal concentration shows that trace metal concentration in urban natural environment is closely related to human activities.

In the past decade, the investment in environmental protection in Shanghai has been increasing, the direct discharge of industrial wastewater has been decreasing gradually, and the rate of meeting the standards has been rising day by day. However, the concentrations of trace metals, nitrogen, phosphorus, and organic carbon in the sediments have not been significantly reduced. In addition to the complex exchange effect between water and sediment, this may be related to the increase of total pollutant discharge and the hysteresis effect of sediment deposition. Based on the above situation, it is preliminarily deduced that the pollution trend of trace metals and other pollutants to the water body (water and sediments) of the Dianshan Lake will not show a significant downward trend in the short term, which needs further study and response. The feature of the behavior and pollution of trace metals, nitrogen, phosphorus, organic carbon, and organochlorine pesticides in the Dianshan Lake are summarized as follows:

1. The distribution, pollution, and temporal and spatial variation of trace metals in the Dianshan Lake, Shanghai

Urban lake sediment is a reservoir of water pollutants, which might record and preserve the pollution situation in different historical periods of the urban. Trace metals in sediments may be re-released to cause "secondary pollution" when environmental conditions change. Based on the study of environmental quality evolution of lake sediment records, the evolution trend of trace metal pollution in the environment of the Dianshan Lake and its surrounding areas is mainly recognized as follows:

The concentrations of trace metal element Cu, Cd, Cr, Pb, Hg, and As in the sediments of the Dianshan Lake are 16.012–60.731 mg/kg, 0.119–4.532 mg/kg, 10.474–57.831 mg/kg, 33.972–83.710 mg/kg, 43.088–145.658 g/kg, and 4.473–15.281 mg/kg. The concentrations of Cd and Pb in the sediment of the Dianshan Lake are higher than those of the surrounding lakes and exceeded the relevant national environmental quality standards, respectively.

The vertical distribution trend of trace metals shows that the concentrations of Cu, Cr, Pb, and Hg increase with the decrease of deposition depth. Most of the maximum concentration elements are distributed on the surface or upper part of the sedimentary column, while the minimum of that are distributed at the bottom or lower part of the sedimentary column. The concentrations of element Cd and As vary greatly with depth. Although the variation regularity of different sample points is different, almost all the elements in the 0–5 cm section of the sedimentary column (roughly equivalent to the time after 1994) show an obvious increasing trend with the depth decreasing, and the increasing rate of most elements is faster. It can be preliminarily judged that during this period, the Dianshan Lake was affected by exogenous pollution, and trace metal pollution was intensifying with time. According to the correlation analysis of sediments, there is no correlation between Cd and other elements in most of the sediments, indicating that Cd may have different sources from other elements.

The potential ecological risk of trace metals in the sediments of the Dianshan Lake is assessed by geo-accumulation index, and the results show that Cu, Cr, Hg, and As are not in pollution, while Pb and Cd are in pollution to varying degrees. Correspondingly, the pollution indices of Cd and Pb increase with the decrease of sedimentary depth, indicating that the pollution degree of Cd and Pb in sediments increases gradually in recent decades, especially in surface sediments over 9 cm (roughly equivalent to the time after 1984).

The distribution of trace metal elements in the Dianshan Lake sediment is:

Arsenic in the sediment is mainly in the unbioavailable speciation (85.46%), the bioavailable speciation (acid-soluble speciation), and the potential effective speciation (reducible and oxidizable speciation) accounting for a lower proportion and has less potential harm to the environment. Pb mainly exists in the latent available speciation (83.28%), the proportions of unbioavailable speciation and bioavailable speciation all are small, and the proportion of bioavailable speciation of Pb increases with the decrease of sediment depth, which has potential harm to the environment. The Cu also exists mainly in the latent effective speciation (45.63%) and has an obvious increasing trend in the sediment above 10 cm. The proportion of the effective speciation is 12.72%, which increases with the decrease of the sediment depth. Therefore, the potential harm of Cu to the environment is relatively great. The results of distribution ratio analysis of secondary and primary phases show that the index has not reached the pollution level, but the pollution has a gradual aggravation trend; Cu is in a mild pollution situation, and the pollution degree is also in obvious gradually aggravating trend; Pb is in a serious pollution situation, but the pollution trend is gradually weakening with time; it has a greater potential ecological risk of the Pb pollution for the Dianshan Lake water quality in a longer period of time.

2. **Coupling relationship and significance of trace metals with N, P, C, and pesticides in sediments of the Dianshan Lake**

(1) **N concentration and evolution of the Dianshan Lake**

In this study, the history and state of eutrophication in the Dianshan Lake are studied and discussed by measuring the nitrogen concentration in its sediments combining with the available data of other nutrients.

The total nitrogen (TN) concentration in the sediments of the Dianshan Lake ranges from 0.437 to 3.804 mg/g. The variation trend of TN concentration in the sediments of the two sampling sites is almost the same. The maximum value appears in the surface layer. The change trend of TN concentration in sediments decreases with the deepening of deposition depth. According to the sedimentation rate of the

Dianshan Lake (about 4 mm/a), the nitrogen concentration might change little and increase slowly from 1960 to 1980 [3]. After the China's reform and opening up (after 1980), the rapid development of economy of the industry and agriculture in the surrounding cities resulted in the input of nitrogen in the Dianshan Lake, and the total nitrogen concentration in the lake began to decrease in the late 90s. This may be related to the fact that people are beginning to realize the seriousness of the environmental problems, to pay attention to environmental protection while developing economy, and even to reduce the discharge of nutrients to the Dianshan Lake.

The results of nitrogen speciation show that more than 95% of the total nitrogen exists as organic nitrogen, and mainly nitrogen enriches in organisms. Among inorganic nitrogen, NH_4^+−N is a large proportion, which may be related to the cations being easily adsorbed by colloids in the system; in contrast, NO_3^-−N is easy to be washed away by water because of its existence in anionic form and that less adsorbed by colloids, and the concentration of NH_4^+−N change is less with very concentration as the increase of sediment depth. The concentration of the NH_4^+−N at the surface sediment is only slightly higher than that of other parts. The correlation between total nitrogen, ammonium nitrogen, and organic nitrogen shows better, and the correlation with organic nitrogen is more obvious than that with ammonium nitrogen. From this fact, it can be demonstrated that the accumulation of nitrogen in the Dianshan Lake may be mainly caused by organic nitrogen accumulation, followed by ammonium nitrogen. Compared with the trend of total nitrogen concentration, the concentrations of ammonium nitrogen and organic nitrogen decrease gradually with the deepening of deposition depth, which is the same as the trend of total nitrogen, while the concentration of nitrate nitrogen is very low and basically unchanged except in the surface–sediment–water critical layer. It is concluded that the total nitrogen concentration in sediments is greatly affected by ammonium nitrogen and inorganic nitrogen, but less affected by nitrate nitrogen concentration.

Compared with the trend of Shanghai's total agricultural output value index, the nitrogen concentration in the Dianshan Lake increases with the development of agriculture and the increasing use of fertilizers and other agricultural materials. It shows that human behavior is constantly changing the natural development process of the Dianshan Lake and accelerating the pace of eutrophication of the lake. With the development of science and technology and the enhancement of environmental protection consciousness of the country and the people, the trend of rapid increase of nitrogen concentration in the Dianshan Lake has slowed down in recent years; the main factor of the occurrence and development of eutrophication in the Dianshan Lake is the influence of human activities around it, and the environmental protection measures on the Dianshan Lake have been strengthened. It is a long-term and important task to reduce the discharge of nutrients and pollutants to the lake and control the pollution in time.

(2) **Phosphorus concentration and evolution of the Dianshan Lake**

Influenced by economic construction, the water quality of urban lakes will change and accelerate the eutrophication process. Based on the records of columnar sediments, the evolution trend and ecological significance of phosphorus and its pollution in the Dianshan Lake are carried out with discussion.

The phosphorus concentration ranges from 0.2776 to 0.5197 mg/g in the 0–25-cm-deep sediments of the Dianshan Lake, with an average value of 0.3335 mg/g. The phosphorus concentration in the surface sediments reaches 0.5197 mg/g. Compared with the total phosphorus concentration and eutrophication degree of some similar lake sediments in China, the total phosphorus concentration in the sediments of the Dianshan Lake is gradually approaching the phosphorus level of eutrophication lake sediments. The characteristics show that the phosphorus concentration in the Dianshan Lake has experienced a slow and fluctuating increase, gradually stabilized, and then increased sharply in the last decade. These conditions warn that if relevant measures cannot be taken artificially to slow down the sharp rise of total phosphorus concentration, in the next few years or a little longer, the Dianshan Lake may be with more serious eutrophication phenomenon.

The concentrations of phosphorus in sediments of the Dianshan Lake are as follows:

Soluble phosphorus concentration is 0.0008-0.0032 mg/g, accounting for about 1% of total phosphorus; iron–aluminum-bound phosphorus concentration is 0.4344–0.0672 mg/g, accounting for about 15% of total phosphorus concentration; occluded phosphorus concentration is 0.0973–0.1247 mg/g, accounting for about 33% of total phosphorus concentration, accounting for the largest proportion of inorganic phosphorus; authigenic calcium-bound phosphorus concentration is 0.0153–0.0441 mg/g, accounting for about 7% of total phosphorus; detrital calcium-bound phosphorus concentration ranges from 0.0456 mg/g to 0.1817 mg/g, accounting for about 21% of total phosphorus; organic phosphorus concentration ranges from 0.6452 mg/g to 0.9070 mg/g, accounting for about 23% of total phosphorus. Compared with other shallow lake sediments in the Yangtze River drainage basin, the concentration of organic phosphorus is lower. Apart from the concentration of occluded phosphorus which decreases first and then increases with the depth deepening, other forms of phosphorus in the sediments of the Dianshan Lake show a trend of decreasing with the depth deepening, and the trend of total phosphorus is basically the same. Exchangeable phosphorus and organic phosphorus change slowly, while iron–aluminum-bound phosphorus, authigenic calcium phosphorus, and detrital calcium phosphorus increase sharply with depth decreasing shallow sediment (0–5 cm). The concentration percentages of exchangeable phosphorus, iron–aluminum-bound phosphorus, and organic phosphorus in the shallow sediments are small, which may be related to the release of soluble phosphorus into water under the influence of water disturbance. The concentration of occluded phosphorus does not change obviously with depth, but its percentage increases with depth, which indicated that occluded phosphorus in the sediments of the Dianshan Lake could be transformed to other forms with the development of the process of phosphorus deposition and the time passing by.

The order of phosphorus forms in the sediments is as follows: detrital calcium-bound phosphorus/De-P > authigenic calcium-bound phosphorus/ACa-P > iron–aluminum-bound phosphorus/Fe/Al-P > organophosphorus/Or-P > soluble phosphorus/Ex-P > occluded phosphorus/Oc-P. The accumulation of phosphorus in the sediment of the Dianshan Lake is mainly related to the accumulation of inorganic phosphorus, of which calcium-bound phosphorus accounts for the largest proportion, followed by iron–aluminum-bound phosphorus. Organic phosphorus is one of the factors that can affect the accumulation of phosphorus in sediments. There is a certain correlation between exchangeable phosphorus and iron–aluminum-bound and calcium-bound phosphorus, and the correlation between exchangeable phosphorus and organic phosphorus is obvious, but the correlation between occluded phosphorus and other forms of phosphorus is not too obvious.

There are obvious correlations between total organic carbon and authigenic calcium, phosphorus and organic phosphorus in sediment of the Dianshan Lake. The significant correlation between authigenic calcium-bound phosphorus and organic phosphorus indicates that the organic phosphorus in the sediment of the Dianshan Lake to some extent mainly comes from the life substances of the lake itself, and the anthropogenic input is not obvious. From the beginning of liberation to the beginning of reform and opening up, the gross output value of industry and agriculture in Shanghai grew slowly. Since the reform and opening up, the development of industry and agriculture in Shanghai has been rapid. It is concluded that the phosphorus pollution in the Dianshan Lake, the accumulation of phosphorus in sediment, and the eutrophication process in the lake are closely related to the development of industry and agriculture in Shanghai.

(3) **The carbon concentration and evolution of the Dianshan Lake**

The acid–base property of sediment of the Dianshan Lake is generally in neutral, and the pH ranges from 6.68 to 7.53. The pH value changes from weak acidity to weak alkalinity from shallow to deep of the sediment. The total organic carbon concentration in sediments of the Dianshan Lake is between 0.659 and 0.221%. The percentage of total organic carbon (TOC) in sediments within 8–25 cm depth fluctuates greatly, which may be related to the fluctuation of annual input intensity of organic matter to sediments under the influence of occasional rainstorm weather during the period of serious soil erosion. The concentration of TOC in sediments of 0–8-cm section increases steadily with depth. This may be related to the strengthening of vegetation protection and the slowing down of soil erosion in recent years.

Total organic carbon and organic phosphorus concentration shows a positive correlation; that is, with the increase of deposition depth, total organic carbon concentration decreased, and organic phosphorus concentration also shows a downward trend. From Org-Or-P correlation fitting line and Or-P/Org variation curve with deposition depth, it can be demonstrated that phosphorus accumulates gradually in the process of organic matter degradation, and the change trend of total phosphorus concentration is consistent with the development trend of Shanghai industry and agriculture.

The variation trend of phosphorus and carbon concentration with time in the sediments of the Dianshan Lake is consistent with the growth trend of Shanghai's total industrial output value, agricultural output value, and per capita GDP. These coupling relationships indicate that the phosphorus pollution in the Dianshan Lake, the accumulation of phosphorus in sediments, the eutrophication process of the lake, and the development of industry and agriculture in Shanghai have obvious links.

(4) **The concentration and evolution of organochlorine pesticides in the Dianshan Lake**

The use of organochlorine pesticides in Shanghai began in the 1950s, and the largest amount use was in the 1970s. The use of organochlorine pesticides was to be forbidden in the 1980s after the state banning on the use of organochlorine pesticides. Two major organochlorine pesticide residues, dichloro-diphenyl-rricgloroethane (DDTs) and hexachloro-cyclohexane (HCHs), in sediment of the Dianshan Lake, are studied with following conclusion obtained:

A. The residues of HCHs and DDTs in sediments of the Dianshan Lake range from 0–59.9940 ng/g to 0–122.3174 ng/g. HCH residue in sediments after the 1980s is lower than the detection limit, while DDT concentration in sediments after the 1980s is stable at a low level. This phenomenon may be related to the fact that HCHs are less soluble than DDTs and have better biodegradability than DDTs.
B. The concentration of HCHs and DDTs in the sediments of the Dianshan Lake has the same trend with the age; that is, in the sediment it is in stable residual state with low value.
C. The concentration of β-HCH monomer in the sediments of the Dianshan Lake is with a relative higher rate in the total HCHs, and it is most stable and difficult to degrade one. The residue of δ-HCH in the sediment decreases with the increase of depth when organochlorine pesticides are used legally, indicating that the degradation ability of δ-HCH is the strongest among the HCHs monomers.
D. After the ban on organochlorine pesticides promulgated (1982), in the sediment of the Dianshan Lake the value of (DDD + DDE)/DDT is less than 1, indicating that there might be still illegal use of organochlorine pesticides around that time. After 2000, (DDD + DDE)/DDT > 1, the situation of the illegal use of DDT improving, the DDT concentration in the sediment continues to the state of reducing by normal continuous degradation or metabolism, leading to the proportion of DDD + DDE increasing.
E. The coupling relationship between organochlorine pesticides in sediments of the Dianshan Lake can be divided into two stages. Before 1980s, the correlation coefficient between HCHs and HCHs + DDTs ($R^2 = 0.8087$) is higher than that between DDTs and HCHs + DDTs ($R^2 = 0.51333$), while the correlation coefficient between DDTs and HCHs + DDTs is 1 after 1980s. It shows that the distribution of HCHs + DDTs residues is closely related to HCHs before 1980s, and the distribution of HCHs + DDTs residues after 1980s is related to DDTs. There is no correlation between HCHs and DDTs residues.

The trend of HCHs, DDTs, and HCHs + DDTs distribution in the Dianshan Lake sediment is the same as that of chronological changes. The residues of HCHs, DDTs, and HCHs + DDTs in sediments are higher in the period of 1950s to 1980s, being from the beginning use to being banned for the use of organochlorine pesticides. Owing to the hysteretic effect of the concentration in the sediment, the use of organochlorine pesticides in large quantities since the 1970s has been evident in the sediments of the mid-1970s. After the ban was issued in 80s, the residues of the organochlorine pesticides began to decline. At present, the residues of HCHs, DDTs, and HCHs + DDTs in sediments are so small that they cannot be detected by instruments.

In conclusion, the growth trend of trace metals, nitrogen, phosphorus, and organic carbon in sediments of the Dianshan Lake is consistent with the development of industry, agriculture, social economy, and population growth in Shanghai. From the early period of liberation to the time of reform and opening up of China, the development of industry, agriculture, social economy, and population growth in Shanghai were relatively slow, and the increasing trend of trace metal concentration in sediments in corresponding periods was also relatively slow. The concentrations of trace metals, nitrogen, phosphorus, organic carbon, and other pollutants in sediments also increased rapidly during the corresponding period. All these phenomena show that the trace metal concentration in the Dianshan Lake system is closely related to the social and economic development of Shanghai, and it is a typical environmental phenomenon that human disturbs the level of substance concentration in the natural system.

6.3 Some Significant Behaviors and Relative Features of Trace Metals in Environment

6.3.1 Major Differences and Relations Between Natural Trace Metal Pollution and Artificial Trace Metal Pollution

The concentrations of this work are trace metal environmental behavior and pollution. As mentioned before, this concentration can be divided into two categories: natural causes and human causes. Therefore, the research object of this research, regional trace metal pollution caused by natural background factors, and urban trace metal pollution caused by synthetic factors belong to two kinds of environmental problems with the same object and different causes. Under this premise, there are differences and connections between the two issues of trace metal pollution.

The connections are mainly manifested in the causes of pollution, the chemical behavior of elements in the environment, and the environmental effects of the pollution bellow:

A. From the point of view of the cause of the pollution, for the both types of pollution, water is important agent and transport media of pollutants in the process of the pollution. Physicochemical properties (pH, temperature, solutes) and transport (flow) of water are the most important factors affecting the formation, diffusion, and even environmental effects of the pollution.

B. The influencing factors, changing rules, and restricting conditions of the adsorption, desorption, and transformation of the contaminated elements are consistent when the pollutants are into the environment.
C. The environmental effects of the two types of pollution, such as the normal exertion of the environmental functions of various environmental media and the toxic effects on organisms, are also consistent.

The differences are mainly manifested in the release or superposition mechanism of pollutants, the spatial variation regulation of pollution, the time effects of pollution, and the development inertia of the pollution trend as follows:

A. The release of trace metal contaminants from natural background to the environment is a natural process, mainly due to weathering or water dissolution and movement from pollutant sources. Human action is only a promoter. The types and intensity of pollutants are mainly determined by natural background and natural processes. The release of trace metal pollutants from cities to the environment is due to human activities to bring or lead to the pollutants into the environment, and in these processes, human action is the dominant factor.
B. Trace metal pollution caused by natural background has the characteristics of large pollution range (regional), and the change of pollutant concentration in environmental medium shows a slow and obvious regularity. Urban trace metal pollution is always with a variety of spatial distribution features due to different pollution sources, but the pollution scope generally is relatively limited and the concentration of pollutants in the medium changes sharply.
C. From the point of view of the development inertia of pollution trend and the time effect of the two types of pollutions, trace metal pollution caused by natural background is a natural phenomenon under the action of natural forces and has a strong natural inertia. Once the pollution is formed, it will last for a long time and is difficult to control. However, urban trace metal pollution is caused by man-made action, and the dispersion of trace metals to the environment is directly controlled by man-made factors, except that the transport of pollutants in the environment, chemical action, and change are affected by natural forces and natural conditions; other factors such as the sustainability of pollution occurrence and spreads, the intensity, and scope of the pollution are also relatively easy to be controlled.

Regional trace metal pollution caused by natural background factors and urban trace metal pollution caused by man-made factors are two extreme situations of trace metal pollution in nature. Objectively, human factors often play an important role in the pollution caused by natural background factors, and trace metal pollution caused by man-made factors often occurs in urban areas. There are also many subtle differences maybe between them. Therefore, the above summary is the basic respects based on some basic facts from this research.

6.3.2 The Simulation Experiments on the Mechanism of Trace Metal Behavior and Pollution

1. Desorption law of trace metal elements in soil

This work shows that the desorption of trace metal elements in soil under surface conditions obeys the following rules under the action of aqueous solution.

$$Y = C_{\mathrm{pH}}\mathrm{e}^{-ax}$$

In the formula above, Y represents the concentration of elements in desorption solution. X stands for desorption time. C_{pH} represents the initial desorption value of elements under certain conditions, which can be measured by experiments (pH, temperature). A is the desorption coefficient of elements under certain conditions (pH value, temperature), which is the intrinsic property of desorption of elements from soil in the solution, and it can be obtained by fitting curve of the experimental data.

2. The law of soil absorption for trace metal elements

Under the common earth surface conditions, the adsorption capacity of soil to metal ions (adsorption capacity and the time required to reach the capacity limit) varies with the number and concentration of metal ions in the solution. When the species of metal ions in the solution decreases and the concentration increases within a time scale of several dozen hours, the adsorption capacity of soil increases accordingly. The adsorption capacity of soil to metal ions in the solution with metal ions is stronger than that of metal ions in multi-element metal solution. The adsorption capacity of metal ions by soil under acidic condition is lower than that under alkaline condition. The adsorption of metal ions by soil under acidic conditions reaches equilibrium earlier than that under alkaline conditions.

3. Regulations of releasing metal particles from solid surface (rocks and minerals)

A. It is demonstrated by the work that under the earth supergene condition, the dissolution amount of metal elements from solid particles in aqueous solution increases with the decrease of pH value and the increase of ambient temperature, and the time required for metal elements in solid particles to reach the release equilibrium in aqueous solution varies with the element, solution pH, and ambient temperature.
B. The effect of pH value and ambient temperature on the release of metal particles from solid particles is mainly manifested in the change of dissolution amount and the time needed to reach the dissolution equilibrium within a short period after the dissolution reaction (the data were obtained in the experiment for 120 h in this study).
C. Under common earth supergene conditions, the release of metal particles from rocks and minerals to the environment (the amount of metal dissolution per unit surface area of the solid used in the experiment) is positively correlated with the

concentration of the element in the solid, and this correlation exists no matter how the environmental conditions (pH) change.

6.4 The Probing and Improving on the Research Methods for the Trace Metals in Environments

As mentioned earlier, there are many shortcomings and defects in speciation analysis in the study of trace metal contamination. Nowadays, all methods are based on operation conditions, specific materials, or the type of compounds defined by reagents. Due to the complexity of the speciation problem itself, it is bound to make different extraction methods in addition to the cumbersome operation process. The comparability of data obtained by procedures is also poor. This work attempts to explore a more convenient, applicable, and efficient method for speciation analysis of trace metals on the basis of previous research results and own laboratory work accumulation.

According to the five trace metal forms classified by Tessier method, i.e., exchangeable, carbonate, Fe–Mn oxide, organic, and residual speciation, the exchangeable trace metals adsorbed on clay minerals and other components in the medium and the carbonate-bound trace metals in carbonate minerals may be released to the solution of the environment when the pH value decreases, and these parts of the trace metals belong to the most easy to be used by biology. Trace metals bound to Fe–Mn oxides, aluminum oxides, and hydroxides, and organic metals chelated with organic matter or bound in sulfides are released only when the environmental Eh value changes. These metals are usually less readily bioavailable. The residual trace metals bound to the lattice of crystalline silicate minerals such as clay are difficult to reenter the solution of the system under normal earth surface conditions; that is, they are difficult to be bioavailable and are generally considered to have no biological activity. Under the earth's surface conditions, pH is the most direct, common, or possible influencing factor for trace metal speciation. Therefore, this work recommends that the exchangeable trace metals and carbonate trace metals defined in Tessier's process are collectively referred to as "easy bioavailable trace metals." The Fe–Mn oxide trace metals and organic trace metals defined in Tessier's process are collectively referred to as "bioavailable trace metals." The residual trace metals defined in Tessier's process are referred to as "unbioavailable trace metals." This classifying method seems to be more suitable or close to the actual situation of the trace metals in environment and is conducive to the understanding and evaluation of trace metal biological effects in practical work. In this study, this classification method is named the "three-step procedure."

The extraction procedure of "three-step procedure" is summarized in Table 6.4.

In the speciation analysis of trace metals, the concentration of two parts is the most important for the pertinence of the problem. One is the concentration of bioavailable speciation, which is a direct indicator of the bioavailability of pollutants, and the

Table 6.4 Schedule of three steps for the sequential extraction of metal speciation in soil and sediment (for 1 g dry sample)

Procedure	Extracted speciation	Extractants	Extraction conditions
1	Easy bioavailable speciation	40 ml 0.11 M CH_3COOH	16 h, room temperature, 4000 rpm 20 min
2	Bioavailable speciation	10 ml 8.8 M H_2O_2, evaporation to dry (using same dosage to act two times), 40 ml 0.5 M $NH_2OH \cdot HCl$, 5 ml 5.5 M NH_4AC	pH2, 16 h, room temperature, 4000 rpm 20 min
3	Non-bioavailable speciation	HCl–HF–$HClO_4$	Digestion (for As and Hg digestion temperature below 100 °C)

other is the concentration of unbioavailable speciation, which is an indirect indicator of the bioavailability of pollutants. Compared with other methods, the "three-step procedure" has obvious advantages of simplicity and clearer environmental significance. The results of the "three-step procedure" analysis for various kinds of samples show that no matter what part of the concentration data are obtained, it is relatively beneficial to the popularization and application of this method.

Compared with the modified BCR method and "three-step procedure," the results show that there is a good consistency between the two methods in the bioavailable metal concentration obtained from analyses for different media. According to the results from the analysis for soil, water sediment, and coal rock, the data of bioavailable speciation concentration of each element from "three-step procedure" in the studied Pb, Zn, Cu, and As are higher in raw coal than that of the modified BCR method, and the concentration of each element from "three-step procedure" is 0.96–31.03% higher than that of the BCR method. The concentrations of the elements in sediment from "three-step procedure" are lower than that of BCR method 7.52–25.72%. The above results show that the error of bioavailable speciation concentration in the two methods is relatively acceptable. The analytical results of bioavailable speciation in different media for elements are different. Among them, the data coincidence between different analytical methods of soil is better than that between sediment and raw coal. Except for Zn and Cu in this work, the datum coincidence of other elements for two important concentration parts of metals, the easy bioavailable and unbioavailable speciation in different methods for sediment is all in a meaningful range, which has a clear significance for exploration and practical application.

According to the similarity of the release conditions of trace metal elements from the medium in nature, the "three-step procedure" classifies the trace metal forms and makes it clear that it is difficult to distinguish for the bioavailabilities of trace metals and the secondary pollution risks in environment.

It should be pointed out that the method of speciation analysis of trace metals is still at the stage of operation definition. Based on the specific conditions of the medium and the complexity of the speciation problem, it is inevitable that there are differences among the data obtained by various methods. The operation procedure and analysis results of the "three-step procedure" method mentioned above show that the method has obvious maneuverability and applicability compared with the current mainstream methods, and it is meaningful and worthy of further exploration and development. It should be said that, limited to the experimental quantity and accuracy of this work, the significance of the revelation and the enlightenment from it should be far more important than the data obtained.

6.5 The Perspectives on This Study

This research involves more concentrations and larger scope. The opinions in this book may only limited to author knowledge and academic view, the range or extent that could not reach to will hope to point out and make it perfection in the future. Moreover, in some aspects of the study, it is found that some phenomena were in inconsistencies with the existing theory or general experience; in addition, there are still many shortcomings in the research methods and ideas. For the newly discovered abnormal problems, on the one hand, it may be caused by some factors in the research work, and on the other hand, there may be people's existing understanding on the problem or the theory itself needs to be improved; in a word, all these need to be further deepened or strengthened in the future research.

The main problems found and opened in this study are as follows:

A. The elements with high background concentration (Cu, Zn, As, Cd, Sb, Pb) developed in the upper reaches of the Hetao Area migrate to the soil and groundwater in the lower reaches. The average concentrations of As, Cd, and Sb in the downstream soil are nearly three times of the background concentration, the highest concentration of As in the diving water is nearly 20 times higher than the national drinking water standard (0.97 ug/L), while the average concentrations of Cu, Zn, and Pb in the soil are only less than two times of the background concentration, and the diving concentration is within the drinking water standard. According to the fact that Cu, Zn, and Pb polymetallic deposits predominate in the upper reaches of the area, the elements with high enrichment in the upper reaches are not the most significant elements migrating downstream. This may be caused by differences in element migration. This difference is a theoretical problem that needs further study in environmental science and element geochemistry.

B. The problem has not yet been clearly explained in the work and needs further confirmation as below. These phenomena, on the one hand, may indicate the need for more in-depth and detailed work to explore the mechanism of action and, on the other hand, may indicate that the environmental significance of the data obtained from existing experimental schemes, such as the speciation analysis procedure proposed by Tessier et al. in 1979, needs to be questioned. All these need a lot of research and practices to confirm.

a. When Al and Zn are released from solid particles to the solution under common supergene conditions, the amount of Al and Zn will decrease as the temperature rising opposite to that of other elements.
b. Among the simulated desorption conditions (pH = 4, pH = 5.6, pH = 7), the trace metal elements in weak acidic solution (pH 5.6) have the highest desorption rate from soil.
c. The desorption amount of Cr and Co in soil of the Hetao Area (Table 6.4) is greater than that of the easy efficient speciation.

C. The working methods and conclusions of the study on trace metal pollution are the results of a case study only. The general practical significance of the study still needs a lot of work to accumulate data or verify its applicability. It is hoped that the conclusions of this study or the suggestions based on the conclusions will be as data to enlighten to continue exploring or applying in the practical work in this field.
D. Because of the relatively large time span of this research work and the fact that environmental problems have changed and evolved in a variety of social realities, it is possible that some changes may occur with the time from that on the basis of the circumstances at that time. To grasp the evolution trend of pollutants and their pollution and to understand their causes is no less important than the significance of specific data, or even more important, in the heavy tasks of environmental protection we are facing.

In particular, the data or research results obtained in this work are based on the conclusions of this work period, which is only of phase significance, but also expected to be continuously enriched, improved, or revised in the development of work.

References

1. Stuer-Lauridsen F. Review of passive accumulation devices for monitoring organic micropollutants in the aquatic environment. Environ Pollut. 2005;136(3):503–24.
2. Van Metre PC, Mahler BJ. Trends in hydrophobic organic contaminants in urban and reference lake sediments across the United States, 1970–2001. Environ Sci Technol. 2005;39(15):5567–74.
3. Yao S, Xue B, Li S. The environmental meaning of the measuring for the depositing velocity of lake sediment, a case study on the Honghu lake, Taihu lake, and the Chaohu lake. Resour Environ Yangtze River Drain Basin. 2006;15(9):569–73.
4. People's Republic of China Environmental Protection Bureau. Surface Water Environmental Quality Standard–GB 3838-2002. 2002.
5. US EPA. Current national recommended water quality criteria. http://www.epa.gov/waterscience/criteria/wqcriteria.html (2006).

6. Luo X, Chen S, Mai B, Zeng Y, Sheng G, Fu J. The distribution and source of PAHs in the surface sediment of the Pearl River and the northern field of the South China Sea. China Environ Sci. 2005;26(4):129–34 (in Chinese).
7. US Geological Survey. Sediment Quality Chemical Criteria. http://www.ecy.wa.gov/programs/tcp/smu/sedchem.htm (2006).
8. White J, Triplett T. Polycyclic aromatic hydrocarbons (PAHs) in the sediments and fish of the Mill River, New Haven, Connecticut, USA. Bull Environ Contam Toxicol. 2002;68:104–10.
9. Tay KL, The SJ, Doe K, Lee K, Jackman P. Histopathological and histochemical biomarker responses of Baltic clam, Macoma balthica, to contaminated Sydney Harbor, Nova Scotia, Canada. Environ Health Perspect. 2003;111(3):273–80.
10. Fung CN, Zheng GJ, Connell DW, Zhang X, Wong HL, Giesy JP, Fang Z, Lam P. Risks posed by trace organic contaminants in coastal sediments in the Pearl River Delta, China. Mar Pollut Bull. 2005;50:1036–49.

MIX
Papier aus verantwortungsvollen Quellen
Paper from responsible sources
FSC® C105338

If you have any concerns about our products,
you can contact us on
ProductSafety@springernature.com

In case Publisher is established outside the EU,
the EU authorized representative is:
Springer Nature Customer Service Center GmbH
Europaplatz 3, 69115 Heidelberg, Germany

Printed by Libri Plureos GmbH
in Hamburg, Germany